THE U.S. FOREST SERVICE: A HISTORY

To Larry—
With best wishes and
hopes for productive reading
Sincerely
Peter Steen

The U.S. Forest Service

A HISTORY

HAROLD K. STEEN

UNIVERSITY OF WASHINGTON PRESS
Seattle and London

LIBRARY OF CONGRESS CATALOGING IN PUBLICATION DATA
Steen, Harold K.
The U. S. Forest Service.

Bibliography: p.
Includes index.
1. United States. Forest Service—History.
SD565.S75 353.008'233 76-15932
ISBN 0-295-95523-6

Preface

THE UNITED STATES FOREST SERVICE of today is receiving mixed reviews. Its supporters point to its many accomplishments within an enormously complex mission and insist that pleasing everyone is an impossibility. Its detractors lament that the old values of conservation that the agency had followed, or even invented for others to adopt, have been abandoned in recent times. Both sides see the historical achievements of the Forest Service as admirable but strongly disagree whether the earlier goals have been lost in the rush of events. It is most timely, then, to examine the Forest Service as it grew from a one-man consultant, working largely at home, in 1876 to an agency with thousands of staff and facilities across the nation one hundred years later.

The Forest Service, the largest bureau in the Department of Agriculture, has three main divisions: National Forest Administration, State and Private Forestry, and Research. National Forest Administration is the most familiar to the general public, dealing as it does with the myriad resources on the vast national forest system. State and Private Forestry is the arm of the Forest Service which is involved with cooperative programs with state forestry agencies, private forest-land owners, and others. Research maintains the Forest Products Laboratory in Madison, Wisconsin, and a network of regional experiment stations in each major forest region. These diverse components assure that the Forest Service is involved to some extent in all forestry activities in the United States.

Not only is the Forest Service large, but its varied and complex mission forces any historian to adjust to the realization that completeness is an elusive goal. Instead, as historians are well aware, analyses of representative issues and events must portray the bureau during its development. It is regrettable that this work could include only infrequent references to the routine, daily lives of Forest Service personnel. The mass of available documents forced a focusing on higher administrative levels and fairly well precluded more than an occasional glance at the real heroes of this story—the ranger and his staff.

Of the many Forest Service traits, esprit de corps is the most outstanding. Pride in one's work and loyalty to the agency did not dissipate after the first old-timers retired; rather, they became part of the Forest Service mystique and a part of the ranger's personality. As a Forest Service personnel specialist once observed, by most frames of reference, the agency has "abused" its people with long hours and modest rewards. Instead of feeling resentment, those who elect a Forest Service career point with wry pride at the discomforts brought on by budgets too small to support their responsibilities.

To many in the field, the most onerous demand has been a transfer every three years or so, usually linked to career advancement. Enforced mobility and pressures to live in agency dwellings have made homeowning a dream unrealized for many until the retirement years. For the convenience of the Forest Service, in certain regions moves are often scheduled for the spring, not during the busy summer field season, so school-age children must leave mid-term or be left behind until vacation.

Summers, the traditional vacation time to spend with family, are in the Forest Service West the time for fires instead. Fire standby is what one does if not actually combatting a going fire—unless, of course, it rains, as the in-joke goes, then the forest supervisor grants time off for a picnic.

In a sense, this book is dedicated to these people in the field who carry out their difficult and routine but still important assignments with pride and dedication. Timber, water, forage, wildlife, and recreation are only some of the forest resources the ranger protects and manages. He lives and raises his family in a small town where the major employers utilize resources found in his district. The ranger meets daily with representatives of special-interest groups and must constantly balance the responsibilities of working for a national, federal agency with the welfare of his home community. His is an assignment to challenge the most talented and energetic. But that story would be vastly too long and complex and perhaps even a little dull.

An elaborate organization serves the ranger, feeding him advice, information, and directives. His district is one of several that constitute a national forest; a dozen or more national forests typically make up one of the nine Forest Service regions. Overall direction emanates from the chief's office in Washington, D.C., but traditional decentralization allows an unusual degree of autonomy to the men in the field. Each administrative level is served by a staff of subject or function specialists.

Although the most familiar figure, the ranger represents only National Forest Administration. The lesser known divisions of State and Private Forestry and Research add dimensions that make the Forest Service an atypical federal bureau. State and Private Forestry operates largely through compatible state agencies rather than directly to the public, making it somewhat different from most comparable federal services. Research provides scientific and technical information, not only to meet the ranger's needs but to solve the forestry problems of the nation and perhaps of the world. In this book, because of time constraints, neither Research nor State and Private Forestry is treated to the extent that it deserves.

This book takes a middle course. It focuses on the chiefs of the Forest Service and their staff and on national policies evolved as part of the conservation movement. It views the Forest Service from the tip of the administrative triangle, rather than from its very broad workaday base. The conservation movement, in which many Forest Service personnel participated in a major way, is described as it specifically relates to bureau history. Broad, national issues, such as public power or reclamation, are treated lightly, if at all; the issues that other authors have already adequately treated are mentioned only in passing, with reference to the appropriate monograph. Since the Forest Service is a public agency, and the relation of an agency to its public is of considerable relevance, changes in Forest Service policy resulting from new directions of national interest are therefore noted.

It seemed inappropriate for a historical work to analyze in detail events of the most recent past. After all, members of other disciplines are better equipped to study issues that are still unresolved. Further, documents vital to the historical process become available over time; most pertinent materials are still in personal or active files, making access difficult if not impossible for the time being. For these reasons, selected events of the 1960s and 1970s are portrayed in chapter 12 only in summary. This final chapter, or epilogue, reflects impressions gained from a range of written materials in combination with experiences as a professional forester and historian. Intensive analysis ends with the passage of the Multiple Use–Sustained Yield Act in 1960, as discussed in chapter 11. The last chapter is written within the framework of testing the success of the 1960 law. There are obviously other equally valid approaches, but the one selected is most relevant to this book. As a result, the administrations of chiefs Edward P. Cliff and John R. McGuire receive substantially less attention than those of their predecessors. It is only a matter of time until historians turn their attention to these affairs.

Acknowledgments

ASSISTANCE IN THIS STUDY has been generous, at times overwhelming. Chief John R. McGuire of the Forest Service provided written authorization for access to official records, and his staff responded with candor to questions about internal processes. It is to its credit that the Forest Service accepted the need for scholarly independence, even though the agency's financial support made this book possible. The late Clifford D. Owsley of the Forest Service History Office, then Frank Harmon provided day-to-day contacts.

At the request of the Forest Service, the following personnel, active and retired, read all or parts of the manuscript, offered much valuable advice, and discovered many errors of fact: Joseph Baker, C. Edward Behre, Howbert W. Bonnett, Arthur A. Brown, Robert E. Buckman, James J. Byrne, Robert Z. Callahan, Craig C. Chandler, W. Ridgley Chapline, Charles A. Connaughton, Edward C. Crafts, Anthony P. Dean, John W. Deinema, E. L. Demmon, Reginald M. DeNio, Ernest E. Draves, Walter L. Dutton, Joseph A. Fitzwater, Clarence L. Forsling, Paul J. Grainger, Arthur W. Greeley, Richard K. Griswold, Dwight Hair, Frank Harmon, Verne L. Harper, Clare W. Hendee, Pieter E. Hoekstra, Raymond M. Housley, Myles R. Howlett, Jay M. Hughes, Floyd Iverson, Edward A. Johnson, Thomas R. Jones, Horace R. Josephson, Edward S. Kotok, Henry L. Lobenstein, Merle S. Lowden, Lennart E. Lundberg, Richard E. McArdle, William McGinnes, John R. McGuire, Russell P. McRorey, Ira J. Mason, L. K. Mays, William E. Murray, Nolan O'Neal, Carl E. Ostrom, Earl S. Peirce, R. Max Peterson, Donald A. Pomerening, Charles E. Randall, Ralph A. Read, Robert S. Rummell, Chester A. Shields, John H. Sieker, Paul Slabaugh, J. Herbert Stone, Herbert C. Storey, Lloyd W. Swift, Donald D. Strode, George Vitas, Alfred A. Wiener, Edwin J. Young, Eliot Zimmerman.

In addition to the above list, the following representatives of a variety of disciplines read all or parts of the manuscript and offered invaluable comments and insights: Vernon Carstensen, Marion Clawson, Henry E. Clepper, Samuel T. Dana, Gordon R. Dodds, Wilmon H. Droze, Grant McConnell, Roderick Nash, Mildred Nelson, Harold P. Newson, Gerald R. Ogden, James Penick, Jr., Harold T. Pinkett, Wayne D. Rasmussen, Arthur C. Ringland, and Warren Rogers. Also, Richard G. Lillard made many thoughtful suggestions while the project was still in the outline stage.

One reviewer must be singled out for special recognition. Ivan Doig carefully read the entire manuscript and prepared detailed critiques of content and style. Many others have offered advice during various stages of the project, and I offer inclusive thanks.

It is unlikely that any similar work has benefited from such intensive review. Of course, I assume full responsibility for all interpretations and for any errors which may remain.

Special thanks are due to Joan Hodgson of the University of California at Santa Cruz Library, who, with stamina and graciousness, retrieved countless documents via interlibrary loan. Through her help, the library resources of the nation were truly at hand. Other librarians and archivists across the country responded in full to detailed requests and offered direct assistance to on-site research. Leland J. Prater cheerfully searched official photograph files for suitable illustrations.

Associates within the Forest History Society offered much help and encouragement. Elwood R. Maunder, executive director, proposed the project initially and offered administrative shelter from outside pressures. Richard C. Davis and Ronald J. Fahl, who were concurrently engaged in archival and bibliographic research, provided a daily trading of information and advice on Forest Service history. Many of this volume's citations resulted from this continuous exchange. Karen L. Burman proofread the manuscript and made editorial suggestions. Rosanne Theiss and Roline Loung saved many hours of research labor.

And far from least, my wife Judy was a major contributor to this effort. She directly participated during the research stages; her unmatched skills at unlocking the mysteries of libraries assured examination of numerous hard-to-find materials. For this specific assistance and for general support I am most grateful.

ABBREVIATIONS USED IN THE FOOTNOTES

AFA	American Forestry Association
AFC	American Forest Congress
AFI	American Forest Institute
ANCA	American National Cattlemen's Association
FHS	Forest History Society
FIC	Forest Industries Council
FRC	Federal Records Center
GPO	Government Printing Office
NASF	National Association of State Foresters
NFPA	National Forest Products Association
NFPC	National Forestry Program Committee
NLMA	National Lumber Manufacturers Association
NRCA	Natural Resources Council of America
RFS	Records of the Forest Service
SAF	Society of American Foresters
USDA	United States Department of Agriculture
USDI	United States Department of the Interior
WFCA	Western Forestry Program Committee

Contents

Illustrations

THE U.S. FOREST SERVICE: A HISTORY

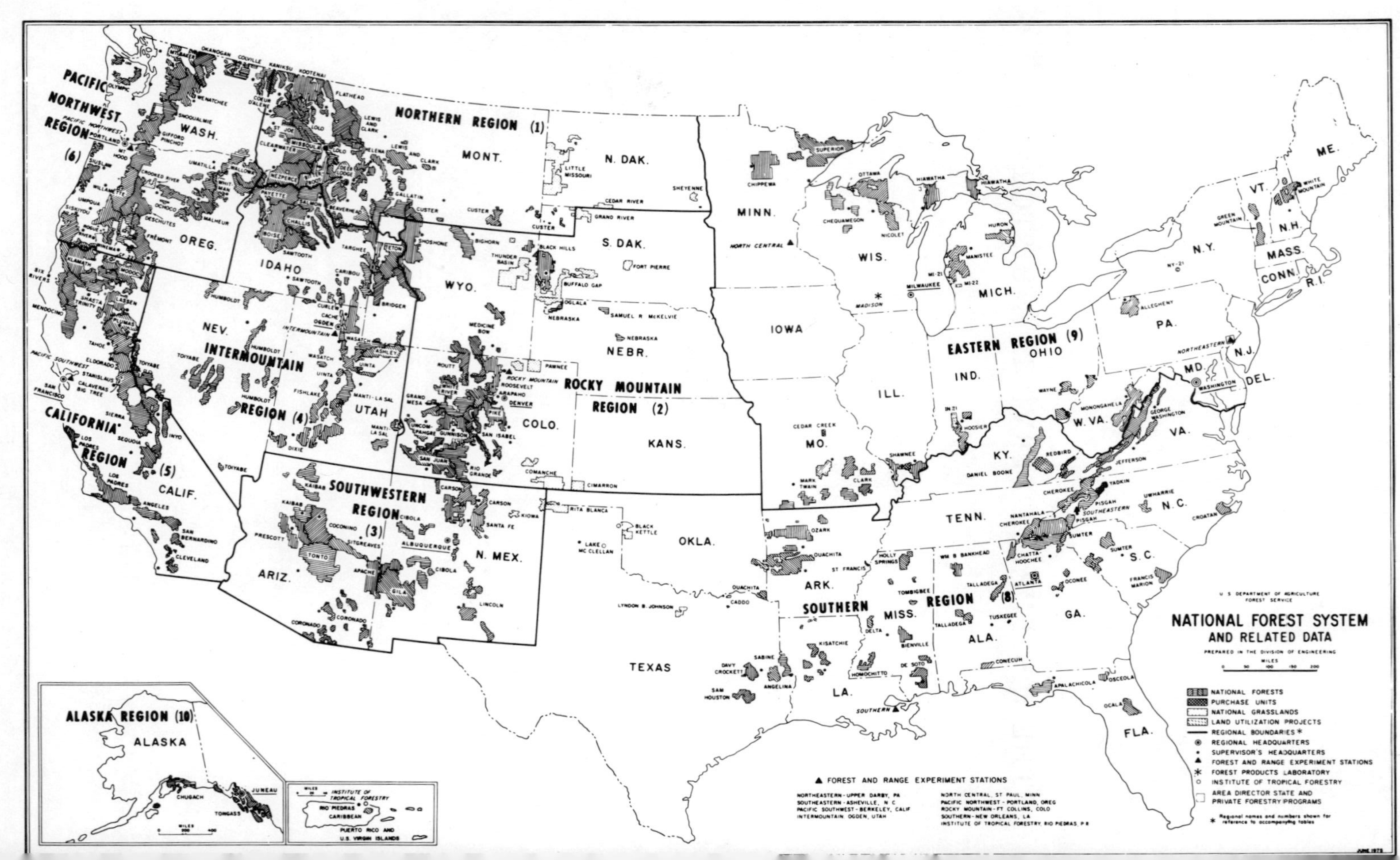
U S DEPARTMENT OF AGRICULTURE
FOREST SERVICE
NATIONAL FOREST SYSTEM
AND RELATED DATA
PREPARED IN THE DIVISION OF ENGINEERING
MILES
0 50 100 150 200
NATIONAL FORESTS
PURCHASE UNITS
NATIONAL GRASSLANDS
LAND UTILIZATION PROJECTS
REGIONAL BOUNDARIES *
REGIONAL HEADQUARTERS
SUPERVISOR'S HEADQUARTERS
FOREST AND RANGE EXPERIMENT STATIONS
FOREST PRODUCTS LABORATORY
INSTITUTE OF TROPICAL FORESTRY
AREA DIRECTOR STATE AND PRIVATE FORESTRY PROGRAMS
* Regional names and numbers shown for reference to accompanying tables
▲ FOREST AND RANGE EXPERIMENT STATIONS
NORTHEASTERN - UPPER DARBY, PA
SOUTHEASTERN - ASHEVILLE, N C
PACIFIC SOUTHWEST - BERKELEY, CALIF
INTERMOUNTAIN OGDEN, UTAH
NORTH CENTRAL, ST PAUL, MINN
PACIFIC NORTHWEST - PORTLAND, OREG
ROCKY MOUNTAIN - FT COLLINS, COLO
SOUTHERN - NEW ORLEANS, LA
INSTITUTE OF TROPICAL FORESTRY, RIO PIEDRAS, P R
PACIFIC NORTHWEST REGION (6)
CALIFORNIA REGION (5)
NORTHERN REGION (1)
INTERMOUNTAIN REGION (4)
SOUTHWESTERN REGION (3)
ROCKY MOUNTAIN REGION (2)
EASTERN REGION (9)
SOUTHERN REGION (8)
ALASKA REGION (10)
WASH.
OREG.
CALIF.
IDAHO
NEV.
MONT.
WYO.
UTAH
COLO.
ARIZ.
N. MEX.
N. DAK.
S. DAK.
NEBR.
KANS.
OKLA.
TEXAS
MINN.
IOWA
MO.
ARK.
LA.
WIS.
ILL.
MICH.
IND.
OHIO
KY.
TENN.
MISS.
ALA.
GA.
FLA.
S.C.
N.C.
VA.
W. VA.
PA.
N.Y.
MD.
DEL.
N.J.
CONN.
R.I.
MASS.
VT.
N.H.
ME.
ALASKA
JUNEAU
CHUGACH
TONGASS
INSTITUTE OF TROPICAL FORESTRY
RIO PIEDRAS
CARIBBEAN
PUERTO RICO AND U.S. VIRGIN ISLANDS
PORTLAND
DENVER
ALBUQUERQUE
MILWAUKEE
ATLANTA
WASHINGTON
SAN FRANCISCO
OGDEN
MADISON
JUNE 1975

CHAPTER I

Forestry in Agriculture: An Accident of History

THE SETTING: Oklahoma, September 16, 1893. By 7:30 that morning crowds of men began to take positions at the starting line, some paying five dollars for a better spot. Waiting for noon to arrive, the assembled thousands sweltered in the heat. Soldiers patrolled the line to see that no one crossed early. Finally, at twelve o'clock, the starter's shots rang out and the men surged forward on horseback, bicycle, on foot, or in horse and buggy, racing for land and a new home.[1] The opening of the Cherokee Strip in Oklahoma marked the last of the great land rushes, epics familiar to every American student today. Land was beginning to seem scarce in this vast nation, and those who wished to claim a piece for themselves were willing to scramble for the right.

In the America of the late 1780s, millions upon millions of wilderness acres stretched west from narrow clusters of population along the navigable waters of the eastern seaboard. This acreage, all the way to the Mississippi River, north to Canada and south nearly to the Gulf of Mexico, had been included in the treaty with Great Britain which followed the Revolutionary War. The coastal collection of former colonies became a national expanse beckoning settlement. Land not surprisingly became a dominant theme of American thought.[2]

Consequently, the policy governing the tremendous sweep of

1. For a graphic account of the land rush, see Everett Dick, *The Lure of the Land: A Social History of the Public Lands from the Articles of Confederation to the New Deal* (Lincoln: University of Nebraska Press, 1970), pp. 289–94.

2. The most comprehensive treatment of American land history is Paul W. Gates, *History of Public Land Law Development* (Washington, D.C.: GPO, 1968).

American land would grow in the next two centuries into a complexity of legislation and administration. Within this land policy, American forest policy was evolving, too. The conservation movement began with concern over forested lands and their water sources. Congressional committees responsible for land sifted many bills, some of which ultimately formed the corpus of conservation law, although ostensibly they dealt with land. When Congress established specialized agencies to administer forestry legislation, it placed them in departments already administering land policy. A history of the United States Forest Service, therefore, begins with a review of public land policy.

The Public Domain

Land lying between the Appalachian Mountains and the Mississippi River came under federal ownership immediately following the Revolutionary War. Some exceptions did exist, as certain of the original states insisted upon retaining parcels of the western frontier of the new nation; but approximately one-half of the country became public domain. The fledgling federal government, facing large military debts and uncertain income, viewed these western lands as an important source of revenue. Tariffs, excise taxes, and land sales were the major federal income for well over a century. In addition to bringing in much-needed revenue from land sales, settlement west of the Appalachians would validate still-tenuous sovereignty and bolster the national economy through expanded commerce.

Even the Continental Congress, with all of the distractions besetting it, had given substantial attention to public land disposition. Two ordinances, in 1785 and 1787, established the rectangular surveying system to be used on public lands, determined procedures by which territories would become states, spelled out the principles of land sales and public reservations, and in general laid down the public land philosophy used to date. Following adoption of the federal constitution in 1789, the United States Congress enacted a lengthy series of laws providing for the eventual transfer of all western lands from public to private ownership. Congressional willingness to experiment is well demonstrated by the many variations found in these laws. All shared two basic principles, however: the land was to be purchased, not bestowed as a gift; and settlement was to be orderly following surveys.

Land sold briskly except during economic slumps; indeed, one can readily judge the nation's fiscal vigor, year by year, by comparing acreages of public lands sold. Most sales were in 160-acre units, or multiples thereof, although at times very large areas were marketed to land companies, which would survey and in turn offer farm-sized parcels to individuals. By mid-nineteenth century the land-disposal system was well tested and offered a range of options to the purchaser. At that point, also, the public domain was vastly larger than when land policies were developed sixty and seventy years earlier. Purchase, negotiation, and war had added the remainder of the continent lying between the Mississippi River and the Pacific Ocean. To deal more effectively with its extensive domain, Congress had created the General Land Office in 1812, and in 1849 the Department of the Interior was established to cope with a myriad of public land responsibilities.

The Civil War, a watershed in American history in many respects, had a decided effect on land policy. Without Southern votes to contend with during the early 1860s, Congress enacted several major laws dealing with the public domain and its disposition. Although land grants to canals and railroads had been given prior to the Civil War, the enormous, precedent-setting grant to the Central Pacific and Union Pacific railroads assured closer ties between East and West while transferring vast acreages to private ownership. An even larger grant to the Northern Pacific Railroad was soon to follow. All told, railroads received over 200 million acres during the last half of the nineteenth century. In return the nation received several transcontinental railroad lines.

Congress had been debating giving land away instead of selling it. The Homestead Act of 1862 enabled those of pioneer spirit to acquire 160 western acres for the price of a filing fee and living on the tract for five years (proving up). Now the farmer could acquire land via purchase under several public laws, buy from a railroad and be close to transportation, or instead pick a spot and enter a homestead claim with the nearest land office. Through such means farmers acquired nearly 550 million acres. In 1850 there had been 1.5 million farms; by 1900 the Bureau of the Census would report 5.75 million farms in the United States. To better serve the farmer, Congress established the Department of Agriculture in 1862, elevating it to cabinet status in 1889.

During 1862 Congress enacted another measure, the Morrill Act, which affected public lands. The act granted to every state 30,000 acres of nonmineral public land for each of its members of Congress as determined by the 1860 census. Two senators and at least one representative in the House assured each state a minimum of 90,000 acres of public land; altogether, nearly 8 million acres of land were made available for sale. Proceeds from sales of such lands were ear-marked for supporting colleges of agriculture and the mechanical arts. Congress amended the act in 1890 to give additional assistance to land-grant colleges. Thus, the Morrill Act, through these schools, provided means to prepare technical specialists needed for agriculture, industry, and ultimately forestry.

"Doing land office business," the thoroughly American catch phrase denoting a runaway work load, came into the language from the hectic heydays of the land agency when administrative paperwork at times lagged several years behind events. That the disposal laws and the survey system worked at all, considering the primitive state of the art and the magnitude of the job, is miraculous. Yet, by all the means available, over 1 billion acres of public land—fully one-half of the entire nation—were transferred to private ownership during the nineteenth century. All things considered, Congress had devised a reasonable and realistic process of land disposal.

Abuses and frauds, the delight of many historians, were bound to occur—and occur they did. To make a show of curbing these frauds and abuses, Congress legislated a number of efforts which were perhaps half-hearted at best; most congressional interest, after all, lay with placing the West in private ownership rather than in impeding the process by intensive policing. During colonial times, Americans had recognized the significance of local wood shortages and enacted ordinances to regulate short-sighted logging. The British Parliament had decreed in 1691 that trees suitable for masts be reserved for the Royal Navy. Although the colonials understood the need to maintain supplies for their own use, they observed the mast reservation order in the breach. Following the Revolutionary War, several federal statutes reimposed restrictions on cutting wood needed by the American navy, with about as little success as the British had before. As early as 1831, Congress forbade removal of timber from the public domain. By 1854 the General Land Office had been given responsibility for protecting the public domain; during the 1880s Division P of the Department of the Interior was established to coordinate protection policies. Much (and perhaps too much) can be made of colorful episodes of land agents battling timber rustlers in the Lake States; or theft and fine agreed upon in advance by logger and federal officer; or dragging a boat on wheels across dry land in order to qualify fraudulently for a swamp land entry.[3]

As far as timbered portions of the public domain were concerned, the problem of disposition was twofold. For the first century of this nation's existence, Congress did not officially recognize timberland in its legislation. With minor exceptions, public land was classified either as agricultural or not suited for agriculture. Congress designed land laws to serve the farmer. Acquiring public land under existing laws usually meant pledging agricultural intent. Any other use was fraudulent by definition, contributing to the impressive statistics showing large-scale abuse. No matter how legitimate lumbering might appear from modern perspective, Congress had only sanctioned farming.

The second reason for frequent flouting of the law may be found in the nature of government and morality on the frontier. Subduers of the wilderness, taking all of the risks entailed, were not overly concerned about what seemed to them to be unrealistic regulations conjured up a thousand miles or more to the east. If a man was apprehended and indicted for some land misuse or another, the accused frequently would be found by a jury of his peers to be guilty of nothing more serious than making the best of a difficult situation. Judge, jury, and defendant were often united against the federal agent, who was considered an outsider. Neither Indians nor "unreasonable" federal laws would impede settling the West.

Whatever judgment one may pass on the seriousness of nineteenth-century land fraud and timber depredation, those directly responsible for administering the law were at least concerned, and at times outraged. Report after annual report of the secretary of the interior and commissioner of the General Land Office from mid-century on carried complaints of theft, misrepresentation, and fraudulent practices. In 1874 Secretary of the Interior Columbus Delano reported "the rapid destruction of timber," especially on public lands. He predicted that all timber of value would soon be removed unless something was done. Protective legislation was "absolutely necessary," the secretary insisted, to protect the public interest.[4]

3. Emphasis on fraud is evident in Roy Robbins, *Our Landed Heritage: The Public Domain, 1776–1936* (Lincoln: University of Nebraska Press, 1942), and especially in Dick, *Lure of the Land*. John Ise, in his classic study, *The United States Forest Policy* (New Haven, Conn.: Yale University Press, 1920), appears outraged by land fraud.

4. USDI, *Annual Report* (1874), p. xvi.

Administrators in the Department of the Interior begged for increased appropriations to obtain the staffing needed if they were to live up to their commitments. They also recommended repeatedly that defective laws be amended or repealed, or, as we have seen, asked for new legislation to deal with the situation. An occasional reformer would attempt verbatim interpretation of the law, causing western outrage to rattle the houses of Congress. It was unfair, claimed those who viewed themselves as victims of unpredictable enforcement, to upset the western way of life with spurts of morality. Congress seemed to agree: increased funding was not forthcoming, laws with obvious loopholes remained in force, and congressmen pressured secretaries of the interior to be less aggressive.

The decade of the 1870s, however, ushered in changes which were highly significant in retrospect but nearly imperceptible at the time. Congress recognized timberland as a legitimate category, enacting the Timber Culture Act in 1873, which elaborated upon the earlier Homestead Act by allowing the settler to substitute planting and cultivating trees for part of the residence requirements. Five years later the Timber and Stone Act authorized sale of nontillable public timberland for personal use. Neither law was motivated by a sense of conservation; each was meant to liberalize timber supplies to meet western needs.

In the main, scientists and other highly educated men were the ones to show concern for future supplies. Congressional thinking for the most part was still at a more elementary level. The versatile and oft-quoted George Perkins Marsh had published his landmark monograph, *Man and Nature*, in 1864 to warn of environmental deterioration and to lecture on the ethics of land use. Annual reports of Interior and Agriculture made frequent reference to Marsh and others, at times reprinting in full a speech or article that carried a conservation message. There were many official warnings, comparing events in the American West to some environmental disaster in Europe, apparently caused by similar abuses.

The states, too, dealt with the resources within their boundaries. Western states owned large amounts of land, acquired upon admittance to the union. All states had responsibilities for private land. Early legislation dealt more with protecting commerce than with conservation, but forest fires, forestation, and trespass received a degree of regulation. In March 1885 the first state—California—created a forest board, which was limited to education and research. Other states followed suit. Not all that impressive, except when viewed in the exploitative context of the times. Organized private concern had appeared only a decade earlier.

Franklin B. Hough

In September 1875 the American Forestry Association held its organizational meeting in Chicago. Called together by John A. Warder, the delegates agreed that collection of statistical information on forest areas, lumber production, reforestation, and botanical descriptions of various tree species was very important. The group continued to meet and merge with other organizations having similar interests. Published proceedings made available to a broad audience the many forestry papers read at each session. Certain names appeared over and over again at the American Forestry Association meetings.[5] One of these names was Franklin B. Hough of Lowville, New York.

In August 1873 Hough, a physician, historian, and statistician, attended the annual meeting of the American Association for the Advancement of Science, held in Portland, Maine. He had been a member of the association for over twenty years. As a statistician, Hough had analyzed census reports and noticed that lumber production was falling off in some areas and building up in others, indicating to him that timber supplies were being exhausted. Hough wondered how long the other remaining supplies would last. In addition to statistical studies, he was very interested in meteorology and rainfall, at a time when it was generally believed that there was a direct relationship between forests and climate. Hough saw forestry as a composite of natural history, geology, mathematics, and physics. He began to focus seriously on the subject when in 1872 he was appointed to a state commission to study the need for a public forest park. He had read and was strongly influenced by Marsh's *Man and Nature*. Now he was to read a paper to his scientific colleagues which reflected his newly formalized interests.[6]

5. For convenience, the name American Forestry Association will be used. Initially, however, there were two groups, the other called the American Forestry Congress. In 1882 the two merged, adopting the name American Forestry Association, which has remained in use. For a full history of the AFA, see Henry Clepper, "Crusade for Conservation: The Centennial History of the American Forestry Association," *American Forests* 81 (Oct. 1975). This history was also issued in 1975 under the same title in book form by the AFA.

6. Edna L. Jacobsen, "Franklin B. Hough: A Pioneer in Scientific Forestry in America," *New York History* 15 (July 1934):311–18; David Lowenthal, ed., *George Perkins Marsh: Versatile Vermonter* (New York: Columbia University Press, 1958), p. 268.

A week before the Portland meeting, Hough noted in his diary: "Began to write a paper on Forestry." He completed it the next day. On August 21 Hough jotted: "Read my paper on Forestry and forest culture in the eve at city hall to a large crowd." The assembled scientists received with favor Hough's paper, entitled "On the Duty of Governments in the Preservation of Forests." The following day they passed a resolution to memorialize Congress "on the importance of promoting the cultivation of timber and the preservation of forests." Hough, Harvard botanist George B. Emerson, and seven others formed a committee to handle the matter. Hough and Emerson were to travel to Washington, D.C., to bring the memorial directly to congressional attention.[7]

In November Hough roughed out the memorial and sent it to Emerson for comment. The following February the two arrived in Washington and went straight to Joseph Henry, director of the Smithsonian Institution, "to talk over the subject of Forestry." They also saw several congressmen and senators, and chatted at length with Commissioner of Agriculture Frederick Watts. The next week another meeting with Watts produced an invitation for Hough and Emerson to meet President Ulysses S. Grant and present their forestry memorial to him personally. They conferred with Grant "for some time about forestry."[8]

After meeting the president, Hough and Emerson visited the Department of the Interior, where they talked to Willis Drummond, commissioner of the General Land Office. During two previous interviews Drummond had shown "lively interest" in their memorial. This time he promised to endorse it and also to secure the endorsement of Secretary of the Interior Delano. Drummond proved as good as his word, recommending to Delano that he support this memorial, which was "indispensably necessary" to halt timber destruction and to provide for reforestation. He included with the memorial a draft joint resolution for congressional consideration. Delano quickly approved

7. Franklin B. Hough, Diaries, Aug. 12, 13, 21, 1873, Franklin B. Hough Papers, New York State Library, Albany. A microfilm copy of the diary is held by the FHS, Santa Cruz, Calif. U.S. Congress, House, "Cultivation of Timber and the Preservation of Forests," Report no. 259, 43 Cong. 1, March 17, 1874. Although Hough corresponded with Emerson on a regular basis, few of the letters have survived in the Hough materials at Albany. Apparently there are no Hough materials included in the Emerson papers held by the Massachusetts Historical Society, Boston.

8. Hough, Diaries, Nov. 20, 1873, Feb. 2, 9, 12, 1874.

and forwarded the memorial and draft resolution to President Grant, who in turn sent it on to Congress.[9]

The memorial declared forest preservation and growth to be of "great practical importance." Predicting timber shortages in the near future, it asked for a law creating "a commission of forestry," appointed by the president and the Senate, to study and report on the amount and distribution of woodlands, the influence of forests upon climate, and European forestry methods. The memorial and a draft resolution were received and referred to the Public Lands committees of the respective houses.[10]

By this time Emerson had returned to Boston, leaving Hough in Washington to deal with Congress. Hough met regularly with congressional supporters and spoke to the House Committee on Public Lands about the merits of his memorial. He had been invited to appear before the Senate committee, too, but the opportunity to speak failed to materialize. When he was not campaigning directly, Hough spent long hours at the Library of Congress, reading all he could find on forestry. One day he determined to find the "origin of the term 'forestry' " and concluded that forestry was "quite new to the language."[11]

On March 4, 1874, Hough met Congressman Mark H. Dunnell of Minnesota, a member of the House Public Lands Committee. From then on Dunnell acted as Hough's champion, and for the next decade they conferred with great frequency about forestry matters. As chairman of the subcommittee reviewing Hough's memorial, Dunnell arranged to have five thousand copies printed for general distribution and drafted for full committee consideration a bill that contained the memorial's essence. The bill provided for presidential appointment, with senatorial approval, of a man "of approved scientific attainments"

9. Ibid., Feb. 9, 12, 16, 18, 1874; Drummond to Delano, Feb. 17, 1874, Delano to Grant, Feb. 18, 1874, and Grant to Senate and House of Representatives, Feb. 19, 1874, copies in "Cultivation of Timber," House Report no. 259, pp. 3–5.

10. "Memorial from the American Association for the Advancement of Science upon the Cultivation of Timber and the Preservation of Forests," reprinted in "Cultivation of Timber," House Report no. 259, pp. 5–6. Henry Clepper believes that the idea of a forestry commission was influenced by creation of a Commission of Fish and Fisheries in 1871; see his *Professional Forestry in the United States* (Baltimore, Md.: Johns Hopkins Press for Resources for the Future, 1971), p. 17.

11. Hough, Diaries, Feb. 27, March 3, Feb. 13, 14, 16, 1874.

who also knew statistical methods and was familiar with forestry. Hough gloated in his diary, "My bill introduced in House."[12]

Waiting for congressional action was excruciating for Hough. Extreme economy was in congressional favor that session, he was advised, and his bill stood little chance. He reported the situation to Emerson in a letter full of "discouragement." Dunnell remained confident, but Hough returned to his room and waited alone. Finally on April 20, Dunnell advised Hough to go home: the bill was dead. Eighteen seventy-four was the same year that Secretary of the Interior Delano pleaded for legislative action to stem destruction of timber on public lands, but that bill died too in a Congress distracted by greater issues.[13]

At home, Hough continued to study forestry during the remainder of 1874 and through 1875, writing papers and presenting a series of lectures at the Lowell Institute of Boston. Hough kept in touch with Dunnell, asking for progress reports on the bill. Showing unusual eloquence, Hough pledged not to "accept *failure* as a *defeat*." He would continue to advance the study of forestry through the AAAS and to collect information. Hough suggested that they might again seek presidential influence. Uncertain, he first crossed out and then reinserted a ringing prediction in a letter to Dunnell, "I am convinced that this is destined to be one of the great questions of the near future and that those who take active interest in it *now*, whether in or out of Congress, will deserve and hereafter secure an honorable place in the Annals of our Forestry."[14]

In January 1876 Dunnell introduced another forestry bill, similar to his earlier one. It, too, made little progress through Congress. After he and Hough exchanged ideas on the subject, Hough journeyed to Washington in February to testify on the bill before the House Committee on Public Lands. "They listened attentively," Hough thought, "but I felt that they regarded the subject with much indifference." Hough was perceptive: the bill bogged down. Even so, Dunnell remained confident and told Hough that he would undoubtedly be

12. Ibid., March 4, 17, 1874. Two years earlier Dunnell had supported creation of Yellowstone National Park, but apparently Hough was not aware of this. See W. Turrentine Jackson, "The Creation of Yellowstone National Park," *Mississippi Valley Historical Review* 29 (Sept. 1942):202–3. The Dunnell collection at the Minnesota Historical Society is fragmentary. No Hough materials are included.

13. Hough, Diaries, March 10, 17, Apr. 3–6, 20, 1874; USDI, *Annual Report* (1874), p. xvi.

14. Hough, Diaries, 1874–75 *passim*; Hough to Dunnell, May 11, Nov. 13, 1874, Hough Papers. For more details, see Andrew Denny Rodgers III, *Bernhard Eduard Fernow: A Story of North American Forestry* (Princeton, N.J.: Princeton University Press, 1951), pp. 36–38.

appointed to carry out the bill's provisions. The Minnesota congressman hoped that the appointment would evolve into something better for Hough. Then in August 1876, to resuscitate the dying campaign, Dunnell made a motion to transfer the substance of the bill stalled in the Public Lands Committee to the general appropriations bill. The rider authorized $2,000 to support a forestry study, obscurely tucked away as seed distribution money for the Department of Agriculture. On August 15, 1876, the appropriations bill received approval, and the commissioner of agriculture was authorized to "appoint a man of approved attainments" to study and report on forest supplies and conditions.[15]

At the time, the significance of the means by which the forestry measure had been passed could not be fully appreciated; but the last-minute, parliamentary tactic shifting the bill from the Public Lands Committee to the agricultural appropriations bill began the century-long tradition of having a forestry agency in the Department of Agriculture.

Hough was delighted at the news. In a lengthy and uncharacteristically detailed diary entry, he explained how he had received urging from Dunnell to pay a personal visit to Commissioner of Agriculture Watts in order to be sure that he got the appointment. Hough set out immediately for Washington, justifying his hasty enthusiasm by explaining: "I said to myself, 'If I go and get the appointment I can exchange views with Mr. Watts fully and begin with a full understanding of what is expected of me. If I fail, I never shall have reason to reflect that I may have lost an opportunity.' "[16]

Hough could well be optimistic, for Watts had supported forestry on previous occasions. The year before, Watts had reported that because of "rapid deforestation" of large areas, "forestry has excited much attention in the United States." The commissioner feared a "timber famine at no distant day" unless appropriate actions were taken.[17]

Hough arrived in Washington to find that Watts was at his home in Carlisle, Pennsylvania. Spencer F. Baird of the Smithsonian joined Hough on the train north, promising full support of his appointment

15. H.R. 1310, 44 Cong. 1; Hough, Diaries, Feb. 24, June 10, 1876; Rodgers, *Fernow*, pp. 38–39. In 1875 Dunnell had tried to get a $25,000 appropriation to construct roads and survey boundaries in national parks, but the effort failed. See H. Duane Hampton, *How the U.S. Cavalry Saved Our National Parks* (Bloomington: Indiana University Press, 1971), p. 37.

16. Hough, Diaries, Aug. 18, 1876.

17. USDA, *Annual Report* (1875), p. 249.

and the accompanying forestry study. At Carlisle, Watts quickly assured Hough that he intended to appoint him as soon as he returned to Washington. He gave Hough complete freedom in making the study and promised to arrange for its publication. Hough realized that the two-thousand-dollar appropriation would demand the "strictest economy." He told Watts that he had expected the bill to be approved in its own right, instead of as a rider to an unrelated subject; but since the project authorized was nearly identical to his original proposal, he was satisfied. They parted cordially.[18]

The next morning, Hough arose early with an "oppressive sense of the magnitude of newly acquired burdens." After three years of advocacy, he had his prize and was humbled. Recovering his composure, he predicted that he would "do credit to myself and the country."[19]

Returning to Washington, Hough called on Commissioner Watts to discuss forestry matters. They had a pleasant chat, agreeing on Hough's plan. Hough decided that no further discussion was needed until he had the report ready to submit. He searched the files of the Department of Agriculture for names of correspondents to contact about forestry, studied a few more items in the Library of Congress, then went home to Lowville to compile his report.[20]

Hough had a head start of at least five years, during which time he had been collecting forestry information. On each trip to Washington, D.C., to lobby for support, he had spent hours at the Library of Congress, reading, translating, or indexing pertinent materials. He had never missed an opportunity to ask a diplomatic legation for pamphlets or reports on forestry conditions in its native lands. In slightly more than a year's time, Hough presented his "Report upon Forestry" to the commissioner of agriculture, as directed by the enabling legislation. An impressed Congress authorized a printing of twenty-five thousand copies.

The report reflected Hough's varied interests and the sort of materials available for quick gathering. He discussed relevant land laws, planting or transplanting trees, soil types, use of wood by railroads and iron manufacturers, problems of insects and fire, meteorology and effects of forests on climate, and the forest resources of many states and other nations.

Hough began his 650-page compilation with a few trial thoughts

18. Hough, Diaries, Aug. 22, 1876.
19. Ibid., Aug. 23, 1876.
20. Ibid., Feb. 24, 29, March 1, 1877.

about underlying conditions affecting forests. Property rights, a sacred American tradition, deterred public intervention in cases of destructive practices occurring on private land. Still, he argued, the government could invoke its right of eminent domain for public welfare. Aware that the remaining public domain was slated for private ownership, he thought another obstacle to the introduction of forestry was the lack of publicly owned land available for reforestation. Even if land were available, he pointed out, there were no trained foresters, since there was no prospect of employment.[21] The situation appeared hopeless—no jurisdiction over private property, no public lands available, and no specialists to carry out a forestry program.

Probably influenced by Hough's study, Watts in his own report reiterated the problem of property rights. "The owner absolutely owns all the rights" beyond the government's right of eminent domain and taxation. Therefore, the commissioner concluded, the forest situation in Europe was only slightly applicable to the United States. Watts did acknowledge, however, that technical or botantical information obtained in Europe could be applicable to American forests.[22]

Hough continued, describing the forest situation in blistering terms, alluding to a pioneer mentality—"little that can be commended and much that can be blamed." The waste of timber shocked Hough, but even so it was faint comfort that overstocked markets brought ruin "upon the greedy." Trespass—a euphemism for thievery—was commonly reported. Enforcement of existing laws and public forest reserves, Hough suggested, would alleviate the problem.[23]

Other anxious voices were heard at the same time. Secretary of the Interior Carl Schurz warned of the enormous depredation of public timber and agreed with Hough that enforcement of existing land laws was vital. Schurz was alarmed by the rate at which forests were being "stripped," estimating that within twenty years timber, along with water, would be in short supply. For those engaged in timber theft, he recommended punishment for both the buyers and sellers of stolen logs. The interior secretary also supported Watts's contention that the government had no jurisdiction over private property.[24]

Early in 1879 Hough received a lump-sum payment of two thousand dollars as compensation for his report. By that time he was already

21. Franklin B. Hough, *Report upon Forestry* (Washington, D.C.: GPO, 1878), pp. 7–9.
22. USDA, *Annual Report* (1878), p. 28.
23. Hough, *Report upon Forestry*, pp. 9, 16.
24. USDI, *Annual Report* (1877), pp. xvi–xvii.

immersed in additional forestry work. He traveled through the Lake States and eastern Canada, observing forest conditions and gathering information for another volume. Rutherford B. Hayes had been sworn in as president in the spring of 1877, naming able but blunt William G. LeDuc to be his commissioner of agriculture. When Hough returned to Washington after his Canadian trip, he presented a memorandum to LeDuc that described his plans for the next report. The commissioner "professed to be satisfied" and planned to use some of the information in his annual report.[25]

Next, Hough vigorously campaigned on Capitol Hill for increased appropriations, with Dunnell either supplying an introduction to a key congressman or intervening with a colleague himself. No doubt this effort enabled LeDuc to secure a $6,000 forestry appropriation, of which he earmarked $2,000 for the West, $1,000 to send Hough to Europe to study forestry conditions, and $3,000 for general forestry purposes. Hough and Dunnell were not satisfied with increased appropriations alone and pressed for statutory permanence for the office. As usual, Dunnell was confident, but Hough "was weary of this waiting from day to day." Repeated lack of a quorum prevented committee action, and Congress adjourned without even providing for the printing of Hough's second report. Both he and Dunnell were disgusted. Hough went home to Lowville.[26]

The next winter Hough returned to Washington and conferred with LeDuc. They had a "friendly visit," with Hough recommending objectives for a forestry commission, should one be established. The two agreed to ask Congress again to fund the printing of Hough's second report. Keeping up constant pressure, by summer they obtained authority for printing, and the second volume of *Report upon Forestry* finally went to press.[27]

This volume, appearing two years after the first, emphasized foreign commerce. Since funds were still limited, Hough again found himself compiling from available sources. Frustrated by variations in reporting systems used by other agencies, his main source of data, Hough raged in his opening remarks that "it would be impossible to

25. Hough, Diaries, Dec. 6, 1878, Jan. 14, 1879; Earle D. Ross, "The United States Department of Agriculture during the Commissionership: A Study in Politics, Administration, and Technology, 1862–1889," *Agricultural History* 20 (Apr. 1946):129–43 (for discussion of Watts's administration see p. 135). LeDuc's report for 1878 praised Hough's efforts, and he asked for $6,000 to support the forestry project (see pp. 27–32).

26. Hough, Diaries, Feb. 18, 19, 24, 25, 26, March 1, June 14, 1879.

27. Ibid., Jan. 28, June 2–5, 1880.

Chief Bernhard E. Fernow at the Division of Forestry exhibit, Columbian Exposition, Chicago, 1893. Courtesy of the National Archives

Student assistants, Bureau of Forestry, in West Virginia, 1903 (Arthur Ringland fourth from right). Courtesy of the U.S. Forest Service

Selective logging in ponderosa pine with piled slash, Black Hills Forest Reserve, South Dakota, 1903. Courtesy of the U.S. Forest Service

Use Book revision committee, 1905. Front row, left to right: Rufus K. Wade, Seth Bullock, Chief Forester Gifford Pinchot, Albert F. Potter. Second row: B. H. Crow, Daniel S. Marshall, R. E. Miller, E. A. Sherman, L. F. Kneipp, E. S. Mainworing. Courtesy of the U.S. Forest Service

excuse the stupidity" that authorized lack of continuity under the guise of economizing.[28] He then launched into his report of export statistics for a large variety of forest products, forestry legislation enacted by states, and the forestry situation in Canada.

The reports seemed to have an effect. In 1881 a Division of Forestry was established in the Department of Agriculture, where previously there had been no formal agency, and Hough, up till then merely a "forestry agent," was named chief. But there were bad times ahead.

Hough's relation with agricultural commissioners Watts and LeDuc had been satisfactory, at least to Hough. There were, however, some indications of a cooling with LeDuc before he was succeeded in July 1881 by George B. Loring, President James A. Garfield's choice for commissioner. The new commissioner was a pompous physician who had long wanted the office. The Garfield administration marked the peak of the spoils system; the wishes of special interest groups received the special consideration they sought.[29]

The two physicians could not get along. After one meeting with Loring which he described as "most unsatisfactory," Hough complained that "he does not seem to comprehend my ideas, or I do not his, and it would be the easiest thing in the world for me to resign, or for him to discharge me."[30]

Dunnell offered to talk with Loring to see if he could smooth things out. He discovered that the commissioner was "extremely prejudiced" against Hough but had no intention of firing him, although he would accept his resignation. Dunnell urged Hough to stay on but assured him that if resignation seemed necessary, the congressman would oppose Loring in Congress. The relationship continued to deteriorate, with Dunnell trying to convince Hough to remain. The forestry chief began to see plots against him from other employees of the Department of Agriculture, especially the chief clerk, and the wear on his nerves is evident in his diary. "Feel very low spirited and all my ambition is gone," he confided; "if Loring remains, I see no other better way for me to do, than to resign and go home."[31]

Hough's spirits sank even lower after Loring appointed Frank P. Baker of Topeka, Kansas, to be a special forestry agent. The appointment was made without Hough's knowledge, and Baker's previous

28. Hough, *Report upon Forestry*, vol. 2 (1880), p. 1.
29. Ross, "U.S. Department of Agriculture during the Commissionership," p. 135.
30. Hough, Diaries, Jan. 19, 1882.
31. Ibid., Jan. 19, 21, 28, June 18, 1882.

contribution to the *Report upon Forestry* had convinced Hough that he was unsuited for the task assigned. With a perverse sense of optimism Hough resolved to wait for the perfect opportunity to make public Baker's "utter incompetence." But worse was in store for Hough's ego. In addition to Baker, Loring had hired another agent, E. W. Ayers. The new forestry agent had preferred an appointment as senatorial staff, but the senator passed him off to the Division of Forestry. Hough sneered that Ayers was "a rebel and a Democrat," a resounding epithet in post-Civil War times.[32]

In spite of his running battle with Loring, Hough remained. With Dunnell's ever-present enthusiasm and support, yet a third Hough volume on forestry cleared the many congressional hurdles and appeared in 1882. It was only half the length of the first two volumes, undoubtedly reflecting the hostile conditions under which it was prepared.

Hough focused on "measures deemed of highest importance" in dealing with problems described before, frequently comparing American conditions to those found in Europe. He called attention to the thoughtless destruction of young timber that was needed to supply future generations, judging it to be "the highest degree of folly." Hough then recommended "*that the principal bodies of timber land still remaining the property of the government . . . be withdrawn from sale or grant*." He proposed that these lands be placed under regulations "*calculated to secure an economical use of the existing timber*." Timber would be cut under lease and young growth would be protected. Hough had in mind a timber lease system comparable to that used in Canada, to be administered by the General Land Office.[33]

Hough had now produced three impressive reports, codifying the forestry knowledge of the western world, compiling statistics on commerce, and recommending appropriate courses of action—all three for a pittance of appropriations.

Hough had done much. His patient compilations brought together the forestry thinking of the time. He worked faithfully with members of Congress to sustain their financial support. Active in the American Association for the Advancement of Science and the American Forestry Association, Hough kept a steady flow of information traveling between scientists and laymen interested in forestry affairs.

32. Ibid., Sept. 16, 24, 1882.
33. Hough, *Report upon Forestry*, vol. 3 (1882), pp. 1, 6, 8, 14.

Similar to his daily association with the Department of Agriculture, Hough's affiliation with the American Forestry Association also suffered at Commissioner Loring's hands. Hough had to travel to the 1882 meeting at his own expense and on his own time. Ironically, Loring also attended, as president of the conservation group. Hough was a frequent participant at the annual sessions, although he did not attend the organizational meeting of the American Forestry Association, held in Chicago during September 1875. The second meeting was held concurrently with the national centennial commemoration at Philadelphia in September 1876. While in Philadelphia, now as a federal forestry official, Hough read once again "On the Duty of Governments in the Preservation of Forests."[34] Also in attendance was a young German forester, Bernhard E. Fernow. Ten years later, Fernow would take over the position in the Department of Agriculture that Hough started. They may have met, but Hough made no mention of Fernow in his diary accounts of the Philadelphia forestry meeting.

At the 1882 AFA session, Hough read a paper entitled "The Forestry of the Future." He predicted that diminishing timber supplies would bring higher lumber prices and offered a detailed economic justification for reforestation. Hough's paper was only one of many read at the 1882 meeting. Growing interest in forest conservation is reflected in the eighty-seven titles presented at the Cincinnati session and the sixty-four additional papers read during the second half of the meeting held in Montreal later the same year.[35]

During 1882 Hough worked on other creditable contributions outside of government auspices. Robert Clarke, a Cincinnati publisher, invited him to write a book about forestry. Hough proposed two volumes, the first elementary in scope and, if sales were adequate, the second a more advanced version. He began work on the first volume immediately and six months later sent the last of the manuscript off to the printer. Five months after the publication of *Elements of Forestry* in July 1882, sales had reached 690.[36] During the same period, Hough and Clarke also produced, with Hough as editor, *The American Journal*

34. Hough, Diaries, Apr. 22, 1882, Sept. 15–21, 1876; Rodgers, *Fernow*, pp. 49–50. Hough had attended a meeting a week earlier of the American Forest Council, held at Sea Grove, Cape Point May, N.J. The council, which was affiliated with the AAAS, asked at the Philadelphia session to merge with the AFA.

35. Franklin B. Hough, "The Forestry of the Future," *The American Journal of Forestry* 1 (Oct. 1882):15–26; AFC, *Proceedings* (1882), *passim*.

36. Hough, Diaries, Jan. 18, 19, July 21, 1882, Jan. 21, 1883.

of Forestry. The first issue appeared in October 1882, but publication was suspended the following September.

Amid such success, the final blow came in 1883 when Loring demoted Hough from chief to "agent of the department," the same title held by Baker and Ayers. Hough remained and continued his forestry studies, although he believed in "the utter impossibility of doing anything to please the Commissioner—or of preparing anything that he would accept." To an overseas colleague he wrote in despair that he doubted Congress would ever take effective action in forestry matters or that any member had even read his reports. The 1884 *Report upon Forestry* contained four studies by Hough dealing with use of wood by railroads, Ohio woodlands, forestry commerce in New Hampshire and West Virginia, and maple sugar production. He died the following year on June 11 at the age of sixty-three.[37]

Hough had predicted to Dunnell, a decade earlier and before any tangible forestry program had received federal support, that "those who take an active interest in it *now* . . . will deserve and hereafter secure an honorable place in the Annals on Forestry."[38] His prediction proved true, as Franklin Hough deservedly won his place in American forest history.

Nathaniel Egleston

If Hough is remembered with favor, his successor, Nathaniel Egleston, is not. Egleston's most outspoken critic referred to him as "one of those failures in life whom the spoils system is constantly catapulting into responsible positions."[39] The record is scanty, but there is evidence to support the substance of such a description.

Most of what we know about Egleston is through his active role in the American Forestry Association, serving on AFA committees and addressing annual meetings. An objective reading of Egleston's speeches in the proceedings fails to reveal the source of his blemished reputation, as his talks seem as perceptive as those of his contem-

37. A composite of diary entries suggests that infirmities of age and need for income caused Hough to stay on, even if reluctantly. Rodgers, *Fernow*, p. 86; Dietrich Brandis, "The Late Franklin B. Hough," *Indian Forester* 11 (Oct. 1885):429; Egleston, *Report upon Forestry*, vol. 4 (1884), *passim*.

38. Hough to Dunnell, May 11, 1874, Hough Papers.

39. Gifford Pinchot, *Breaking New Ground* (1947; reprint ed., Seattle: University of Washington Press, 1972), p. 135.

poraries. But his *Report upon Forestry* does show an unevenness and lack of focus which contrast with his predecessor's efforts.[40]

Egleston's relations with the commissioner of agriculture, Norman J. Colman, indicate his unsuitability for an administrative position. In June 1885, a month after Colman was sworn in, Egleston confided to Robert W. Furnas, a former Nebraska governor for whom he had obtained a position as forestry agent, that he had yet to get an appointment to talk about the role of the Division of Forestry. Egleston wanted to stay on as chief, but he did not "exactly like to ask." His concern was well founded.

Colman asked for his resignation, then returned it after holding it a month. Yet the situation failed to improve. Egleston waited in vain for Colman to give him an assignment, feeling too insecure to offer a plan of his own. Colman hired two agents without Egleston's knowledge and without outlining what the new men should do. Befuddled by indecision and uncertainty, Egleston meekly waited to be fired. He lingered in anguished limbo for three years until relieved by a professional forester, Bernhard E. Fernow, in 1886.[41]

During the decade of Hough's and Egleston's administrations, the Division of Forestry, although essentially a one-man operation without statutory permanence, established itself in the Department of Agriculture. The four-volume *Report upon Forestry* gave official recognition to forestry conditions in the United States and compared them with the situation in other nations. The commissioner of agriculture, as well as the commissioner of the General Land Office, learned the value of this information. So, too, did congressional committees dealing with agriculture, the public domain, and appropriations. The bureau made no progress in curbing forestry abuses, the responsibility for which lay with the Department of the Interior; but it did create a niche for itself in the government bureaucracy. Not a bad showing, considering the mood of the times. It would take two more decades before Congress was ready to approve an effective administrative forestry agency.

40. Egleston, *Report upon Forestry*, vol. 4.

41. Egleston to Furnas, May 11, June 16, July 20, Aug. 12, 1885, roll 5, Robert J. Furnas Papers, Nebraska Historical Society, Lincoln.

CHAPTER II

The Watershed of the Nineties

THE DECADE of the 1860s was a turning point for American land policy—a vital era of land grants to transcontinental railroads and institutions of higher learning, free homesteads for the farmer, and creation of the Department of Agriculture. So, too, was the decade of the nineties.

Americans had become interested in reform and were questioning institutions that they had always accepted. Prior to the nineties, America was predominantly rural, but the nation was now turning industrial and urban, with concurrent changes in population, technology, and economy.[1] The decade, given the misnomer, the Gay Nineties, actually saw the Sherman Anti-Trust Act, debates on free silver, labor riots, Coxey's Army, and the Spanish-American War, which marked the apex of American imperialism. What little gaiety there was seems to be accounted for in the three months of the 1893 Columbian Exposition in Chicago. Even there a solemn note was sounded, when a young professor earnestly offered an explanation for much of what was happening.

On the evening of July 12, 1893, Professor Frederick Jackson Turner of Wisconsin, age thirty-one, faced his learned audience of historians and read the paper he had been working on right up to his time on the program, entitled "The Significance of the Frontier in

1. Henry Steele Commager, *The American Mind: An Interpretation of American Thought and Character since the 1880's* (New Haven, Conn.: Yale University Press, 1959), p. 41; Harold U. Faulkner, *Politics, Reform and Expansion, 1890–1900* (New York: Harper and Row, Torchbook ed., 1959), p. xiii.

American History." Turner's thesis, that America's democratic institutions owed much of their identity to the western frontier, was provocative enough to open "a controversy that was large enough to command the attention of his peers for four generations."[2]

Turner concluded, perhaps prematurely, from figures in the 1890 census that western settlement at last was dense enough to eliminate officially a continuous north-south line demarking the frontier. After explaining how key American traits were related to the existence of a frontier, Turner predicted major changes in the national thought. Termination of the frontier meant a lessening of cheap resources; Americans would have to learn to adjust their economic, political, and daily lives to a new kind of world.[3] Oklahomans dashing into the Cherokee Strip, only months later, offered prompt support for Turner's prediction.

Elsewhere on the exhibit grounds the lumber and forestry building housed exhibits designed to appeal to a much broader audience than the one that had listened to Frederick Jackson Turner. A mighty colonnade of tree trunks, one from each state, symbolized the federal structure of American government. As chief of the Division of Forestry, U.S. Department of Agriculture, Bernhard E. Fernow manned his agency's exhibit, chatting with visitors about the theme on display—the development and teaching of forestry as a science.[4]

At this Chicago exposition celebrating the four-hundredth anniversary of Columbus' discovery of North America, both Turner and Fernow looked at conditions they believed to be uniquely American. At least from certain vantage points, both were correct.

Third Chief, First Professional

Bernhard Eduard Fernow had replaced Nathaniel Egleston as chief of the Division of Forestry on March 15, 1886. On June 30, Congress gave full statutory recognition to the division, removing it from subjection to the whim of a commissioner of agriculture. Although Congress provided stability, it was still parsimonious, even though it was concurrently wrestling with huge budgetary surpluses. The state of

2. Ray Allen Billington, *Frederick Jackson Turner: Historian, Scholar, Teacher* (New York: Oxford University Press, 1973), p. 127; Richard Hofstadter, *The Progressive Historians: Turner, Beard, Parrington* (New York: Alfred A. Knopf, 1968), p. 164.

3. Billington, *Frederick Jackson Turner*, p. 128.

4. Andrew Denny Rodgers III, *Bernhard Eduard Fernow: A Story of North American Forestry* (Princeton, N.J.: Princeton University Press, 1951), pp. 198–200.

New York, recognizing the efforts of Hough and others, had appropriated fifteen thousand dollars for forestry that year; but funds for Fernow's federal agency totaled only ten thousand dollars. Responsibility for forested regions of the public domain still lay with the Department of the Interior.

Fernow had first come to the United States to attend the American Forestry Association meeting held in Philadelphia during the national centennial celebration. Fernow then stayed on after marrying his American sweetheart. Having held a German forestry license since 1869, he began work as a forestry consultant to a Pennsylvania firm. Proceedings of the American Forestry Association show his increasing importance as a figure in forestry affairs. There is much evidence that he associated with both Hough and Egleston for these annual sessions. Fernow's abilities did not go unnoticed, and Abram A. Hewitt, prominent leader of Democratic politics, recommended him to President Grover Cleveland as Egleston's successor, even though it was well known that Fernow was a Republican. As far as Fernow was concerned, Egleston bore him no ill will but instead was relieved to be free of the responsibilities. "We have been and continue to be on the best of terms."[5]

While Fernow was getting settled at the Department of Agriculture, public forest lands seemed under siege. General Land Office Commissioner William Sparks had denounced timber frauds in northern California. A lumber company had openly used farmers, sailors—any available person—to file under the notorious Timber and Stone Act. These benign conspirators would sell their claims to the company for a modest sum and go about their business. A year before, Sparks, in desperation, had suspended all entries in order to gain control of the situation. One of his agents reported that the going rate for dummy entrymen ranged from $50 to $125; you could buy a witness for $25. The same agent estimated that three-quarters of the claims filed with him were fraudulent; a more optimistic colleague guessed that one-half was a better figure.[6]

As Land Office staff watched reports of fraud and depredation pour in, a pattern emerged. When the price of timber increased, so did

5. Rodgers, *Fernow*, p. 108; Fernow to Edgar T. Ensign, March 24, 1886, Letters Sent by the Division of Forestry, 1886–99, Record Group 95 (hereafter cited as RG 95–2), RFS, National Archives.

6. John Ise, *The United States Forest Policy* (New Haven, Conn.: Yale University Press, 1920), pp. 74–75; USDI, *Annual Report* (1886), pp. 95, 200, 213.

timber fraud. Unfortunately for the bogus entrymen, the price of timber frequently dropped before they could clear the claim. Agents for Interior's Division P investigated thousands of fraud and trespass cases every year but were unable to stem the tide. As John Ise has put it, fraud was a frontier way of life.[7]

Despite widespread complacency about timberland problems, efforts for corrective legislation did continue. One of Fernow's first assignments was to draft an enforcement bill for Senator Eugene Hale of Maine. As was frequently the case, Fernow acted in his capacity as an officer of the American Forestry Association, not as chief of the Division of Forestry. The bill elicited opposition instead of support. Kansas Senator Preston B. Plumb, usually a friend of forest protection, objected to having the bill referred to the Committee on Agriculture and Forestry. After all, the bill did provide for creation of a commissioner of forests in Interior. Fernow probably cared little which committee heard the bill. He saw the difficulties involved in getting control of the forests away from the General Land Office and assessed the administrative capability of his department as inadequate for the management task. At the time, Agriculture lacked cabinet rank.[8]

The Hale bill died in committee, but it did provide a blueprint for later legislation. Hale had proposed to suspend entry on all federal forest land until it could be examined and classified. The secretary of the interior could recommend to the president which lands should be reserved. The commissioner of the General Land Office would appoint rangers and make regulations for logging, grazing, and other uses on the reservations.[9]

Whatever other defects members of Congress may have seen in the Hale bill, it was too ambitious for the times. To propose both forest reserves and the means to administer them was unrealistic. Progress would have to travel a much longer and indirect route. There had been repeated attempts to deal with public forest lands, but never adequate congressional support.[10] The Department of the Interior

7. USDI, *Annual Report* (1888), p. 85. For example, in 1889, 55 agents spent 30 man-years investigating 3,307 cases. Five hundred eighty-one of these cases were for timber trespass, valued at 3 to 6 million dollars. USDI, *Annual Report* (1889), pp. 275–80; Ise, *Forest Policy*, p. 79.

8. Rodgers, *Fernow*, pp. 115–16; Jenks Cameron, *The Development of Governmental Forest Control in the United States* (Baltimore, Md.: Johns Hopkins Press, 1928), pp. 203–4.

9. Rodgers, *Fernow*, p. 115; Cameron, *Development of Governmental Forest Control*, p. 220.

10. For a listing of over 220 forestry bills introduced between 1872 and 1897, see AFA, *Proceedings* (1897), pp. 43–64.

was limited to policing trespass and investigating fraudulent entries under a myriad of laws. Then, as with the 1876 seed-distribution rider that authorized a forestry agent in the Department of Agriculture, Congress almost accidentally made a major advance toward protecting forest lands.

Creation of Forest Reserves

Much of the original documentation has been lost for what is now called the Forest Reserve Act of 1891. It is unfortunate that one of the most important legislative actions in the history of conservation is so obscure. Section 24 of this law authorized the president to reserve certain forest lands from the public domain.[11] The reservation clause had a lengthy list of precedents, but in the main the idea began two years earlier. In April 1889 the law committee of the American Forestry Association, consisting of Fernow, Egleston, and Edward A. Bowers of the General Land Office, met with President Benjamin Harrison. Egleston as spokesman presented a petition advocating adoption of an efficient forest policy. The president was cordial but took no noticeable action. The following year, after Fernow's prompting, the American Forestry Association memorialized Congress to make reservations and to provide a commission to administer them.[12]

The same American Forestry Association law committee then made an appointment to see Secretary of the Interior John W. Noble. Fernow, Bowers, and Egleston were joined by others, including John Wesley Powell of the Geological Survey. Years later, a prejudiced Fernow recollected Powell dominating the meeting, trying to convince Noble "that the best thing to do for the Rocky Mountain forests was to burn them down." Fernow used the brief time remaining to impress upon the secretary his responsibility to protect the public domain. Accounts vary as to who said what, but it is generally accepted that as a result of the meeting, Noble personally intervened with a congressional conference committee at the eleventh hour to get Section 24 added.[13]

11. 26 Stat. 1095 (March 3, 1891), Section 24: "That the President of the United States may, from time to time, set apart and reserve, in any State or Territory having public land bearing forests, in any part of the public lands wholly or in part covered with timber or undergrowth, whether of commercial value or not, as public reservations, and the President shall, by public proclamation, declare the establishment of such reservations and the limits thereof."

12. Rodgers, *Fernow*, pp. 142–43; *Congressional Record*, 51 Cong. 1, pt. 3, pp. 2537–38.

13. Rodgers, *Fernow*, pp. 154–55, 199. Fernow's version of the meeting is contrary to Pow-

This presidential authorization to create forest reserves was added in a House/Senate conference committee and not referred back to the originating committees for their consideration. Historians have given much attention to this deviation from standard procedure. That Section 24 became law of the land improperly has also been well emphasized, as well as the fact that Congress passed this most important bill without being aware of its content. These views distort the legislative history of the law of 1891, so important to the development of the Forest Service.[14]

The main purpose of the bill was to revise a series of land laws. Debates in Congress on this subject had been common during the 1880s. Mark Dunnell, who had returned to Congress in 1888 after a three-term absence, opposed attaching forest reserve legislation to general land reform, as he believed that forestry was important enough to warrant its own measure. Too, he was upset that the Timber-Culture Act of 1873, which he had originally introduced in the House, was one of the laws scheduled for repeal. It is not clear from the record, but apparently Dunnell remained forestry's champion, even though he opposed the addition of Section 24 as a rider. Others on the Public Lands Committee overrode Dunnell's objections and the clause stayed.[15] In retrospect, Dunnell's tactics at least made sure that his congressional colleagues were aware of—even familiar with—the substance of Section 24. It is likely, though, that few, if any, could imagine the impact of what was to follow.

President Harrison wasted no time in exercising his new powers. He first set aside the Yellowstone Forest Reserve. By the end of 1892,

ell's; see Carroll Lane Fenton and Mildred Ann Fenton, *The Story of the Great Geologists* (Garden City, N.Y.: Doubleday, 1945), pp. 232 ff. Most historians have accepted the Fenton version, but see Herbert D. Kirkland, "The American Forests, 1864–1898: A Trend toward Conservation," Ph.D. diss., University of Florida, 1971, pp. 171–75. Kirkland casts doubt upon Noble's specific role and whether Fernow was even aware that the amendment was under consideration. The papers of Secretary Noble have yet to be located; if they do exist, they might possibly shed additional light on his role. R. E. Pettigrew on the Senate Public Lands Committee remembered immodestly and perhaps inaccurately thirty years later that it was he who added Section 24 in conference, leading "to one of the most bitterly fought parliamentary struggles in which I have ever participated" (Pettigrew, *Triumphant Plutocracy: The Story of American Public Life from 1870 to 1920* [New York: Academy Press, 1921], pp. 11–12). In a rather garbled account, Robert Underwood Johnson takes credit for alerting Noble to the need to protect forest land and describes Bowers as the probable author of Section 24. Johnson, *Remembered Yesterdays* (Boston: Little, Brown, 1923), pp. 293–94.

14. Ise, *United States Forest Policy*, p. 117; Samuel T. Dana, *Forest and Range Policy* (New York: McGraw-Hill, 1956), p. 101.

15. For detailed descriptions of events in Congress, see Kirkland, "American Forests," pp. 140–87, and Joseph A. Miller, "Congress and the Origins of Conservation: Natural Resources Policy, 1865–1900," Ph.D. diss., University of Minnesota, 1973, pp. 230–38.

mostly to protect water supplies, he had created fifteen reserves containing over 13 million acres. His successor, President Cleveland, added 5 million acres more and then stopped. Until Congress provided the means to protect the forest reserves, Cleveland said that he would set aside no more. After all, without protection, the reserves fared no better than unreserved lands in the public domain.

Protective legislation for the reserves was promptly advocated. In his 1891 report, General Land Office Commissioner Thomas H. Carter pointed out the need for managing the new reserves, as did the American Forestry Association and others. Secretary Noble "urgently recommended that Congress take proper action to have the reservations . . . established as national parks" or to be granted to the states for public use. The American Forestry Association, having made detailed recommendations for areas to be reserved, asked for "a wise and just system" that would be scrupulously and rigorously enforced. The conservationists feared that the administrative procedure of requiring cutting permits from the Land Office would do little to protect the reserves.[16]

When Grover Cleveland returned to the White House in 1893 after a four-year absence, he named Hoke Smith as his secretary of the interior, bringing a mind more imaginative than Noble's to that office. Smith had actively sought the appointment, and Cleveland, with some misgivings, consented. The president believed that Smith at least would take a hard line against raiders of the public domain. Although Noble had been influential in getting the reserves established, he had viewed them as simply augmenting the national park system. Smith immediately recommended legislation to provide a comprehensive forestry system and creation of a forestry commission to advise the commissioner of the General Land Office.[17]

Support swelled both within and without government for legislation to deal specifically with administration of the forest reserves. Within two weeks of the 1891 law's passage, Fernow was advising that his agency would cooperate with the American Forestry Association on implementation. He saw need for data on proposed reserves and new regulations for the Department of the Interior to regulate timber cutting. In his annual report, Fernow explained that more national

16. USDI, *Annual Report* (1891), pp. xiv, 55; AFA, *Proceedings* (1891), pp. 16–18.

17. Rexford Guy Tugwell, *Grover Cleveland* (New York: Macmillan, 1968), p. 191; USDI, *Annual Report* (1893), p. ix.

parks were not the intent of this law; the goals were protection of public property and production of revenue. He reminded his readers that the Division of Forestry had no jurisdiction over public forests; the General Land Office administered what regulations there were. Fernow also supported Cleveland's nomination of J. Sterling Morton to be secretary of agriculture. Traditionally, secretaries became president of the American Forestry Association, and Morton was no exception. He had acknowledged before his senatorial confirmation that as secretary he could do much to advance the interests of forestry.[18]

In March 1892 Senator Algernon S. Paddock, chairman of the Committee on Agriculture and Forestry, introduced a bill, "For the Protection and Administration of the Public Forest Reservations." The bill was later reintroduced, having been expanded in committee to provide for withdrawing all public timberlands and placing them under military protection, and also for returning agricultural land to the public domain for disposal under existing legislation. Paddock's bill gave the responsibility of administering the reserves to the Department of Agriculture. Fernow enthusiastically favored the bill, even though he correctly surmised that it asked for too much. He would settle for less.[19]

Concurrently with the Paddock proposal, Congress was considering H.R. 119, the sort of bill Fernow believed to be more realistic. Congressman Thomas R. McRae, chairman of the House Committee on Public Lands and a member of the American Forestry Association, introduced his bill, "To Protect Forest Reservations," in 1893. Later he would give Fernow credit for convincing him that forestry meant use of forests, not reservation from use. McRae's bill strongly resembled the Hale bill of 1888 drafted by Fernow, which provided for sale of timber to the highest bidder. In McRae's version, the Department of the Interior would administer the reserves.[20]

Secretary Smith, Fernow, and the American Forestry Association supported McRae. Opposition came from the West, against both opening the reserves to logging, thus jeopardizing water supplies, and impeding the miner and stockman. There was strong sentiment for providing free timber to settlers; sales would be a dangerous prece-

18. Fernow to George H. Parsons, March 24, 1891, RG 95–2, RFS; Division of Forestry, *Annual Report* (1891), p. 224; *Annual Report* (1893), p. 31; James C. Olson, *J. Sterling Morton* (Lincoln: University of Nebraska Press, 1942), pp. 350–51.

19. Kirkland, "American Forests," pp. 201–3; USDI, *Annual Report* (1895), pp. cxxviii-cxxix.

20. Kirkland, "American Forests," pp. 204–8; AFA, *Proceedings* (1895), p. 62.

dent. McRae revised his proposal so settlers could get free timber and brought it back to the floor of the House, this time with a favorable Public Land Committee report. A motion passed 117 to 54, but lacking a quorum, this effort failed, too. McRae's third version ended western opposition by allowing mining on the reserves. The 159-to-53 vote in fact reflected strong western support.[21]

When H.R. 119 was referred to the Senate, Henry M. Teller of Colorado, who had been secretary of the interior, 1882–85, substituted his own version. Teller wanted to limit the standards for reserve creation to water protection only, believing that timber supply was not a proper justification. He moved his proposal quietly through the Senate, but both his and McRae's bills died in conference.[22] Although Congress would eventually adopt the main feature of McRae's proposal, it would happen by less straightforward means.

Washington's Birthday Reserves

As early as 1889, the American Forestry Association had advocated a special commission to study public timberlands and recommend how they should be treated. In fact, naturalist John Muir and Secretary of the Interior Carl Schurz had proposed such a commission in the 1870s, but this earlier recommendation went unheeded. Support for the idea increased as the reserves remained unprotected. Fernow could see no reason for further study. It was readily apparent that the reserves needed protection, so why waste time with a study? Instead, all energy should be aimed at getting legislation. But within the American Forestry Association developed a strong backing for a commission.[23]

In June 1895 Wolcott Gibbs, Charles S. Sargent, and Gifford Pinchot met at Sargent's home in Boston. Pinchot was a young forester with boundless enthusiasm, Sargent an eminent botanist, and Gibbs president of the National Academy of Sciences. The trio decided to use the organization to bypass Congress, which was reluctant to establish a forestry commission. At a meeting of the American Forestry Association that fall, the idea of a commission gained support despite Fernow's protests that further study was unnecessary. The time was ripe for action. Fernow insisted that a commission would be a back-

21. Kirkland, "American Forests," pp. 208–12; Ise, *United States Forest Policy*, pp. 124–26.
22. Ise, *United States Forest Policy*, pp. 127–28.
23. USDI, *Annual Report* (1889), p. xxxvi; Kirkland, "American Forests," p. 212.

ward step. Pinchot and journalist Robert U. Johnson, however, succeeded in carrying the meeting and won a resolution favoring a forestry commission. Fernow fought back, and at the next meeting of the association's executive committee, he managed to have a statement inserted into the proceedings branding such action as "prejudicial to the passage of the definite legislation now before Congress."[24] Failing to achieve more than this token of opposition, Fernow then decided to support the commission.

Needed was a letter signed by Secretary Hoke Smith requesting the National Academy of Sciences to appoint a forestry commission. In November 1895 Pinchot noted in his diary that he had drafted such a letter. Another draft, dated December 1895 and unquestionably in Fernow's handwriting, contains the same phraseology as in the letter signed by Smith the following February. It lends strong support to Fernow's later contention that he, through the American Forestry Association, was instrumental in creating the commission. Pinchot, too, takes full credit. In any case, the important fact is that Smith signed a formal request to the National Academy of Sciences on February 15, 1896.[25]

The letter asked the academy to study the forests on the public domain and report back in time to send the information to Congress during the current session. The commission was to determine whether fire protection and permanent forests were practical on the public domain; estimate the influence of forests on climate, soil, and water; and recommend specific legislation. Gibbs explained that the assignment was the largest the academy had ever received from the government and that the report would be impossible to complete before Congress adjourned. Gibbs named to the commission Alexander Agassiz of Harvard, Army engineer Henry L. Abbott, William H. Brewer of Yale, geologist Arnold Hague, and Pinchot, the only

24. Gifford Pinchot, Diaries, June 5, Sept. 4, 1895, Gifford Pinchot Papers, Library of Congress (microfilm copy of diary at FHS, Santa Cruz, Calif.); S. B. Sutton, *Charles Sprague Sargent and the Arnold Arboretum* (Cambridge, Mass.: Harvard University Press, 1970), p. 159; AFA, *Proceedings*, special 1895 meeting, p. 35; *Proceedings* (1896), p. 41.

25. Pinchot, Diaries, Nov. 18, 1895; Smith to Gibbs, Fernow draft in FHS clipping file dated December 1895. Curiously, a notation not in Fernow's handwriting at the top of the draft reads: "Original draft of letter for Sec'y Hoke Smith, calling on Academy of Science for advice Dec. 1895." This notation was probably added after the fact in order to confirm Fernow's claim of primacy. For Pinchot's version see *Breaking New Ground* (1947; reprint ed., Seattle: University of Washington Press, 1972), pp. 86–90; see also Rodgers, *Fernow*, p. 220. Robert Underwood Johnson noted his own role as key in convincing Smith to write the letters in *Remembered Yesterdays*, p. 297.

one who was not a member of the academy. Armed with a twenty-five-thousand-dollar appropriation, the commission headed west to study the forest reserves.[26]

Sargent was awed by his commission's assignment. The reserves already included 20 million acres, and he suspected that local residents would be uncooperative. He admitted to having "more on my hands than I can manage" but immodestly pointed out that there was no one else with his knowledge to head the project. He glumly predicted that Congress would ignore the final report, anyway. Pinchot did not share Sargent's opinion, noting that the chairman was "utterly without plan or capacity to decide on plans submitted." As committee work progressed, Pinchot's disenchantment grew.[27]

Fernow's biographer justifies leaving the chief of the Division of Forestry off the commission, as it would have meant Fernow might sway the report to validate his own policies.[28] Since the commission was to study the reserves in Interior and Fernow worked in Agriculture, his exclusion does seem unjustified. Fernow himself was bitter.

A frequent respondent, Abbot Kinney, wrote to Fernow, reporting on the commission's brief inspection of the forest reserves in southern California. He wanted to know why Fernow was not a member. The Californian regretted also that the group had been unable to see examples of major flood damage in a reservoir. With sarcasm, Fernow explained that the committeemen had "imagination enough to describe the condition from reports of others sufficiently well for the sake of argument to secure legislation." Fernow then told Kinney that he had been instrumental in the commission's creation, having written the letter for Hoke Smith's signature. He pointed out that this information was an "inside history" of the executive committee of the American Forestry Association. "Nevertheless, I have been neither consulted nor in any way asked to contribute my share, nor recognized in my existence as the representative of the Government on this question." But, he philosophized, "such is life, and such are people." What counted to Fernow were the ultimate results. He expected that

26. Copy of Smith to Gibbs, Feb. 15, 1896, in Rodgers, *Fernow*, p. 220; Kirkland, "American Forests," pp. 222–23; Sutton, *Sargent*, p. 159. Johnson takes credit for obtaining the appropriation by his influence with Joseph Cannon, chairman of the House Appropriations Committee. Johnson used George P. Marsh's *Man and Nature* to convince Cannon (*Remembered Yesterdays*, pp. 297–98).

27. Sargent to Thiselton-Dyer, quoted in Sutton, *Sargent*, p. 160; Pinchot, Diaries, May 16, 1896, and *passim*.

28. Rodgers, *Fernow*, p. 221.

the recommendations would not vary much from the programs advanced by the American Forestry Association over the years.[29]

But they did. Fernow and the American Forestry Association had accepted western opposition to forest reserves as a political reality. Therefore, they supported a slow, gradual reservation program to avoid triggering strong protests. The commission, however, with the full support of the new secretary of the interior, David R. Francis, recommended the immediate creation of thirteen forest reserves covering 21 million acres. Gibbs's letter of transmittal to President Grover Cleveland suggested that Washington's birthday, 1897, would be an appropriate date for the proclamations. President Cleveland obliged and set off the furor Fernow had feared.[30]

Furor indeed. Five days later, Fernow wrote NAS Commissioner Arnold Hague that he was not surprised by the "howl" raised over the president's action, and he predicted more. As to the thirteen reserves, Fernow added, "I want to claim a good share of the credit in this for the Forestry Association as having paved the way towards making it possible to secure not only reservations but the committee itself."[31]

The howl Fernow described grew louder. Typical of western outrage was a Seattle Chamber of Commerce memorial to Congress. The northwesterners fumed that they were being treated as a "mere dependency" and their further economic development was being prohibited by the "gratuitous suggestions of three irresponsible strangers [the NAS committee], after a flying visit of a couple of days. . . ." The whole episode was a "galling insult to [our] local sovereignty." Pulling out all stops, the chamber raged that "King George had never attempted so high-handed an invasion upon the rights" of Americans.[32]

Scarcely a week after Cleveland's precipitous act, Fernow told a colleague that the situation "has changed most unexpectedly and most seriously." To another he lamented, "Alas! Our forward steps have

29. Kinney to Fernow, Oct. 7, 1896, Letters Received by the Division of Forestry, 1888–99, Record Group 95 (hereafter cited as RG 95–1), RFS; Fernow to Kinney, Oct. 9, 1896, and Fernow to J. D. Barrett, Jan. 9, 1897, RG 95–2, RFS.

30. Cameron, *Development of Governmental Control*, p. 207; Rodgers, *Fernow*, p. 223; Sutton, *Sargent*, pp. 163–64; Francis to Sargent, Oct. 2, 1896, box 578, Pinchot Papers. Fernow via the AFA acknowledged at this point that a sudden increase in the reserve system could dramatize the need for regulatory legislation and prompt congressional action; see AFA, *Proceedings* (1897), p. 31.

31. Fernow to Hague, Feb. 27, 1897, RG 95–2, RFS.

32. Memorials from the Seattle Chamber of Commerce and others are reprinted in U.S. Congress, Sen. Doc. 68, 55 Cong. 1, May 6, 1897.

frequently to be taken back. . . ." He described Cleveland's act as "injudicious" because it made no provision for managing the reserves. The proclamation had "stirred up such an antagonism as we have never had before."[33]

Sundry Civil Bill

Buried under an avalanche of protests, congressmen moved to appease constituents. An amendment to the Sundry Civil Bill restoring the entire area to the public domain passed the Senate. Fernow labored behind the scenes and got the House conference committee to insert an amendment giving the secretary of the interior authority to establish a division to protect and manage the forest reserves, with the right to sell timber—something he had tried for years to achieve. His effort only partly succeeded: President Cleveland refused to sign the bill because of other defects.[34]

Secretary of the Interior Francis left us with his recollections of Cleveland's refusal. The appropriations measure cleared Congress and arrived at the White House on the day of McKinley's inauguration. The outgoing president asked each member of the cabinet to comment. Francis, second to last (being senior only to Secretary of Agriculture Morton), pointed out that the measure revoked the proclamation that had created the 21 million acres of reserves. At that moment a messenger interrupted to announce that McKinley had arrived. Cleveland hesitated, then threw the Sundry Civil Appropriations measure on the floor saying, "I'll be damned if I sign the bill."[35]

Cleveland's pocket veto of the appropriations measure left the government without funds for the new fiscal year. McKinley quickly called Congress into extra session on March 15. Forces concerned about forest reserves rallied. The opposition made full use of its momentum, but the defection of Senator Richard Pettigrew of South Dakota eventually turned the tide in favor of the reserves. This powerful member of the Senate Public Lands Committee had been a staunch foe of Cleveland's "Washington's Birthday" reserves, but

33. Fernow to W. G. Steel, March 1, 1897, RG 95–2, RFS; Fernow to H. G. deLotbiniere, March 6, 1897, ibid.

34. Fernow to H. G. deLotbiniere, March 6, 1897, ibid.

35. Walter B. Stevens, "When a Missourian Forced a Special Session of Congress," *Missouri Historical Revue* 23 (Nov. 1928): 46–47. Another account, as related to Robert Johnson, quotes Cleveland as saying, "I will veto the whole damned Sundry Civil Bill" (*Remembered Yesterdays*, p. 300).

Charles D. Walcott, director of the U.S. Geological Survey, won him over.[36] Walcott persuaded Pettigrew to sponsor an amendment to the new Sundry Civil Appropriations bill. Walcott drafted the amendment, modeling it after McRae's much-battered H.R. 119. The amendment specified the criteria for reserve designation—water protection and timber production—and excluded mineral and agricultural land. Also, settlers could have free timber and stone.

Then Walcott convinced Pettigrew of the need for another clause, suspending the new reserves for nine months. The suspension clause was a clever tactic to overcome western demands for total elimination. Under this clause, those who had entered the designated reserves could transfer their claim, within nine months, to other parts of the public domain. By this so-called lieu-selection process they could select new tracts in lieu of the original claim.[37]

Walcott met with NAS committee members and others to plan strategy. After convincing McKinley's newly appointed secretary of the interior, Cornelius Bliss, they approached the president. McKinley, not wanting to alienate any congressmen this early in his administration and facing much greater and more important demands to go to war with Spain, was cordial and strongly supportive of the forest reserves. It would have been easier for him to rescind Cleveland's besieged proclamations, but McKinley agreed to let them stand, although he could not offer open support.[38]

The Pettigrew amendment to the Sundry Civil Appropriations bill won handily in the Senate, with strong western backing. That the measure was favored by opponents of conservation probably resulted from western concern over unmanaged, locked-up reserves. Assurance that the reserves would be open for use eliminated the main reason for opposition. Ironically, McRae, whose H.R. 119 had been the pattern for the Pettigrew amendment, fought acceptance in the House. He believed that opening the reserves would jeopardize their

36. Pettigrew, *Triumphant Plutocracy*, p. 207; Charles D. Walcott, "Forestry Legislation of 1897," n.d., copy in box 73, SAF Records, FHS. This is apparently a draft of a memo Walcott wrote to John Ise on April 2, 1919; the original is in the Pinchot Papers. Ise also uses the substance of the memo in his *United States Forest Policy*, pp. 132–33.

37. Ise, *United States Forest Policy*, pp. 132 ff.; Walcott memo, "Forestry Legislation of 1897," SAF Records; Pettigrew, *Triumphant Plutocracy*, p. 208.

38. Pinchot, Diaries, March 5, 1897; Sutton, *Sargent*, pp. 166–69; Walcott memo, "Forestry Legislation of 1897," SAF Records. Over twenty years later, Sargent claimed to have convinced McKinley to let the reserves stand. Sargent to Henry S. Graves, Sept. 7, 1921, Henry Graves Papers, Yale University. The author is indebted to Henry Clepper for calling this letter to his attention.

flood control capacity. The House adopted the measure, and a conference committee ironed out differences. The only important conference change was a modest reduction in the lieu-selection generosity. With congressional approval, President McKinley signed the bill on June 4, 1897.[39] Thus the third major piece of forestry legislation moved through Congress as an amendment, never having had to surmount the full legislative process.

The victory belonged to many. Fernow certainly deserves major credit. The American Forestry Association provided its good offices, bringing together foresters, legislators, and others concerned about the reserves. Sargent, Pinchot, and the National Academy of Sciences commission were able to bring presidential intervention into a congressional stalemate. Several members of Congress—McRae, Pettigrew, Paddock—made contributions of their own. The achievement itself may be judged on its own merits.

What some historians call the Pettigrew Amendment turned out to be the basis of federal forest reserve management for sixty-three years until supplemented in 1960 by the Multiple Use–Sustained Yield Act. The law authorized the U.S. Geological Survey to examine the forest reserves. It stipulated that no reserve could be established "except to improve and protect the forest within the reservation, or for the purpose of securing favorable conditions of water flows, and to furnish a continuous supply of timber for the use and necessities of citizens of the United States. . . ." Further, the secretary of the interior was directed to make rules and regulations for the protection of the reserves. Perhaps most significant, in terms of what Fernow, Pinchot, and others had sought, was the authorization of the sale of mature or dead timber. Timber selected for sale had to be appraised, advertised, sold at or above appraised value, "marked and designated" prior to cutting, and supervised during cutting. Within three weeks, field agents of the General Land Office had been instructed how to proceed under the new law.[40]

Two of the most important legislative events in the history of the Forest Service, the laws of 1891 and 1897, took place during the decade of the nineties. The nation now had forest reserves and the

39. Ise, *United States Forest Policy*, pp. 137–38.

40. 30 Stat. 34–36 (June 4, 1897). The Multiple-Use Act of 1960, 74 Stat. 215, is considered to be supplemental to the 1897 law. In 1973, *Izaak Walton* v. *Butz*, which was decided in favor of the plaintiffs, declared clearcutting practices of the U.S. Forest Service on the Monongahela National Forest to be in violation of the intent of the 1897 law. General Land Office Circular, June 30, 1897, in USDI, *Annual Report* (1897), pp. cix-cxiv.

means to protect and manage them. More adjustments would take place, but the basic elements of federal forestry now were intact. Dwelling on these two laws, however, leaves an unbalanced view of Forest Service history. Obviously, most of the agency's energies have been devoted to nonlegislative tasks.

Fernow as Scientist and Administrator

As pointed out above, shortly after Fernow had taken over as chief of the Division of Forestry in the spring of 1886, the agency was granted full statutory status but was denied adequate funding. Fernow had been hesitant to accept the federal appointment, writing a colleague, Edgar T. Ensign, that he had accepted the position "after all." The position seemed political rather than technical, and he feared that he would be out of his element. For ten years he had worked at a variety of consulting jobs. Now he was looking for "useful" things to do, but first he had to file a report in a few months. Fernow offered to pay his associate's expenses for gathering information on and making a map of federal timberlands in Colorado.[41]

Fernow furiously searched forestry literature for information. Three days after having taken office, he wrote to V. M. Spalding that he wanted to place the division on a "more scientific and systematic basis," but priorities dictated a report by July 1. He hinted that perhaps Spalding's work on white pine could be far enough along for inclusion. To another correspondent, Fernow outlined his hopes that the division would bring into closer connection all agricultural schools, forestry associations, and horticultural societies. He closed with a plea for data on federal forest lands in Minnesota and offered to pay travel expenses.[42]

Fernow's hurried efforts produced a surprisingly sound statement of forestry principles. He made the usual references to the influence of forests on streamflow and climate and noted the exhaustibility of supply. Eastern forests, he predicted, would be depleted in fifty years, and even the stupendous western forests would ultimately share the same fate.[43] Fernow's most important contribution, one that introduced modern professional forestry concepts, dealt with economics.

41. Rodgers, *Fernow*, pp. 108–9; Fernow to Ensign, March 18, 1886, RG 95–2, RFS.

42. Fernow to Spalding, March 17, 1886, RG 95–2, RFS; Fernow to J. Fletcher, Williams, March 22, 1886, ibid.

43. USDA, *Annual Report* (1886), pp. 151–58. In 1879 Secretary of the Interior Schurz had estimated a twenty-year supply nationally.

To Fernow, the basic deficiency in handling forest resources was the failure to distinguish between interest and capital. Sustained yield, after all, was based upon the notion of harvesting annual or periodic growth increments—the interest. Overcutting occurred when the resource base (the capital) was depleted to a point where the desired growth could not be maintained. Cutting into forest capital reserves would in the long run lead to exhaustion.[44]

Fernow perceived overcutting as unnecessary as well as undesirable. Oversupply of lumber was the problem; solve it, and the threat of exhaustion would disappear while the lumbermen's profits would increase. Utopia? Not at all. Railroad land grants had made available vast amounts of low-cost timberland for speculative purchases. The flood of low-cost timber yielded low-cost lumber, forcing prices lower and lower. Lower prices meant that the lumberman had to sell—and therefore produce—more lumber to meet fixed costs, feeding even more to an already glutted market. Fernow believed that the lumber industry was unique in that the laws of supply and demand worked only to its detriment. The industry was unable to cope with a fluctuating marketplace and needed "the fostering care of a far-seeing governmental policy."[45]

Fernow offered no governmental panacea for industrial problems, in recognition of the traditional American reverence for private property rights. He did propose, however, that government timber be withheld as a means to alleviate oversupply. He hoped that examples of good forestry on federal reserves would be adequate to start the industry on the right track, obviating a potential need for government intervention.[46] Although oversimplified in many respects, the report was an amazing effort for the thirty-five-year-old chief who had spent only four months on the job.

That out of the way, Fernow wrote to the commissioner of agriculture proposing changes in the division's organization, now that it had statutory permanence. Listing Egleston as one of his two assistants and the eight clerks shown on the division roster, he complained that six of the clerks were not under his jurisdiction, one was sick, and the eighth only worked part-time. Fernow therefore asked that all of the clerks be taken off his payroll. He apologized to his superior for grumbling about clerical inefficiency in the staff he had inherited, but

44. Ibid., p. 155.
45. Ibid., p. 168.
46. Ibid., p. 166.

not one of them knew botany or forestry, had command of a foreign language, was skilled in research technique, or could operate a typewriter. He also asked that either of his assistants be removed; he did not care if Egleston stayed.[47]

Fernow's requests were modest, indeed. He asked only for "a small plot of ground"—perhaps the Bureau of Animal Husbandry could spare some space at its station. For a clerk, he wanted a recent college graduate botanist, who could be "a young lady with knowledge of typewriting." This botanist clerk would be paid $900 per year. His total budget was $8,000: $3,500 for the division headquarters and $4,500 to support special agents studying western forests.[48]

Fernow divided the work of the division into four categories: general and statistical, economic, forest botany, and forestry proper. Under general and statistical he included studies of the forests of Colorado and California by Edgar Ensign and Abbot Kinney, as well as a review of the scientific arguments on the effects of forests on climate. Economic studies would focus on the wood-using industries—charcoal, iron, mining, lumber, and railroads—and "especially those directly controlling forest property." Fernow listed thirteen timber types or species deserving investigation under biological studies, indicating that commercial value was obviously of prime importance. Finally, he proposed to publish manuals on tree seed, nurseries, and thinning, which he thought would be "exceedingly valuable" for educational purposes. In sum, his was an ambitious program for an understaffed division with a budget of only eight thousand dollars.[49] His own house in order, Fernow looked at the broader problems of American forestry.

In 1886, the same year Fernow became chief of the Division of Forestry in the Department of Agriculture, Edward A. Bowers joined the Department of the Interior as an inspector of public lands. The two were well acquainted, having been leaders in the American Forestry Association. Bowers asked Fernow to propose policies for government-owned forest lands.

To Fernow, the forest was a valuable national property exposed to "reckless and shameful deterioration and depredation." Opposition to reform could stem only from ignorance or "people not wholly disin-

47. Fernow to Norman J. Colman, July 1, 1886, RG 95–2, RFS.

48. Fernow to Colman, July 1, 1886, RG 95–2, RFS. Apparently Colman had reserved for departmental use $2,000 of the $10,000 appropriation.

49. Ibid.

terested in the thieveries upon the public domain." To him the issue was simple: "*How* shall we preserve for legitimate and economic use" the remnants of the public domain? Sustained yield was Fernow's answer to Bowers.

He calculated that federally owned timber was worth conservatively 280 million dollars. At 5 percent, annual income from public forests would be at least 14 million dollars. Certainly as a straight business investment the government could well afford to set up an effective management program—a program Fernow modeled after his earlier Prussian experience. First, withdraw all forest land from sale. Land found suitable for agriculture could be restored to the public domain later. Second, create an enforcement bureau in the Department of the Interior, probably in the General Land Office. The bureau would have a central headquarters with district offices convenient to forested areas. Each district office would use inspectors to assure compliance with policy by rangers (Fernow's title for those responsible for the smallest administrative units). Policy would emanate from the central office; staff would be required to make field inspections at least once annually. Under his plan, timber prices and sales would be based on local conditions, and local demand should receive "primary consideration."[50]

Fernow made little headway either reforming Interior or inspiring his own Department of Agriculture. Sadly he came to the conclusion "that under present conditions no practical work will be done and we might as well satisfy ourselves, that all we can do is talk." Perhaps it was just as well, he thought, for much specific information about American forests was needed "before we can even judiciously suggest" a correct management system.[51]

Fernow saw a need to advance forestry on more than the governmental front. At the opening ceremonies of the Pennsylvania Forestry Association in 1886, he had made suggestions on how to promote forestry locally, while waiting for the national movement to coalesce into strong, federal programs. To ask a lumberman to cut fewer trees would be futile, since he was absorbed in supplying a sawmill. The most productive target, according to Fernow, would be the farmer who owned a small woodlot. The forester told his audience that the individual farmer would show little concern for forest de-

50. Fernow to Bowers, Dec. 19, 1886, ibid. Fernow's proposed forestry agency bears striking resemblance to the modern Forest Service.

51. Fernow to J. G. Kern, March 10, 1887, ibid.; Fernow to Kinney, Apr. 6, 1887, ibid.

struction in general; the message would have to deal with his own property. Forestry manuals would not suffice: a farmer might read a brief, clearly written article but not a manual. The most fruitful method would be for a "competent plain spoken man" to address the farmer face to face. Fernow encouraged the association to raise funds in order to support "travelling teachers." He predicted that the teachers would build support for remedial forestry legislation, needed in Pennsylvania and in many other states.[52]

Despite some frustrations, Fernow sat at his desk with a sense of achievement. He had been able to reorganize his staff satisfactorily. By his second year, he had hired two field agents, and Egleston was helping in the office. Each earned fifteen hundred dollars per year. Fernow was especially pleased with his new assistant, George B. Sudworth, and he recommended giving him a 20 percent increase in salary. Fernow's peace of mind had markedly improved over the previous year.[53]

Obviously gaining confidence after being on the job for several years, Fernow told Assistant Secretary of Agriculture Edwin Willets that the Division of Forestry should have executive responsibilities, that is, it should be managing forests owned or controlled by the federal government. His office was mainly educational, since without forests it could act only as a "bureau of information and advice." If Fernow could have known that two years later the first forest reserves would be established, he might have pressed more vigorously the justification for his agency to manage forests. Instead, he seemingly resigned himself to his educational fate and delineated for his superior how he gathered forestry information and which groups used it.[54]

Fernow could be caustic when his agency or its work received less than praise. One critic's name was "burnt into my record of knownothings" for "slandering" certain research projects. Fernow acted in a similar vein when he clashed with Henry Gannett of the Geological Survey over the need for a stronger federal forest policy. In an April 1, 1893, article carried by a Washington, D.C., newspaper, Gannett claimed that the relationship between forests and climate, soil, and water was little proven. He added that forests, although diminished from the original amount, were growing faster than they were being cut. To him the " 'laissez faire' policy seems to be the best." Public

52. Copy of speech in Bernhard E. Fernow Papers, Cornell University.
53. Fernow to Colman, June 29, 1887, RG 95–2, RFS.
54. Fernow to Willets, Sept. 25, 1889, Fernow Papers.

interest in forestry should be limited to improving transportation systems to aid forest commerce.[55]

Fernow was aboard the steamship *Aller*, bound for Germany to assist his mother country in planning its exhibit for the 1893 Chicago exposition, when a shipboard friend showed him Gannett's article. Another friend sent a letter describing it as "stupid." Fernow agreed and counterattacked publicly in his annual report. He acknowledged that the relation of forests to climate had been exaggerated and that much more information was needed on relations with soil and water. But then, branding Gannett's presentation as a dangerous collection of half-truths, Fernow charged that "any observant logger" could quickly spot the inaccuracies. He was discouraged that a man in a high position like Gannett would ignore his responsibility for careful reporting and dismissed the geographer's preference for "laissez faire" government as unwarranted.[56]

Fernow was disappointed when Gannett failed to utilize fully the resources of the 1890 census in gathering forestry statistics. Ten years earlier, Charles Sargent had made an impressive contribution to forestry in the 1880 census, showing the potential of the national inventory at the beginning of each decade. Their spat continued when Gannett's Division of Geography and Forestry acquired functions overlapping those of Fernow's agency.[57] Resolution of differences between the two departments would not occur until after the resignation of Fernow, whose successor would be a great admirer of Gannett.

Not all of Fernow's problems were external. His Prussian temperament and scientific training made it difficult for him to accept the indignities of being a minor bureaucrat in a department that had not achieved cabinet status until 1889.

Fernow had welcomed Secretary of Agriculture J. Sterling Morton to office, but he later had second thoughts. In January 1894 Morton sent a memo to all of his division chiefs, instructing them to keep a daily time record for all employees and submit it to the chief clerk. Absences, the secretary ordered, were to be charged to annual leave. He added that "fabrication" would result in dismissal. Instead of assigning a clerk to this task and going about his work, Fernow elected

55. Fernow to [?], draft of letter [1890] in FHS clipping file; Washington, D.C., *Evening Post*, Apr. 1, 1893.

56. Rodgers, *Fernow*, p. 200; E. P. Martin to Fernow, March 3, 1893, RG 95-1, RFS; Fernow to Morton, Apr. 5, 1893, reprinted in USDA, *Annual Report* (1893), pp. 313–15.

57. Rodgers, *Fernow*, pp. 200–1; USDA, *Annual Report* (1890), p. 193.

to be offended. He expressed his "great regret" to Morton that he evidently intended "to reduce the chiefs of divisions to the levels of clerks and time servers." Fernow protested that division chiefs were finally achieving a degree of respect, and he resented the "insinuation" that he was not giving all the time and energy he could spare to his work. Fernow found the secretary's suggestion that a chief might falsify his reports "so degrading that no self-respecting man will allow it to go without protest. . . ."[58]

As intemperate as this response might seem, it represented an effort at self-control on Fernow's part. His first reaction to Morton's order had been to demand that Morton "consider at once my resignation . . . as I do not desire to hold a position as clerk subject to the dictation of any indiscreet underling that may from time to time be invested with such unheard of authority."[59] He reconsidered and sent a milder version.

Fernow continued to protest what he believed to be a reduction in status. An exasperated Morton asked the attorney general to rule on the relation between the chief clerk of the department and the chiefs of divisions. Hoping to settle the matter, the secretary told Fernow that he had confidence in the chief forester's ability to "accommodate himself to the attorney general's interpretation cheerfully and manfully." But Fernow was not satisfied with the interpretation and pleaded with Morton to make a final ruling. Fernow acknowledged Morton's authority to subordinate an officer to a clerk but could not believe that he really intended to do so. Fernow wrote the secretary that surely he did not mean to "degrade" him and "thereby compel the retirement" of Fernow and other division chiefs. Morton, showing great patience, responded that the attorney general's interpretation had clearly specified the relations between the chief clerk and division chiefs. He asked Fernow to specify ambiguities in the interpretation, but the forester let the matter rest.[60]

Despite his petty disputes over recognition and status, Fernow carried on an admirable technical program. As we have seen, he began his term as chief of the Division of Forestry by inviting men already engaged in forestry studies to participate in the federal program. In

58. Morton to Division Chiefs, Jan. 1, 1894, and Fernow to Morton, n.d., Fernow Papers, FHS.

59. Fernow to Morton, n.d., draft in Fernow Papers, FHS.

60. Morton to Fernow, Feb. 1, 1894, Fernow Papers, FHS; Fernow to Morton, Feb. 4, 1894, ibid.; Morton to Fernow, Feb. 5, 1894, ibid.

addition to the work accomplished by these field agents, Fernow and his staff produced many worthwhile contributions. To use his own terms, during Fernow's twelve-year administration the Division of Forestry published approximately six thousand pages of technical material with a total appropriation of $230,000, working out to be about twenty-four dollars per page. Not a bad price considering the values involved, calculated Fernow.[61]

Publications of the division covered a range of topics, reflecting Fernow's particular interests and the important uses of wood during that period. The drain on forests by railroads was the subject of several monographs; timber physics (the mechanical properties of wood, a favorite of Fernow) occupied many pages of the six thousand total. Fernow's 1893 study on the relation of forests to climate and water supply, a valuable contribution to the subject, included a history of rainmaking lore and experimentation. He had no wish to get involved with weather modification, but Congress insisted. When he found that a favored rainmaking prescription was to shake water loose from clouds with cannon fire, Fernow managed to involve the Army Signal Corps and returned to research he deemed more suitable.[62]

There were other important contributors to scientific forestry literature. Edgar Ensign reported on the forests of Colorado, Abbot Kinney on southern California, and Filibert Roth on the forests of Wisconsin. V. M. Spalding produced a valuable monograph on eastern white pine, Charles Mohr on the pines of southeastern United States, and George Sudworth on tree nomenclature. Sudworth would continue to produce important dendrological works. Frederick V. Coville made seminal contributions toward understanding the forest range. This representative list could be longer, but it is impressive enough in abbreviated form.

Within the Department of Agriculture, however, not all were satisfied with Fernow's accomplishments. Over the years he had been saddled with secretaries disinterested in substantive investigations, preferring instead that the Division of Forestry send seed packets to mollify congressional constituencies or engage in rainmaking experiments. Fernow's timber-testing studies, painstakingly conducted, drew criticism because he published only after acquiring large quan-

61. B. E. Fernow, *Report upon the Forestry Investigations of the Department of Agriculture, 1877–1898*, U.S. Congress, House Doc. no. 181, 55 Cong. 3, 1899, p. 7.

62. For a complete list of Forestry Division publications, see Fernow, *Report upon Forestry Investigations*, pp. 40–44. Rodgers, *Fernow*, pp. 148–49, reviews the rainmaking episode.

tities of data. Publish more quickly, he was told by Morton's assistant secretary. Morton himself, although he had supported the project, eventually thought an extensive timber-testing study to have been impractical experimentation.[63]

No longer secretary, Morton got in one last jibe at Fernow, referring to him as one who was "presumed to know something of the theories of European forestry." Fernow heard of this criticism and was naturally hurt. He told Morton that he had known him to be "thoroughly inconsiderate, injudicious and irresponsible" but, using a Prussian twist of phrase, had always believed him to be "fair, just and sincere." Fernow was grieved to be mistaken. Typically, he offered Morton the opportunity to apologize. Receiving none, he knew "what to think of you and this, I suppose, ends the matter."[64]

Fernow received another slight when he summarized his efforts of the preceding twelve years, a substantial contribution to professional literature in its own right. In his letter of transmittal to the voluminous report, the new secretary of agriculture, James Wilson, called special attention to the fact that Fernow's successor was working in "distinctly different channels." He added, "These plans meet with my full approval."[65]

This successor who was charting new courses was, of course, Gifford Pinchot. His selection must have surprised Fernow, who as late as mid-April 1898 believed that his assistant, Charles Keffer, would be named to replace him.

Meanwhile Fernow had drafted a bill for the New York legislature to establish a forestry school at Cornell University with a 30,000-acre experimental forest, and on April 15, 1898, he was elected director of the new college. He was jubilant. "It is my intention to take the timber physics work with me and I hope by and by it will thrive to the glory of another institution than the one in which I have tried in vain to bring it to recognition."[66]

At Cornell, Fernow began the first professional forestry school in America in the fall of 1898. Within a few years, controversy prompted the New York State governor to veto the program. Whether Fernow's

63. Charles Dabney to Fernow, Nov. 30, 1896, RG 95–1, RFS; Morton to R. L. O'Brien, April 25, 1898, Fernow Papers, FHS.

64. Fernow to Morton, May 10, 1898, Fernow Papers, FHS.

65. Wilson to the President, Jan. 24, 1899, printed in Fernow, *Report upon Forestry Investigations*, p. 3.

66. Fernow to Charles W. Garfield, Apr. 18, 1898, RG 95–2, RFS; Fernow to W. M. Hays, Apr. 19, 1898, ibid.; Fernow to S. T. Nelley, Apr. 18, 1898, ibid.

application of forestry to state lands caused the cancellation is still a matter of debate. In 1907 Fernow moved to the University of Toronto as head of its newly organized forestry faculty. He retired from teaching in 1919. Between 1903 and 1916 he was editor of *Forestry Quarterly*. In 1917 the *Quarterly* merged with the *Proceedings* of the Society of American Foresters to become the *Journal of Forestry*, of which Fernow was editor-in-chief until 1923. In addition to his editorial achievements, he published over 250 articles and bulletins and three books. His *Economics of Forestry*, which appeared in 1902, clearly demonstrates a sophisticated grasp of modern forestry concepts. *A Brief History of Forestry*, in three editions, although dated in some respects, has yet to be fully replaced.[67]

Fernow's influence in key legislation cannot be disparaged, nor can his contributions to technical forestry subjects. He was a cultured, highly educated scientist, out of place in rustic America with its partisan politics and spoils system. His oversensitivity to real or imagined slights adversely affected his ability to function as chief of the Division of Forestry. But despite accusations that he advocated adoption of European forestry methods in America, Fernow understood the forestry needs of the time and set out to fulfill them.

Fernow has suffered much neglect and abuse at the hands of those wishing to give his successor credit for nearly every early advance in American forestry. Pinchot contributed significantly to his own reputation by diminishing Fernow's. At his retirement from teaching, Fernow must have been pleased to receive widespread praise. Two letters in particular were especially apt. Forestry Professor Ralph C. Bryant of Yale University wrote: "No other man has been such a potent force in the advancement of forestry in this country and the wonderful foundation laid by you will always endure." Raphael Zon, a fellow immigrant whom Fernow had fostered, was unabashed in his praise: "You have been more than a teacher of forestry; . . . you were a leader of life."[68] Fernow died on February 6, 1923.

67. B. E. Fernow, *Economics of Forestry* (New York: Thomas Y. Crowell, 1902); Fernow, *A Brief History of Forestry in Europe, the United States and Other Countries* (Toronto: University of Toronto Press, 1907, 1911). He also wrote *Care of Trees* (New York: Henry Holt, 1910).

68. R. C. Bryant to Fernow, Sept. 17, 1919, Fernow Papers, Cornell University; Raphael Zon to Fernow, Aug. 15, 1919, ibid.

CHAPTER III

Growth and Cooperation

WHEN GIFFORD PINCHOT succeeded Bernhard Fernow as chief of the Division of Forestry in 1898, he was no stranger either to the man or the agency. The two first met in 1889 when Pinchot was a student at Yale. In order to learn something of forestry, he offered to work for Fernow at no salary. Fernow accepted, advising the young man to make forestry his secondary, not primary, course of study. Then the chief reconsidered and invited him to continue with forestry. Back at Yale, Pinchot launched into a term paper on American lumber supply and within two months he had decided that "forestry is my meat."[1]

That June Pinchot graduated and went off to study European forestry policies and practices. Fernow invited him to cut his travels short and return to be his assistant chief. The job would be largely office work, paying $1,600 per year. Pinchot would have to "work like a beaver to make up for deficiencies in knowledge and experience." Sorely tempted, Pinchot turned to Dietrich Brandis, a distinguished German forester, for counsel. Following Brandis' advice, Pinchot accepted but asked for the appointment to be deferred for a year. The next day he received a letter from Charles S. Sargent, who advised against accepting, because in his opinion the Division of Forestry and Fernow did not have satisfactory standing in American forestry.[2]

1. Gifford Pinchot, Diaries, Jan. 5, Feb. 25, March 11, 1889, Gifford Pinchot Papers, Library of Congress. Pinchot provides a description of his undergraduate education in his autobiography, *Breaking New Ground* (1947; reprint ed., Seattle: University of Washington Press, 1972, pp. 3–4. His father had first suggested forestry as a career in 1885 (*Breaking New Ground*, p. 2).

2. Gifford Pinchot to James W. Pinchot, Aug. 3, 1890, box 10, Pinchot Papers; Pinchot, Diaries, Aug. 1, 1890.

Pinchot stayed on in Europe, enrolling in L'Ecole Nationale Forestière at Nancy, France. He studied formally, then toured European forests under Brandis' wing. After thirteen months, Pinchot was impatient to return home and put some of his new knowledge to work. He arrived in time to attend the December 1890 meeting of the American Forestry Association, where he reported on European forest policy.[3]

Pinchot met with Fernow, who was still holding the assistant chief position open for him. He got the impression that Fernow preferred being the head of a scientific bureau rather than of one which actually administered forests. Pinchot thought that Fernow needed his "connections and acquaintances" to obtain forestry legislation. They agreed to travel through the southern states to observe forest conditions. Pinchot, regarding Fernow as thin-skinned "but of sterling worth," quickly lost patience with his "endless self-appreciation at the expense of others." He decided that Fernow was a "very queer man." Pinchot nonetheless planned a brief trip to Europe, with a return in July 1892 to become Fernow's assistant. He stayed in the United States, however, as an opportunity soon appeared to practice forestry in another way.[4]

Near Asheville, North Carolina, George Vanderbilt's Biltmore Estate sprawled across thousands of acres of forested and cutover land. Through Frederick Law Olmsted, who was landscaping the estate, Pinchot was hired to be the resident forester. It was Pinchot's duty to manage these lands profitably and develop a forested estate worthy of Vanderbilt's wealth.

Fernow was disappointed that the young man had decided against public service. To Pinchot, however, the Vanderbilt estate offered an opportunity to implement Brandis' suggestion that demonstration forestry was vital. If landowners could see examples of good forestry practices and also see that forestry was a profitable venture, then they would voluntarily adopt approved management methods. Pinchot told Fernow that he was sorry he had decided not to continue with the Division of Forestry: "You have been so kind toward me in many ways. . . ."[5]

For three years Pinchot managed the Vanderbilt estate forests. He

3. Pinchot, *Breaking New Ground*, pp. 10–22.

4. Pinchot, Diaries, Jan. 1, 1891, Dec. 31, 1890, Feb. 3, 1891, March 14, 1891.

5. Pinchot to Fernow, Jan. 31, 1892, Bernhard E. Fernow Papers, FHS, Santa Cruz, Calif.; Pinchot, Diaries, Feb. 5, 1892.

Forestry crew in the field, Santa Fe National Forest, New Mexico, 1907. Courtesy of the U.S. Forest Service

Field barber shop, Carson National Forest, New Mexico, 1911. Courtesy of the Forest History Society

Inspecting the Carson National Forest, New Mexico, 1911. Left to right: Aldo Leopold, deputy forest supervisor; Ira T. Yarnell, forest assistant; H. C. Hall, forest supervisor. Courtesy of the Forest History Society

Ranger saddling up in the Uncompahgre National Forest, Colorado, 1915. Courtesy of the U.S. Forest Service

thought his efforts successful, in terms of both improving the forests with an eye on profit and furthering his own education. As with his forestry education in Europe, however, Pinchot was impatient for greater challenges. He branched out into forestry consulting, developing a management plan for a 40,000-acre tract in New York State.[6] Active in the affairs of the American Forestry Association, he kept in close touch with Wolcott Gibbs and Charles Sargent and soon was involved in plans to create a forestry commission within the National Academy of Sciences, which would prove to be so important to the law of 1897.

Pinchot's personality held fascinating quirks. John Muir, founder of the Sierra Club in 1892, was one of his companions during portions of the western trek for the NAS commission. Muir's journal entry for August 30, 1896, describing their excursion through the Crater Lake country of Oregon, reads: "Heavy rain during the night. All slept in the tent except Pinchot." Muir, of course, was no city-bred weakling himself. One of his most famous exploits was to lash himself to the top of a Douglas fir to experience the full fury of a Sierra storm. Pinchot's insistence upon exposing himself to the elements was typical. Born to wealth, he prided himself on physical prowess. His diary entries commonly refer to his lack of stiffness following a day on horseback. Marksmanship was a source of pride; Pinchot proudly recorded three bullseyes out of five shots with a handgun at fifty yards. When during his western travels he met artist Thomas Hill, renowned for capturing scenes of the Sierra, Pinchot remembered not Hill's paintings but the "splendid" grizzly skin he had in his studio.[7]

Virtually obsessed with a drive to achieve, Pinchot would chastise himself for relaxing when there was anything to be done. "Wasted evening reading" appears in his diary at the end of an account describing a full day of meetings. He seemed nearly as concerned about how he appeared to others. A frequent entry following social events was, "made an ass out of myself." Morbidity was a trait, too. When his fiancée, Laura Houghteling, died after a long illness in 1894, Pinchot's grief continued almost without end. As the years went by, he still felt

6. Pinchot, *Breaking New Ground*, pp. 47–48. For an appraisal critical of Pinchot's craft as a forester, see Carl Alwin Schenck, *The Biltmore Story: Recollection of the Beginnings of Forestry in the United States* (Saint Paul: American Forest History Foundation of the Minnesota Historical Society, 1955 [reprinted as *Birth of Forestry in America: Biltmore Forest School, 1898–1913*, by FHS, 1974]).

7. John Muir, *John of the Mountains: The Unpublished Journals*, ed. Linne Marsh Wolfe (Boston: Houghton Mifflin, 1938), p. 356; Pinchot, Diaries, May 7, 1891.

Laura's presence: good days were when "My Lady" seemed close, "blind days" when his mental picture of her was dim. Two decades later he was still noting the anniversary of her death. Finally at age forty-nine, Pinchot married Cornelia Bryce.[8]

Confidential Agent

After the National Academy of Sciences report had prompted President Cleveland's precipitous proclamation reserving 21 million acres, which were suspended for nine months, McKinley's secretary of the interior wanted to know more about the reserves. Cornelius Bliss asked Pinchot to make a study of the reserves, but Pinchot balked at the title of special agent. Having served on the NAS commission, Pinchot preferred to do the job without fanfare. As confidential forest agent, following instructions he wrote for himself, he was to examine and report on the suspended reserves and propose a structure for a forest bureau in Interior.[9]

In his report Pinchot claimed that the most important fact about the reserves was a "profound change now taking place in public opinion." He believed that the initial protest against Cleveland's "Washington's Birthday" reserves had died down. The new regulations for use had been instrumental in the softening of public opinion. He described problems stemming from imprecise boundaries, fire danger, stream flow, and mineral lands. Pinchot suggested that it would be wise to make agricultural lands included within reserve boundaries available to farmers. He supported retention of existing reserves but felt that exchanges and reopening for entry were needed to deal with agricultural, mineral, and railroad lands. Pinchot saw the administration of forest lands as being significantly different from other types of government work. A forest's longevity, its susceptibility to fire and "injudicious handling," and the need for management distant from centers of government were the key elements of difference. He strongly recommended adoption of three goals: (1) permanent tenure of forest land, (2) continuity of management, and (3) the permanent employment of technically trained foresters. This last point, professionalism, was crucial. Without trained foresters "the high standard of fidelity,

8. Pinchot, Diaries, 1894, *passim*; M. Nelson McGeary, *Gifford Pinchot: Forester–Politician* (Princeton, N.J.: Princeton University Press, 1960), pp. 32–33, 250.

9. Pinchot, *Breaking New Ground*, p. 123; Diaries, June 5, 16, 1897.

honesty, and ability" required to manage the forests would be lacking.[10]

Secretary Bliss had specifically instructed Pinchot to propose an organization to take charge of the reserves. The need to begin modestly prompted Pinchot's suggestion of an initial seventy-thousand-dollar budget, one-half of which would be used for salaries. The organization would be headed by a chief forester, who would supervise seven districts. District foresters would live on or near the reserves under their supervision and make frequent inspection trips.[11] This basic organization plan is strikingly similar to the agency that eventually appeared in the Department of Agriculture and equally similar to the organization Fernow described twelve years before. Pinchot and Fernow shared the belief that the wide spectrum of natural and market conditions found in the various forested regions demanded on-site supervision and that uniform rules administered at great distances would not work well.

Concurrently with Pinchot's assignment, Bliss ordered a more detailed examination and description of all the reserves, suspended or not. The rider to the 1897 Sundry Civil Appropriations Act had authorized the U.S. Geological Survey to make the study, under instructions from the secretary of the interior. A Division of Geography and Forestry was set up to carry out the inventory and mapping. Henry Gannett, Fernow's old antagonist, who had once claimed that the forests were not in danger, headed the new division. Fernow expected to be asked to work on the survey and turned down a chance to study the forests in Arizona to avoid a conflict. Assignment to survey the reserves eluded Fernow as disappointingly as had membership on Sargent's National Academy of Sciences commission. Fernow noted the irony of having Gannett inventory the reserves and thus provide the best data to show that the geographer had been wrong ten years earlier.[12] Despite Fernow's ruffled feathers, Gannett's surveys were of high quality and provided basic information necessary for effective management of the reserves. The three quarto volumes, plus an atlas, are impressive even today.

10. U.S. Congress, Senate, *Surveys of the Forest Reserves*, Sen. Doc. 189, 55 Cong. 2, March 15, 1898, pp. 35–39, 43; Pinchot, *Breaking New Ground*, p. 129.

11. Senate, *Surveys of the Forest Reserves*, p. 43.

12. Fernow to James Toumey, May 18, 1897, Letters Sent by the Division of Forestry, 1886–99, Record Group 95 (hereafter cited as RG 95–2), RFS, National Archives; B. E. Fernow, "Outlook of the Timber Supply in the United States," *Forestry Quarterly* 1 (Jan. 1903): 42.

Aiming a barb in Fernow's general direction, Gannett observed that for nearly a generation, agitation for protecting forests had grown in some persons to become "almost a religion." Tongue in cheek, he thought it strange that there was "practically no knowledge to serve as a basis for such a cult." In a more serious vein, Gannett looked at the forest from two perspectives, as a supply of lumber and "as a physical factor with effects upon climate, erosion, and the flow of streams." As a lumber supply, forests, except for their long maturity time, were analogous to agricultural crops. Timber supplies could be renewed, unlike mined resources. Explaining that forest study until then had been largely botanical ("the geographic and economic sides of the question have received very little attention"), he described the forests and their associated industries state by state and estimated the reliability of his figures. Gannett emphasized that his data were only estimates, and estimates of the same area could vary from man to man. Changing utilization standards and improved inventory techniques accounted for this variation. Because of variation, Gannett predicted that "twenty-five years hence [an inventory] will doubtless show twice as much timber in the same area as if made today."[13]

The volumes provided much more than statistical accumulations about sawmills and timber volumes. Included, too, were descriptions of topography, climate, agricultural potential, and geology. Abundant plates and hundreds of large-scale, color-coded maps enabled the reader to visualize written descriptions. The study was completed in time for the 1900 census, which showed 1,390 billion board feet of timber in America. As Gannett had predicted, this amount was substantially greater than earlier estimates, such as Sargent's 856 billion board feet, excluding Douglas fir and ponderosa pine, twenty years earlier.[14] Future estimates could be larger yet, but policymakers now had a realistic view of conditions on the forest reserves and forest land generally.

Chief of the Division

Although he could not match Fernow's technical forestry knowledge, Pinchot by now was well versed for the time and a logical

13. Henry Gannett, *Nineteenth Annual Report of the United States Geological Survey to the Secretary of the Interior*, pt. 5: *The Forest Reserves* (Washington, D.C.: GPO, 1899), pp. 1–2, 14–15.
14. Samuel T. Dana, *Forest and Range Policy* (New York: McGraw-Hill, 1956), p. 192.

choice to be Fernow's replacement. He had been able to make two excursions through western forests, once for the National Academy of Sciences commission and once for Bliss. He had studied forestry in Europe, managed Vanderbilt's private estate, and worked as a consulting forester. He was active in the American Forestry Association and knew the leaders in the field. On July 1, 1898, at the invitation of Secretary of Agriculture James Wilson, Gifford Pinchot became chief of the Division of Forestry. Charles Keffer, Fernow's assistant chief, stayed on for a short time to aid the transition.[15]

Pinchot at first had refused Wilson's offer, but after consulting with Charles Walcott, Hague, Gannett, and William H. Brewer of Yale, all of whom advocated acceptance, he reconsidered. When his close friend and associate, Henry S. Graves, agreed to be assistant chief, Pinchot took on the new position. He would be chief of one of eighteen divisions in the Department of Agriculture. He began with a staff of sixty, out of a total department of nearly 2,500.[16]

Division headquarters was two rooms on the third floor of the Department of Agriculture building, and a small space in the attic. A dozen men crowded their desks into the largest room. On top of Assistant Chief Graves's desk a big sign ordered "Silence." Pinchot's small office was only semiprivate, as anyone coming down from the attic could see through the transom what was going on.[17] The modest surroundings failed to dampen the enthusiasm of Pinchot and his new staff.

Wilson had offered Pinchot freedom to operate and staff the agency as he saw fit. The secretary introduced Pinchot to the public in his first annual report by saying, "Mr. Pinchot is planning to introduce better methods of handling forest lands in public and private ownership, the private owners paying the expenses of Department agents who give instructions." After only three months in office, Pinchot

15. Keffer to G. Fred Schwartz, July 11, 1898, RG 95–2, RFS.

16. Pinchot, Diaries, May 10, 11, 12, 17, 1898; Harold U. Faulkner, *Politics, Reform and Expansion, 1890–1900* (New York: Harper and Row, Torchbook ed., 1963), p. 64. His fifth day on the job, Pinchot requested that his title be changed from chief to forester, as there were many chiefs in Washington but only one forester. Wilson soon complied and the head of the agency was the forester until 1935, when chief was readopted.

17. Graves's official title was superintendent of working plans. Ralph S. Hosmer, "Early Days in Forest School and Forest Service," *Forest History* 16 (Oct. 1972): 8; Bureau of Forestry, *Annual Report* (1901), p. 326. In 1901 the agency moved to the seventh floor of the Atlantic Building on F Street.

issued Division of Forestry Circular Number 21 to implement Wilson's promise.[18]

Circular 21 offered technical advice to lumber companies, giving consideration "to the present interest of the owner" and to forest protection. The program was immediately popular. Redwood lumbermen responded at once by sending one thousand dollars and promising free transportation and board for federal forestry agents. There were 122 other applications for assistance from 35 states, involving over 1.5 million acres of private land. Every year applications for assistance under Circular 21 came in; by 1905 Pinchot reported that nearly 11 million acres of private lands had been or were being studied.[19]

A good example of Circular 21 in operation is the Kirby Lumber Company in Texas, which applied for assistance in 1902 and contributed 1.2 million acres to the 11 million total. The Kirby Company, formed in 1901 through the consolidation of fourteen smaller firms, held cutting rights to Houston Oil Company lands, which contained 80 percent of the longleaf pine in Texas. To Kirby, it would be good business as well as good public relations to market the mature timber and still maintain the forest's capacity for growth.[20]

Thomas H. Sherrard of the Bureau of Forestry led a fifty-man data-gathering party into the Kirby forests. Divided into four groups, the men lived in camps provided by the company. Sleeping on cots in two-man tents, the men had to supply their own blankets, perhaps in keeping with lumber camp traditions. Wood-frame offices were available for map preparation and the compilation of weekly reports. For the working plan itself, the crews divided their studies into five parts. Kirby needed to know the growth rate of longleaf pine; appraised valuation of the vast stands; growth characteristics of the forests; degree of danger from fire, wind, insects, and disease; and of course they needed good maps.[21] With these data, Kirby could make rational

18. Pinchot, *Breaking New Ground*, pp. 135–36; USDA, *Annual Report* (1898), p. x; Pinchot, *Practical Assistance to Farmers, Lumbermen, and Others in Handling Forest Lands*, U.S. Division of Forestry Circular no. 21 (hereafter cited as Circular 21) (Washington, D.C.; GPO, 1898).

19. Circular 21, p. 1; Chief, Division of Forestry, *Annual Report* (1899), pp. 93, 96; *Annual Report* (1905), p. 210. For another account see Harold T. Pinkett, *Gifford Pinchot: Private and Public Forester* (Urbana: University of Illinois Press, 1970), pp. 47–50.

20. Houston *Daily Post*, Dec. 22, 1902.

21. Ibid.; Overton Price and Thomas H. Sherrard, "Report upon the Preliminary Examination of the Tract of the Kirby Lumber Company and the Houston Oil Company," unpublished typescript, 19[03?].

plans to manage the forests profitably and still protect the resource—forestry's two basic tenets.

Field work needed to prepare working plans was completed in Maine, New York, Michigan, South Carolina, and Tennessee in 1902, the same year Sherrard began work on the Kirby holdings. Pinchot billed the landowners $9,040.86 to pay the expenses of the field crews. As with Kirby, the cooperating firms received recommendations on minimum tree size for logging, which trees to have as a seed source to assure reforestation, a system of marking for cutting that would leave the residual forest healthy and increase its value, and a fire-protection plan.[22]

The cooperative program continued until 1909. Pinchot justified the effort in terms of public interest: since lumbering was the nation's fourth largest industry, then obviously its maintenance was vital to the national economy. He assured his constituents that he was "not expending public money to benefit private interests." To prove the point he had his associate, Overton Price, calculate how the bureau's money was spent. Price's report supported Pinchot's contention that expenditures on public lands increased over fourfold while money for cooperation with industry dropped by a third between 1902 and 1904. In 1907 Pinchot reported that 75 percent of the cooperators had "adopted the plans laid down for them."[23]

Opposition to public help for industry did not by itself cause this program to be gradually phased out. Much more important was the dramatic increase in the bureau's responsibility for public lands following the 1905 transfer of the forest reserves from the Department of the Interior to the Department of Agriculture. Also, the industry, distracted by a faltering economy, found it less appropriate to propose long-term investments. In retrospect, perhaps the greatest value of the program was not the number of acres brought under some degree of management but the cooperative experience gained by the agency and landowners.

Cooperation, of course, was not restricted to the lumber industry. Federal foresters worked with state agencies and other federal

22. USDA, *Annual Report* (1902), pp. 113–14.

23. Ibid. (1903), p. XLVI; Forest Service, *Annual Report* (1907), p. 29; Overton Price to Pinchot, Jan. 13, 1904, General Correspondence of the Forest Service, 1898–1908, Record Group 95 (hereafter cited as RG 95–3), RFS. In 1900 Henry Graves left the Division of Forestry to be the dean of the newly founded forestry school at Yale; Price became second in command. The Pinchot family had been instrumental in funding the school.

bureaus. Nor was cooperation limited to managing forests; forestation was also of interest. Circular 22, issued in 1899, explained a tree-planting program parallel to the forest management assistance offered under Circular 21. The Forest Service would provide the plan, but the landowner had to furnish seeds or trees and the manpower to do the planting. By 1904, 334 plans provided planting schedules for over thirteen thousand acres—a relatively small area, perhaps, but the cooperative process itself added to a building tradition.[24]

Cooperation with Interior

Pinchot was limited to cooperative projects and publishing technical reports and bulletins, at least within the framework of the Department of Agriculture. His agency was in Agriculture, but the forest reserves still lay under the jurisdiction of the Department of the Interior. If he wished to manage public forests, he had to establish some sort of working relationship between the two departments.

Since 1891 the Department of the Interior had been faced with the responsibility of administering the forest reserves. The secretary had advocated increased authorization to manage and protect the reserves, as had many others. Until the 1897 Sundry Civil Appropriations Act provided the enabling authorization, Interior handled the reserves pretty much like other timbered land in the public domain. Agents from Division P tried to combat trespass and fraud, both off and on the reserves. Statistics reveal how futile such efforts had been.

After 1897 the Department of the Interior turned more and more to the Division of Forestry in the Department of Agriculture. When Secretary Bliss appointed Pinchot confidential agent and asked him to examine the forest reserves and recommend a structure for a management agency, Pinchot's design contained the basic elements found in today's Forest Service. By 1901 the Agriculture and Interior secretaries had worked out a formal agreement on the reserves. Interior would continue to patrol the reserves and enforce laws and also provide the routine office work. The foresters in Agriculture were responsible for examining the reserves, making all technical decisions, and administering whatever plans they developed. Interior would not pay for these considerable services but pledged prompt and cheerful

24. Division of Forestry, *Annual Report* (1899), p. 97; Bureau of Forestry, *Annual Report* (1904), pp. 188–90.

assistance from its field agents. Pinchot would report on the reserves directly to the secretary of the interior. Not the ideal solution, Secretary of the Interior Ethan A. Hitchcock noted, but a satisfactory one—"by far the best solution of the problem which can be reached this year."[25]

Pinchot's ego caused unpleasant moments. He wanted to sign correspondence concerning the reserves, but instead was allowed only to initial the paperwork, under the signature of General Land Office Commissioner Binger Hermann. The forester bridled at the apparent slight, and President Roosevelt had to intervene. In a terse, personal letter, Roosevelt told Pinchot of his recent meeting with Secretary Hitchcock. He had been assured that Hermann "absolutely and without reservation acquiesces" in giving Pinchot a free hand on reserve policies. Roosevelt promised his forester that he would "have exactly the same freedom as though you were the independent head of the bureau." The matter of initials versus full signature seemed "utterly unimportant" to the president. Results, not procedures, were vital, and Roosevelt lectured Pinchot on the dangers of making issues out of trivial matters. Finally, he held out bait that Pinchot could not ignore: "The condition of affairs is not expected to be permanent. It is expected to be a transitional stage. . . ."[26]

Chastised, Pinchot explained that he had believed being able only to initial correspondence would reduce his effectiveness, but the president's specific assurances would greatly strengthen his position. Also, he told Roosevelt that he was gratified by the promise of impending changes, as Secretary Hitchcock had led him to believe no immediate reorganization was planned. Pinchot then wrote Hitchcock to acknowledge that he would accept the initialing procedure but pointedly summarized Roosevelt's commitments. The president had promised him, Pinchot emphasized, "the same freedom as though I were the independent head of the Bureau, without one shadow of difference." He also pointed to the presidential assurance of a transitional situation "leading to a permanent establishment."[27]

Two weeks later, Pinchot followed up with a long letter to Hitchcock, outlining the "principles and practices . . . that should govern the administration of the National Forest Reserves." Pinchot saw an

25. Hitchcock to James W. Wadsworth, Jan. 9, 1901, box 580, Pinchot Papers.
26. Theodore Roosevelt to Pinchot, Oct. 18, 1901, ibid.
27. Pinchot to Roosevelt, and Pinchot to Hitchcock, Oct. 19, 1901, ibid.

urgent need to extend the forest reserve system in a major way that winter. To wait even until spring would mean potential reserve sites might be pre-empted in the usual rush of land claims following the snow melt. He also recommended exerting every effort to gain the good will of those living on or near the reserves, for their help would be essential in fighting and preventing forest fires. Pinchot also outlined specific policies on grazing and timber management.

Pinchot advocated more flexibility in management, as the reserves represented diverse conditions. To him, uniform rules were destructive. A major point was that field work, not office reports, should be given priority. Office personnel should "interfere" as little as possible with the important work of the agency. And finally, Pinchot insisted that "every effort should be made to create an esprit de corps"—good work should be followed by promotion, and men in the field should know that the department stood behind them.[28]

Early in 1902 the Department of the Interior issued a manual on administrative procedures and policies for the reserves. Its content and style strongly suggest that Pinchot wrote it, although Pinchot gives credit to E. T. Allen. Protection of timber and regulation of water, referring to the 1897 law, were listed as the two principal reasons for reserves. Timber was to be grown on land unfit for agriculture, and water flow was to be influenced by slowing snow melt with shading, windbreaks to cut drying action, reduction of erosion, and maintenance of the absorptive qualities of soil. The manual admonished: "Keeping in mind the object and purpose of the reserves and their forests, it is clear that the first and foremost duty of every forest officer is to care for the forest, and every act, every decision he is called upon to make should be guided by the thought, will it improve and extend the forest?"[29]

The manual explained that farming on reserve lands better suited for agriculture was desirable; that prospecting and mining were not prohibited; that roads, trails, and irrigation canals could be built by permit only; and that schools and churches could be constructed on public land. Grazing, the rangers were reminded, could be forbidden if damage to the reserve was probable. Regulations prohibited grazing until after it had been shown that no damage would occur. Although

28. Pinchot to Hitchcock, Oct. 31, 1901, drawer 283, Research Compilation File, 1897–1935, Record Group 95, RFS; letter reprinted in Pinchot, *Breaking New Ground*, pp. 194–95.

29. USDI, *Forest Reserve Manual for the Information and Use of Forest Officers* (Washington, D.C.: GPO, 1902), p. 3; Pinchot, *Breaking New Ground*, p. 264.

cattlemen applied for grazing permits individually, sheepmen who were members of a woolgrowers' association could receive their allotment from the association. Sheep had to be kept in flocks on the federal range, but cattle were allowed to roam at large.[30]

The bulk of the manual dealt with timber management. Significantly, the section began with regulations on free use of timber. Free use for legitimate petitioners was a tradition in public land policy, both as a means to reduce western opposition and to stimulate settlement in the region. Free use was "a privilege, and not a right." Settlers and others who lived on or near reserves were eligible, but corporations, sawmills, or other "large establishments" were not.[31]

Sale procedures were described in detail. The two basic criteria were that removal of timber would be limited to that which would benefit the forest ("or at least not [be] detrimental") and that "local demand will have preference." This latter policy bears Pinchot's imprint, as it was central to his philosophy. The sale itself was to be initiated by the purchaser, who had to locate and describe the timber. Next, the government forester would inspect the area and, if appropriate, mark the trees or area to be cut. Then the sale was advertised and the public invited to bid. The original applicant, who did all of the basic work setting up the sale, bid along with his competitors. The highest bidder purchased the timber at the bid price. Those who logged without benefit of permit were guilty of trespass. Willful trespassers had to pay the gross value of the timber, but those guilty of unintentional trespasses were allowed to deduct expenses incurred during logging. As willful trespass was obviously difficult to prove, most of those apprehended with unmarked timber paid the going rate. Admittedly not much of a risk: if caught, you paid the purchase price; if not, the timber was free.[32]

Punishment for grazing trespass was ambiguous. The manual stated that any person herding stock without permit was "guilty of violating the rules of forest reserves. . . ." Apparently the forest officer could only report grazing trespass to his superior and use persuasion against any repeated offense, such as threatening to not award a permit. Considering the fact that the trespasser, by definition, was a stockman who preferred operating without a permit, this threat was weak.

30. USDI, *Forest Reserve Manual*, pp. 4–12. Woolgrowers' associations later lost their right to make allotments; see USDI, *Annual Report* (1903), p. 320.

31. USDI, *Forest Reserve Manual*, p. 12.

32. Ibid., pp. 14–26.

When an officer proposed accepting a fine for grazing trespass, however, Pinchot refused because he believed that "it would not be best to establish a precedent of condoning a trespass by accepting a small payment for the grazing thereby obtained." Such a procedure, he reasoned, would only encourage future trespass. Outright prohibition of grazing ended the matter.[33]

Division R

For a brief period, the Department of the Interior tried to develop its capability to manage the reserves. The General Land Office was to administer funds appropriated by the 1897 Sundry Civil Appropriations law. Filibert Roth, who began with the Division of Forestry in 1892 and moved to Cornell with Fernow in 1898, became chief of Division R—the Forestry Division—on November 15, 1901. Roth was joined by five men from Pinchot's Bureau of Forestry, three for the field and two in the office. Pinchot saw Division R as an opportunity to manage the reserves without being held directly responsible.[34]

Even with Roth as chief of the division, Pinchot still wielded direct control, as well as running his own agency in Agriculture. Writing for Hermann's signature, he crackled orders to Division R men in the field. He cautioned one about a report that showed too many different men working under a free use wood permit. He scolded: "It appears, also, that you have no correct information as a forest officer should have, and would have with diligent and proper oversight of the cutting going on in his jurisdiction. . . ." Another officer was warned about appointing the son of the largest cattleman in the district to be a ranger. Pinchot admonished still another about imposing improper charges for services and ordered him to make immediate refunds.[35]

Locally, reserve supervisors were absorbed with preparing routine vouchers on fire-fighting costs, issuing grazing permits, and making

33. Ibid., p. 26; Binger Hermann to D. S. Marshall, Jan. 14, 1903 (Pinchot wrote the letters for Hermann's signature), Records of Division R, Record Group 49 (hereafter cited as RG 49), USDI, National Archives. Pinchot may have been alluding to a precedent whereby federal officers condoned systematic theft of public timber by arranging with the lumberman for routine payment of the pre-agreed fine. See Ivan Doig, "John J. McGilvra and Timber Trespass," *Forest History* 13 (January 1970), pp. 6–17.

34. James Pinchot, Nov. 1, 1902, box 581, Pinchot Papers. For a brief account of the work of Division R see Filibert Roth, "Administration of U.S. Forest Reserves," 2 pts., *Forestry and Irrigation* 7 (June 1902): 191–93, and (July 1902): 241–44.

35. Binger Hermann to Harrison White, Jan. 15, 1903, and Hermann to Willis M. Slosson, Jan. 28, 1903, RG 49, USDI.

timber sales. Ranger-in-charge Leon F. Kneipp (later to be an assistant chief) of the Pecos Forest Reserve notified an applicant that the commissioner had refused to issue a grazing permit, except during the grazing season of May through December. "You will therefore remove your stock at once" or lose all privileges. On another matter, Kneipp told his superior of the extremely hazardous fire season and complained that it was impossible to patrol the Pecos with only three rangers. He eyed the Gila Reserve with its thirteen rangers and suggested that redistribution was merited.[36]

Division R failed to satisfy Pinchot. In his view, Roth struggled vainly against "land office routine, political stupidity, and wrong-headed points of view." No one, according to the outspoken chief of forestry, could overcome the "tide of crookedness and incompetence" found in the Land Office. Other less dramatic reasons contributed as well to the problems of achieving a professional career in Division R. One of Pinchot's men on leave with Roth was worried that his seniority in the Department of Agriculture was in jeopardy. Those who stayed behind with Pinchot seemed to advance more quickly. In any event, Roth resigned on February 6, 1903, to begin an exemplary academic career at the University of Michigan. To Pinchot, Roth's departure was one more piece of evidence arguing for Interior's withdrawal from the forestry business.[37]

Student Assistants

To facilitate management of the forest reserves for the Department of the Interior, and the work of his own bureau, such as Circular 21 tasks, Pinchot inaugurated a program to stretch meager budgets. Since there were few foresters and fewer forestry schools, trained men were scarce. By using college students Pinchot could recruit and train the specialists soon needed to deal with America's forestry problems. The student assistant program turned out to be one of his most successful.

36. Kneipp to H. K. Kelley, Feb. 29, 1904, Field Office Records of Forest Supervisors, 1898–1904, Record Group 95, RFS; Kneipp to Commissioner, June 7, 1904, ibid. Two accounts of Division R rangers are Harold D. Langille, "Mostly Division 'R' Days," *Oregon Historical Quarterly* 57 (Dec. 1956): 301–13; and Len Shoemaker, *Saga of a Forest Ranger* (Boulder: University of Colorado Press, 1958).

37. Pinchot, *Breaking New Ground*, pp. 196–97; E. T. Allen to Roth, Aug. 3, 1902, drawer 2, Records of the Office of the Chief, Record Group 95, RFS. The close relationship between Division R and the Bureau of Forestry is well demonstrated by the official records of the two agencies. Both contain copies of correspondence issued by each other.

Using students enabled Pinchot to acquire manpower for a modest sum. Their work provided on-the-job training at a time when the opportunity to study forestry formally was very limited and when practical experience was valued more highly than the understanding of theory. As one of Pinchot's staff put it, there was need to "get the Harvard rubbed off the students before they came in contact with the loggers."[38]

Arthur Ringland remembers going to work for Pinchot as a student assistant in 1900. He first learned of forestry by reading about Fernow's school at Cornell. Other students had been attracted to Pinchot's program by an article in the *Saturday Evening Post*. Their enthusiasm surmounted the skimpy pay of twenty-five dollars per month. High spirits were manifested in a number of ways. For example, seizing on the initials of Pinchot's likable chief clerk, Otto J. J. Luebkert, the students quickly dubbed him Jumping Jesus.

Ringland was one of sixty-nine student assistants appointed during that second year of the program; Pinchot had appointed thirty-three a year earlier. By 1902 nearly three hundred students had signed on. Many, like Ringland, Coert DuBois, and Ferdinand A. Silcox, went on to higher positions in the Forest Service. The Old Guard, as this early group of students and their supervisors came to be called, formed the cadre and the esprit de corps upon which Pinchot staked so much. The program was almost too popular, and Pinchot considered limiting enrollment to three seasons. Also, the student assistants would have to be promoted through examination like everyone else. Civil Service rosters would be a safeguard against political appointments.[39]

Some students received office assignments, but most worked in the West after an orientation in Washington, D.C. The low pay was further diminished by requiring the student to pay his own travel expenses. A student assigned to E. T. Allen in the Priest River Forest Reserve in Idaho and Washington learned that he was "expected to assist in whatever work is assigned. . . ." Working for the Bureau of Forestry was not summer school; it was hard and sometimes monotonous but would give an opportunity to see practical forestry.

38. E. T. Allen to Henry Graves, July 13, 1900, drawer 2, RG 95–3, RFS.

39. Arthur C. Ringland, *Conserving Human and Natural Resources*, Oral History Interview by Amelia R. Fry et al., 1970, Regional Oral History Office, Bancroft Library, pp. 2–5; Minutes of the Service Committee, Apr. 18, 1903, RG 95–8, RFS; Coert DuBois, *Trail Blazers* (Stonington, Conn.: Stonington Publishing Co., 1957), p. 23.

The warning was prophetic. The following month the student severely cut his foot twenty-five miles from the nearest road in the back country of Idaho. Allen had to hike out for needles and thread to sew up the wound because the young man could not be moved. Pinchot must have been pleased when he learned that the student's greatest worry was that his parents would find out.[40]

Monotony and hardship could spawn disciplinary problems, and Pinchot met the matter with customary dispatch. Less than a month after receiving his assignment, and before he even reached the field, one student opened and read a stern letter. If he wished to remain a student assistant, Pinchot warned, "it will be necessary for you to do your work cheerfully and well and to avoid complaint of the conditions and management under which you are working." He was told to remember that working for the Bureau of Forestry was a privilege for those worthy of it.[41]

Pinchot also moved vigorously to keep any taint of political favoritism from his agency, making it distinctly different, to his mind, from the corrupt General Land Office. He may also have been remembering the limitations of Nathaniel Egleston, a predecessor appointed under a flagrant spoils system. He held firm against making even one political appointment because he was sure that if the word got out, every favor-seeker would come running. Pinchot's resolve was sorely tested by Colorado Congressman Franklin E. Brooks, who told Pinchot that Simon Guggenheim wanted a summer job for a friend. The chief explained that he had three times more applicants than openings and he hired the most qualified first. He doubted that there would be a place for Guggenheim's friend. Brooks pleaded to give the man a chance, even if he had to work for no pay. The congressman added, "Simon Guggenheim is a very large figure on our horizon just at this time, and he is in a position to do the Department [of Agriculture] a great deal of good." But Pinchot insisted that it was impossible to help the mining magnate, and that settled the matter.[42]

Professionalism was vital. Civil Service would help; so would giving students on-the-job training, as would more forestry schools. Fernow

40. Overton Price to H. J. Brown, June 13, 1905, drawer 2, RG 95–3, RFS; E. T. Allen to Price, July 29, 1901, ibid.

41. Pinchot to Preston Brooks, June 12, 1902, drawer 8, RG 95–3, RFS.

42. Pinchot to Ralph Hosmer, July 13, 1901, box 575, Pinchot Papers; Franklin E. Brooks to Pinchot, May 6, 1905, RG 95–3, RFS; Pinchot to Brooks, May 9, 1905, RG 95–3, RFS; Brooks to Overton Price, May 18, 1905, RG 95–3, RFS; Pinchot to Brooks, May 23, 1905, RG 95–3, RFS.

had a four-year curriculum at Cornell; Carl A. Schenck offered a one-year practical course at the Biltmore Forest School on the Vanderbilt estate in North Carolina; and the Pinchot family underwrote a master's program at Yale. Still needed was a means of gathering foresters together. In November 1900, at Pinchot's invitation, a small group consisting of Henry S. Graves, Overton Price, E. T. Allen, William L. Hall, Ralph S. Hosmer, Thomas H. Sherrard, and Pinchot formed the Society of American Foresters. It was soon nicknamed "The Baked Apple Club" for the gingerbread and baked apples that Pinchot served at subsequent meetings in his home. Students attended these weekly sessions, supplementing their $25 monthly salary with Pinchot's gingerbread and mingling with professionals and distinguished guests. Seventy-five years later, the Society of American Foresters could boast of over twenty thousand members in sections and chapters throughout the nation.[43]

Pinchot and his staff had lesser issues to resolve, too. Office routine for the rapidly growing agency was discussed at weekly meetings. One week the Service Committee, as chief and staff were called, ruled that the American flag should fly over the head man's tent in the field. At the same meeting they decided that the mail clerk should arrive early enough to finish sorting by 9:00 A.M. Three sessions in a row weighed the merits of smoking in the office, but here consensus eluded Pinchot's advisers. Frederick E. Olmsted once recommended to the group that carbon copies of outgoing correspondence be initialed; two years later, he complained that the initialing regulation was observed only in the breach.[44]

Of course, some office procedures showed innovation, such as Pinchot's filing scheme. The method of storing file folders vertically instead of horizontally, allowing the eye to scan more easily, was given wide publicity at the 1893 exposition in Chicago. By 1900 Pinchot began using vertical files. He expanded this program after 1905 to incorporate forest reserve records that were received from the Department of the Interior. Within a few years other government agencies adopted the Forest Service filing scheme. Pinchot's whole system of records management was gradually, over twenty years, adopted

43. Pinchot, *Breaking New Ground*, pp. 150–52; Ringland, *Conserving Natural Resources*, pp. 6–9.

44. Minutes of the Service Committee, May 2, 8, 1903, March 24, 1904, Apr. 18, 1906, box 112, RG 95–8, RFS.

throughout the government.[45] As the instance of the vertical filing system shows, throughout his tenure as chief, Pinchot's administrative skills were everywhere evident—and his talents were continuously tested. There was no moratorium on problems or issues, offering him time to perfect his agency's administrative machinery. Timber management problems he met with Circular 21 and cooperative programs with the General Land Office. Conditions of the western range, too, demanded early attention.

Rangers on the Range

The range, in some respects offering more pressing problems than timber resources, enabled Pinchot to establish important precedents in public land management. At times it was necessary to distinguish between sheep and cattle when establishing regulations; other times grazing could be treated in its broadest context. In 1898 the secretary of the interior prohibited grazing in forest reserves, except those of Washington and Oregon. This exemption was based on the belief that rainfall in the Pacific Northwest supported vegetation of sufficient quantity to withstand grazing pressures. Studies had suggested, too, that grazing did not injure forest trees and indeed might reduce fire danger by removing fuel, although this favorable view of grazing was not widely accepted. Afraid that local opposition could result in his firing, one General Land Office ranger cautioned Pinchot, "Please do not use my name if you cannot find me a better job."[46]

Stockmen agreed that studies were necessary but did not want to face a moratorium on grazing until after all facts were in. Secretary of the Interior Hitchcock granted use of the "accustomed ranges" and asked Pinchot to study the situation. The stockmen said that they would accept the findings.[47]

Pinchot probably wrote the letter for his signature, but Hitchcock asked Agriculture Secretary Wilson to prepare reports on grazing on forest reserves. The Interior secretary wanted to know the importance

45. Harold T. Pinkett, "The Forest Service: Trail Blazer in Record-keeping Methods," *American Archivist* 22 (Oct. 1959): 419–26.

46. S. B. Ormsby to Cornelius Bliss, Nov. 6, 1897, box 577, Pinchot Papers; C. P. Dodd to Pinchot, Nov. 13, 1900, ibid. Approximately 25 percent of Pinchot's 1897 report on the forest reserves as a special agent of the Interior Department concerned sheep grazing on the Cascade Forest Reserve of Oregon. Frederick V. Coville made the grazing study.

47. E. S. Gosney to Hitchcock, Feb. 28, 1901, copy in FHS clipping file.

of the grazing industry in each reserve—the effect of grazing on fire potential and natural regeneration of forests, the relation of grazing to water supplies, and the difference between sheep and cattle grazing.

Pinchot admonished his field agents to be "impartial": there were no "preconceived ideas to establish or to defend" and "no interest to favor except the public interest." Pinchot quoted Wilson (who sounds very much like Pinchot): "Every question of grazing should be decided . . . on its own merits. As a rule, grazing should be regulated, not prohibited. . . ." Forage that could be harvested by grazing should not be wasted. He cautioned the investigators that there were "radical differences of honest opinion" about grazing and that the final report had to be based upon personal observations. Some differences of opinion were not so honest: Pinchot learned that one agent had been offered a ten-thousand-dollar bribe, apparently from cattlemen, to report that sheep grazing was detrimental. Forest officers sometimes were less than objective, too. Arthur Ringland claims with a grin that his orders were, "If you meet a sheep-herding so-and-so on the trail with a broken leg, break the other leg and go on."[48]

Studies, reports, and opinions on grazing abounded. To the secretary of the interior, the grazing problem was the most perplexing in reserve management. Not only was the degree of damage from grazing disputed, but methods of allotting range were controversial, too. When the secretary ordered a reduction in the number of sheep grazing on the Manti Forest Reserve to one-eighth of the existing number, sheepmen asked for a gradual reduction of 25 percent each year. The concept of allotment itself was not debated, for stockmen saw that open ranges required a quota system, because no incentive existed for any one owner to reduce his herd. Overgrazing would be inevitable.[49]

Pinchot worked vigorously to deal with grazing. He authorized studies, attended livestock meetings, and wrote rules; but his most productive act was to bring Albert Potter into federal service. An Arizona range man, Potter brought both knowledge and credibility to the Bureau of Forestry. He quickly began to make order out of grazing chaos.

In 1900 Potter had called on Pinchot to discuss range problems in

48. "Government Investigation of Grazing in Forest Reserves," undated newspaper clipping; Gosney to Pinchot, n.d., handwritten note, FHS clipping file; Ringland, *Conserving Natural Resources*, p. 35.

49. USDI, *Annual Report* (1903), p. 322; unidentified clipping from Salt Lake City, Sept. 17, 1903, FHS clipping file; unidentified clipping, March 5, 1902, FHS clipping file.

the Southwest. Pinchot readily agreed with the stockmen that decisions should be made after field examination and not based on a generalized instruction. Later that year, Pinchot met with Potter in Arizona to study the range firsthand. The story goes that Pinchot made a lasting, favorable impression on Potter and his companions by drinking scum-encrusted water. The easterner passed western initiation rites with flying colors. Here was a man they could deal with.[50]

Now working for Pinchot, Potter went to the stockmen themselves for the answer. He sent questionnaires dealing with fifteen aspects of the range, mainly asking about use and methods of improving conditions. Fourteen hundred stockmen responded, most of them with care, Potter believed. Most important to Pinchot, the stockmen voted five to one in favor of "reasonable" government regulation of grazing. Potter proposed that grazing permits be issued for no more than ten years and that the values involved be determined and an appropriate fee levied. He summarized his views to Pinchot: (1) proper use will be allowed, (2) carrying capacity of range will determine allotment size, (3) permits will be granted equitably, and (4) the regulations will be flexible.[51]

To Pinchot, prospects were brighter. Potter's knowledge and orderly approach were paying off. Stockmen favored regulation and accepted the idea of allotments. Meanwhile, Pinchot asked Overton Price to calculate how much revenue the government could expect from grazing fees. These monies, added to timber sale receipts, perhaps would make the reserves self-supporting. Pinchot warned Price that he wanted to show the figures to important members of Congress and that he might "have to make definite promises regarding it, therefore, it wants to be exceedingly careful."[52] Revenues versus expenditures captured Pinchot's attention for several years; making his agency show a profit was a goal he was slow to abandon.

50. Paul H. Roberts, *Hoof Prints on Forest Ranges: The Early Years of National Forest Range Administration* (San Antonio, Texas: Naylor Co., 1963), pp. 28–29.

51. U.S. Congress, Senate, *Report of the Public Lands Commission*, Sen. Doc. 189, 58 Cong. 3, 1905, appendix p. 5, pp. 8–24; Potter to Pinchot, Dec. 10, 1903, box 577, Pinchot Papers. Clarence S. Forsling, a range specialist who was assistant chief for Research during the 1930s, believes that Potter's job had been made easier by the continuous range wars between cattlemen and sheepmen. Both sides saw regulation as a solution to their differences. Forsling also points out that competitive bidding on forage was discarded, because it would have jeopardized the necessary continuity for stock production. Graziers most dependent upon federal range were given first priority when issuing permits. Forsling to Frank Harmon, July 30, 1975, U.S. Forest Service History Office files.

52. Pinchot to Price, Jan. 19, 1904, Box 49, RG 95–3, RFS.

Much had happened during the first half-dozen years since Pinchot became chief of the Division of Forestry in 1898. The size of the agency had burgeoned from a score of men to over five hundred.[53] Budgets had doubled and tripled. The agency's increasing stature was duly acknowledged in 1901 by its designation as the Bureau of Forestry. Millions of acres of privately owned timber had been examined through the cooperative Circular 21 program, and the forest reserves had felt the first, faltering touches of a professional hand. Much had happened, indeed, but the transfer of the forest reserves in 1905 from the Department of the Interior to the Department of Agriculture would mark a dramatic shift in programs. With that shift the modern Forest Service would arrive.

53. USDA, *Annual Report* (1905), p. 204. Pinchot reported as of June 30, 1905, 6 inspectors, 2 superintendents, 49 supervisors, 5 rangers in charge, 379 rangers, 87 guards, 5 laborers, and 5 forest assistants. Added to this list would be his Washington office staff.

CHAPTER IV

Pinchot, the President, and the Conservation Movement

Everyone waited; many in the crowd dabbed at their faces. It was warm in the Temple of Music at Buffalo's Pan American Exposition. The excitement of waiting for President McKinley to appear made it seem even warmer. Finally at 4:00 P.M. McKinley began to greet the public. Seven minutes later, Leon Czolgosz sprang forward and fired twice through a handkerchief hiding a small revolver. The president lingered for nine days. Ten days after the assassin's shots, on September 14, 1901, the presidential oath was administered to Vice-president Theodore Roosevelt. Then, as Senator Mark Hanna bitterly remarked in private, "that damned cowboy" was in the White House.[1]

Roosevelt was a fascinating man. He was young, only forty-two, at the time of his inauguration. Americans would have to search back to John Quincy Adams to find a president with as broad an intellect and such diverse interests. Harvard-educated, Roosevelt bustled about writing books on American history, enjoying art, engaging in rigorous hunting trips. To overcome youthful frailties, he went West and engaged in strenuous ranch activities. His zestful affection for riding horses became a near legend, accounting for Hanna's cowboy allusion. In keeping with his image, Roosevelt once wrote to a friend that to kill a grizzly with a hunting knife "would be great sport."[2] His swashbuckling charge up San Juan Hill during the Spanish-American

1. Margaret Leech, *In the Days of McKinley* (New York: Harper and Bros., 1959), pp. 592–93; George E. Mowry, *The Era of Theodore Roosevelt and the Birth of Modern America, 1900–1912* (New York: Harper Torchbook, 1962), p. 106.
2. Mowry, *Era of Theodore Roosevelt*, pp. 106–10.

War—"The Bloody Good War"—had assured him and his Roughriders a place in romantic history. Eight years later, Roosevelt would help end the Russo-Japanese War, for which he won the Nobel Peace Prize. This impetuous president believed in asserting power to the limit, not restricting himself to duties specifically authorized.

This attitude and Roosevelt's interest in natural resources and the out-of-doors provided an excellent setting for Gifford Pinchot. He, too, was impetuous—and young, having celebrated his thirty-sixth birthday only a month before Roosevelt became president. Roosevelt, several years after he left the presidency, said of Gifford Pinchot: "I believe it is but just to say that among the many, many public officials who under my administration rendered literally invaluable service to the people of the United States, he, on the whole, stood first."[3] The warm relationship between the president of the United States and the chief of the Forest Service provided an unusual opportunity for a bureau chief to convert his plans promptly into reality. This Roosevelt–Pinchot relationship, made possible by assassination, without question is one of the most significant in the history of the Forest Service.

Roosevelt had been elected governor of New York in 1898, the same year that Pinchot became chief of the Division of Forestry. The next year, Pinchot, while in the Adirondacks examining a Circular 21 application, called on the governor. They were already acquainted, as Roosevelt had nominated Pinchot for membership to the Boone and Crockett Club, a hunting group that Roosevelt had helped to found. The tall forester and the stocky governor hit it off. They wrestled for sport, and Pinchot was soon on his back. Boxing, however, was another matter, and Pinchot's longer arms gave him the "honor of knocking the future President of the United States off his very solid pins." After Roosevelt became president, Pinchot would be his frequent companion on the tennis court or on horseback. Pinchot enjoyed telling how he, the president, and the French ambassador stripped to the buff and swam in Rock Creek Park in Washington, D.C. For some outings, Pinchot would slip a .38 special into his pocket, enabling him to act as presidential bodyguard after a joyous time eluding Secret Service agents.[4] (This latter sport in retrospect

3. Theodore Roosevelt, *Theodore Roosevelt: An Autobiography* (New York: Charles Scribner's Sons, 1926), p. 394.

4. Pinchot, *Breaking New Ground* (1947; reprint ed., Seattle: University of Washington Press, 1972), p. 145 and *passim*.

seems foolish, considering that Roosevelt had become president via an assassin's bullet.)

The Forest Reserves

Three months after becoming president, Roosevelt was telling Congress that the forest reserves belonged not within the Department of the Interior but in the Department of Agriculture, under Pinchot's Bureau of Forestry.[5] The promptness of Roosevelt's action reflects the high degree of Pinchot's influence on the president.

Pinchot had not been the first to see the incongruity of having the forest reserves in the Department of the Interior and the foresters in the Department of Agriculture. The obvious solution was consolidation, but the question was in which department. Early in his administration, Fernow had bowed to the realities of the situation and recommended an agency in Interior. Toward the end of his term, however, Fernow had tried to have the reserves transferred to Agriculture. That proposal met with slight interest but no support, either from the secretary of agriculture or from Congress.[6]

Even before Pinchot replaced Fernow he, too, began striving to bring foresters and the forests together. Pinchot's first efforts echoed Fernow's in trying to establish a forestry agency in the Department of the Interior. When Arnold Hague and Charles D. Walcott advised that westerners would probably oppose formation of such an agency, Pinchot agreed to wait. If the attempt to establish Division R to carry out the 1897 Sundry Civil Appropriations Act should fail, however, he would make every effort to establish a forestry bureau with himself at its head. Then Secretary of Agriculture James Wilson offered him Fernow's job.[7] His acceptance committed him to develop the Division of Forestry into a management agency within Agriculture. As with Dunnell's last-minute tactics in 1876, Pinchot's support for a forestry agency in the Department of Agriculture was more a matter of expediency than preference.

As early as 1899, Pinchot arranged to have the subject of transferring the reserves brought up at a cabinet meeting. Secretary of State Elihu Root introduced the topic. He had known Pinchot and had been

5. Roosevelt, Message to Congress, 57 Cong. 1, Dec. 1901.

6. Charles Keffer to Secretary of Agriculture, Aug. 31, 1897, Letters Sent by the Division of Forestry, Record Group 95, RFS, National Archives; Keffer to Fernow, July 13, 12, 1897, ibid.

7. Pinchot, Diaries, Jan. 5, 13, Apr. 22, May 10, 1898, Gifford Pinchot Papers, Library of Congress.

aware of forest conservation needs since the New York State constitutional convention in 1894. At that time Pinchot had asked Root, as presiding officer of the convention, to help defeat a proposed amendment that would prohibit logging in Adirondack Park, but the amendment was adopted. Now, as chief, Pinchot also solicited support from Wilson, who reported that Interior Secretary Hitchcock, as well as President McKinley, was supportive. Binger Hermann, who as commissioner of the General Land Office held control of the reserves, was opposed.[8]

Hitchcock asked an aide to study the issue. The resulting memorandum, entitled "Some Reasons Why the Care of the Government Forest Reserves Should Be Transferred from the Land Office . . . ," argued for giving the authority to the Department of Agriculture on the basis that forests were crops. Pinchot conferred with the secretary and others. Frederick H. Newell advised against moving the forestry bureau to Interior. The reserves, he insisted, must come to Agriculture. There the matter stood until Roosevelt, following up on his December 1901 message to Congress, asked Hitchcock to request formally technical advice from Pinchot. Hitchcock did more than that: his annual report supported transfer of the reserves.[9]

The House considered a transfer bill in the late spring of 1902. The bill would shift the reserves to Agriculture and authorize making them fish and game preserves. When it passed the House Public Lands Committee by an eight to five vote, Pinchot saw a chance for victory. The committee favored transfer of reserves after boundary questions had been dealt with. In other words, the reserves would be created in Interior, then transferred to Agriculture for administration. The committee acknowledged the technical possibility of jurisdictional disputes under this arrangement but "as a practical question" thought there would not be "any difficulty whatsoever." The group also suggested that Pinchot's agency could be utilized more effectively than in the past.[10]

8. Pinchot, *Breaking New Ground*, p. 183; Mary C. Leger, "A Study of the Career of Ethan Allen Hitchcock," Ph.D. diss., University of New York, 1971, p. 109; Pinchot, Diaries, Nov. 6, 9, 1899, Jan. 22, 1900.

9. J. A. Holmes, "Some Reasons Why the Care of the Government Forest Reserves Should Be Transferred from the Land Office . . . ," Dec. 1899, copy in box 580, Pinchot Papers; Pinchot to James Pinchot, Dec. 19, 1899, box 581, ibid.; Pinchot, Diaries, Nov. 9, 10, 29, 1899; Roosevelt, *Autobiography*, p. 400; USDI, *Annual Report* (1901), p. lxviii.

10. H. R. 11536, 57 Cong. 1; Pinchot to Graves, March 13, 1902, box 575, Pinchot Papers; U.S. Congress, House Report to Accompany H.R. 11536, no. 968, 57 Cong. 1, pp. 2–3.

A minority report opposed the transfer on the grounds of inefficiency and potential interdepartmental friction. As to the majority view toward better utilizing Pinchot, the minority countered with the logical proposal that the forestry bureau in Agriculture could be reduced in size or transferred to Interior. More telling were the arguments that every acre of the reserves was potentially mineral in character, and the proposed dilution of jurisdiction between Agriculture and Interior could increase the difficulty in settling disputes. Too, settlers would be oppressed by needless complexities as they made legitimate claims. The minority report concluded by admitting that "mistakes had been made" by Interior but it was "undisputed" that "much good has been accomplished." Pinchot's bureau should be able to find "a large field of useful employment" studying "the questions of forestry" on private as well as public lands. There was no reason for these studies to interfere with "the policing and administration of the forest reserves by the Interior Department." Views of the minority carried the day. "Transfer bill beaten today by speech of Joe Cannon," a disappointed Pinchot wired an associate.[11]

Speaker Cannon, a powerful force that Pinchot would have to reckon with on several occasions, had opposed the transfer from the start. His opposition remained firm, but as a political realist he eventually bowed to increasing pressures and stopped blocking congressional action. In the meantime, Pinchot remained patient and optimistic. Shortly after the 1902 defeat he told Newell that everything was all right in the Bureau of Forestry, except for the transfer. He hoped that William A. Richards, who favored transfer, would soon replace Hermann as commissioner. Pinchot still hoped for passage of the transfer during the 1902 congressional session.[12]

Cannon received support from western congressmen who saw the Department of Agriculture as being concerned purely with agriculture. Obviously, settlers, speculators, and industrialists were better off if the reserves remained under Interior jurisdiction. Then spectacular scandals in the Land Office reduced western opposition, and

11. House Report no. 968, pp. 1–6; Leger, "Hitchcock," p. 111; Overton Price to Pinchot, June 10, 1902, General Correspondence of the Forest Service, 1898–1906, Record Group 95 (hereafter cited as RG 95–3), RFS; Pinchot to Price, June 11, 1902, ibid.; Pinchot, Diaries, June 10, 1902.

12. Pinchot, Diaries, Jan. 29, 1900; Frank W. Mondell to Pinchot, Dec. 9, 1904, and Pinchot to Newell, Aug. 30, 1902, box 580, Pinchot Papers. Robert U. Johnson claims that it was he who persuaded Cannon to change his mind and support the transfer in *Remembered Yesterdays* (Boston: Little, Brown, 1923), p. 299.

Pinchot reminded Roosevelt of his promise to advocate transfer if he had western backing. Congressman Frank Mondell of the Public Lands Committee was hopeful. Roosevelt reacted by telling Congress that the forest reserves were beyond the experimental stage and should be transferred to the Bureau of Forestry. Hitchcock openly acknowledged that the move to Agriculture "would be productive of better administrative results." In the Far West, the decade-old Sierra Club supported Pinchot by writing to each member of the California congressional delegation.[13]

Transferred at Last

Crucial to the relaxation of congressional opposition was the attitude of Senator Alfred B. Kittredge of South Dakota. One of Kittredge's powerful constituents, the Homestake Mining Company, remained opposed to a transfer until assured that timber on reserves in South Dakota, which the company depended on as a source of mine timbers, could not be sold out of the state. When Pinchot agreed to intervene personally and gain departmental assurances to protect the mining company's timber supply, the final congressional opposition melted.

The American Forestry Association had supported shifting the reserves, beginning with a resolution in 1899. Throughout Pinchot's campaign, the conservation group had worked quietly and effectively, keeping its members informed and building congressional favor. Many stockmen's groups also advocated transfer.[14] The combined efforts bore fruit: on February 1, 1905, President Roosevelt gave final approval.

Sixty-three million acres of forest reserves were now in Agriculture. On July 1, 1905, the Bureau of Forestry was renamed the United States Forest Service, to reflect, in Pinchot's words, that his agency was committed to service.[15] Two years later the reserves were re-

13. Leger, "Hitchcock," p. 115; Pinchot to Roosevelt, Sept. 21, 1903, box 581, Pinchot Papers; Roosevelt, Message to Congress, 58 Cong. 2, Dec. 1903; USDI, *Annual Report* (1902), p. 2; William Colby to Congressmen, March 23, 1904, copy in box 580, Pinchot Papers.

14. Samuel P. Hays, *Conservation and the Gospel of Efficiency: The Progressive Conservation Movement, 1890–1920* (Cambridge, Mass : Harvard University Press, 1959), pp. 43–44; AFC, *Proceedings* (1905), p. 449.

15. Pinchot, Diaries, Oct. 31, 1904; Pinchot, *Breaking New Ground*, pp. 254–62; Forest Service, *Annual Report* (1905), p. 199. Between February 1 and July 1, 1905, additions increased the reserves from 63 to 86 million acres.

named national forests, because to Pinchot the term reserve suggested that these federal forests were to be held inviolate. They were not.

Now the reserves were his. Pinchot had circled at their administrative periphery for almost seven years. Before becoming chief forester, he had examined them both as a member of the National Academy of Sciences commission and as a special agent of the Department of the Interior. Since 1898 he had provided administrative advice at his own suggestion to the General Land Office. No longer would Pinchot have to resort to cumbersome procedures, with the Land Office commissioner as an intermediary. He began at once to form the theory and the machinery to administer those many millions of remote acres.

To set down a policy blueprint for operation of the Forest Service, Pinchot wrote a brief letter to himself for Secretary Wilson's signature. The agency quickly absorbed the principles listed in this letter, and even today its imprint is clearly visible throughout all fundamental policies. Use is not contrary to conservation. Decisions on use would consider needs of local industries first. When in doubt, where conflicting interests must be reconciled, the question would "always be decided from the standpoint of the greatest good of the greatest number in the long run."[16]

This last statement has been frequently pointed to as proof that Pinchot advocated the utilitarian philosophy of Jeremy Bentham. Perhaps so, since Pinchot received a broad, general education at Yale during the late 1880s and he could well have fallen under Bentham's influence. Pinchot credits W J McGee, an associate on many occasions, for advancing the basic philosophy.[17] In any regard, "the greatest good of the greatest number," whatever its philosophical shortcomings to support decision-making may be, still is commonly used by Forest Service personnel as a guide to their daily decisions.

On the matter of administration, Pinchot wrote Associate Forester Overton Price only months following the transfer from Interior that he was "very seriously dissatisfied" with the operation of the forest reserves. He saw repeated failure to send supplies, or mistakes in those that were sent. To Pinchot, this situation was "absolutely intolerable." He instructed Price to make it his personal responsibility at once to see to it that no requisition of any kind went unanswered more

16. Letter quoted in Pinchot, *Breaking New Ground*, pp. 261–62.

17. Pinchot, *Breaking New Ground*, p. 359. McGee never used periods following his initials; printers called him No Stop McGee.

than twenty-four hours without an acknowledgment. Price was authorized to pull in as many men from the field as needed. Pinchot emphasized that he was to see improvement "no matter what is to be sacrificed to get it done." The only mission that had higher priority was examining land to aid in the creation of new forest reserves.[18]

Pinchot's likable associate forester fit well the job of attending to administrative problems, and Pinchot praised him highly for his talents. Price's correspondence files show his method. Succinct, to-the-point letters and directives briskly flowed from his desk to field personnel. Questions received direct, one-paragraph answers. Orders were unmistakable: he told one ranger to stop gambling, another to stop drinking, still another to be concise. Rangers and forest supervisors frequently read that their letters were too lengthy or too detailed or failed to answer one of Price's pointed questions.[19]

Pinchot and Price hammered at the need for administrative efficiency. All unfinished business and letters must be kept on desk tops overnight—no more putting these materials in drawers where they might be forgotten. Letters had to be answered within thirty-six hours of receipt, or a promise file made to insure an eventual reply, with an explanation for the delay. Field men were ordered to use clear language and to avoid unwieldy words and laborious sentences. Pinchot himself frequently edited outgoing correspondence and returned marked-up copy to the sender for revision. To him, few occasions justified a letter longer than one page. Pinchot eventually softened this latter order when he discovered "the tendency to make letters and telegrams brief at the expense of courtesy."[20]

If he badgered for efficiency, Pinchot exerted even greater effort to support the men on the line. He impressed upon his office staff "the necessity of doing everything possible for the men in the field. . . ."[21] Decentralization of the administrative process, as suggested in Wilson's 1905 letter of instruction, gave forest officers

18. Pinchot to Price, Aug. 25, drawer 49, RG 95–3, RFS.

19. "Overton had more to do with the organization and reorganization of the Service than I had, and therefore its efficiency. . . . He was far more than my assistant. He was my associate." Pinchot, *Breaking New Ground*, p. 302. Price's files in the RFS have suffered heavy water damage, but fragments are still readable. Price committed suicide in 1914.

20. Service Order no. 89, Sept. 22, 1905, copy in drawer 164, Records of the Office of the Chief, 1908–47, Record Group 95 (hereafter cited as RG 95–4), RFS; Ralph S. Hosmer, "Early Days in Forest School and Forest Service," *Forest History* 16 (Oct. 1972): 8; Minutes of the Service Committee, Apr. 11, June 30, 1906, box 112, RG 95–8, RFS; Service Order no. 106, July 19, 1906.

21. Minutes of the Service Committee, May 22, 1903, box 112, RG 95–8, RFS.

distant from the Washington, D.C., headquarters a status rare in federal bureaus.

The first steps in decentralization occurred early. Fernow in 1886 had proposed a dispersed administrative structure; Pinchot drafted a similar plan in 1898. In 1901 Price saw an ever-increasing work load falling on his chief, as well as on himself, resulting in too much office time and too little in the field. Price proposed a reorganization plan, which was subsequently adopted. A redistribution of staff responsibilities left Pinchot free to deal with the cooperative program, the Division of Forestry in Interior, legal processes, and editorial matters. Price looked after forest inventory, silviculture, and personnel. Division chiefs handled extension work outside the agency, products, finance, forest management, and dendrology.[22] The forest reserves, of course, then were still under the Department of the Interior but were administered according to Pinchot's scheme, which gave substantial authority to the local forest supervisors.

Pinchot's success in getting the forest reserves transferred to Agriculture caused an overwhelming increase in his administrative work load. Within two years of the transfer, he formulated a new organizational plan for reserves, which by 1907 included over 150 million acres, and submitted the new version to Secretary Wilson. Pinchot had early set up three district offices to ease the administrative burden, but further adjustments were necessary. Creation of six districts would cause no disruption in work because he had long planned for this contingency. As the new district foresters Pinchot in 1908 named William B. Greeley, Arthur Ringland, Smith Riley, Clyde Leavitt, Frederick E. Olmsted, and E. T. Allen. As an added concession to the West, Forest Service receipts would be handled by western banks, improving their cash flow.[23]

Pinchot had spelled out in the 1905 letter his insistence that the man on the ground was the best judge of the situation. He pointedly reiterated this view to his immediate staff, emphasizing that as much work as possible must be done in the field. Forest supervisors were in-

22. Minutes of the Service Committee, May 22, 1903, ibid. Coert DuBois asserts that growing antagonism between field agents and James B. Adams, Pinchot's chief clerk, over the need to revise reports was a major factor prompting reorganization. See DuBois, *Trail Blazers* (Stonington, Conn.: Stonington Publishing Co., 1957), pp. 65–66.

23. Minutes of the Service Committee, Apr. 3, 1907, box 112, RG 95–8, RFS; Service Order no. 96, Feb. 28, 1906, copy in drawer 164, RG 95–4, RFS; Samuel T. Dana, *Forest and Range Policy* (New York: McGraw-Hill, 1956), p. 393; news release, n.d. (also Oct. 7, 1908), copies in FHS clipping file.

structed to increase efficiency by entrusting their rangers with more administrative work. Such delegation would develop their executive ability; after all, a ranger must feel responsible for his district. The public should be taught to turn to the ranger, not the supervisor. The supervisor remained accountable for all that happened on his forest, but the ranger had to take charge of his district. The district foresters, too, had to accept more autonomy. They had been referring an unacceptable number of problems to the Washington office, problems better solved by the men in the field.[24]

A decentralized organization must have some control, some uniform standards, however. Roth's Division of Forestry in the Department of the Interior had produced a manual. Now that the forests were in Agriculture, Pinchot wanted a new version. This one would be smaller, designed to fit into a ranger's shirt pocket. He dubbed it the *Use Book*, because the forests were for use. Henry Graves, forestry dean at Yale, congratulated him on his choice of name. Pinchot insisted that each supervisor obtain an adequate supply of *Use Books* to meet the rangers' needs.[25]

The *Use Book*

Four and one-quarter by six and three-quarters inches, containing 142 pages, the first *Use Book* of regulations and instructions took effect on July 1, 1905. Increases in Forest Service size and responsibility later would cause the *Use Book* to explode into the multivolume looseleaf encyclopedia of procedures lining the office bookshelves of today's administrator. During the early years, however, the ranger could slip everything he needed to know into one pocket, his "fixin's" in another, and patrol his district on horseback.

Many of the concepts and philosophies expressed then are still in

24. Minutes of the Service Committee, Nov. 27, 1907, box 112, RG 95–8, RFS; C. S. Chapman to Supervisors, Nov. 29, 1907, drawer 164, RG 95–4, RFS; Service Order no. 122, Jan. 17, 1907, drawer 164, RG 95–4, RFS; Minutes of the Service Committee, Dec. 30, 1908, box 112, RG 95–8, RFS. Nomenclature is confusing. Since 1905 the Forest Service has had four administrative levels: the headquarters in Washington, D.C.; the district office (renamed regional office in 1930); national forests; and ranger districts. The head of the Forest Service is called chief, except between 1898 and 1935, when the title "forester" was used. District (or regional) foresters are the second-ranking line officers in the agency. Each region is divided into forests, directed by supervisors. The basic administrative unit is the district, headed by a ranger, the fourth-level line office. Each of the four levels is served by staff, which under some circumstances has been given line authority.

25. Henry Graves to Price, Aug. 7, 1905, drawer 36, RG 95–3, RFS; Minutes of the Service Committee, Apr. 11, 1906, box 112, RG 95–8, RFS.

evidence. The book stated the tenets of Forest Service policy "in a nutshell," as Pinchot would have said. "Forest reserves are for the purpose of preserving a perpetual supply of timber for home industries, preventing destruction of the forest cover which regulates the flow of streams, and protecting local residents from unfair competition in the use of forest and range. They are patrolled and protected, at Government expense, for the benefit of the Community and home builder." Pinchot gave Frederick Olmsted credit for overseeing preparation of the *Use Book*, which "represented the best judgment and experience of the whole service."[26]

Significantly, the first regulation assured all legitimate occupants of the forest reserves that the Forest Service would protect their rights even to the extent of giving them preferential treatment. To lead off with this promise demonstrates Pinchot's great concern for gaining local support for his agency. The next seven regulations dealt with permit procedures and free use permits, both detailing the needs and rights of those living on or near the reserves. In general, residents could take specified amounts of materials for their own use without charge, a privilege still granted today.[27]

Grazing and timber received substantial space. The *Use Book* explained that the object of grazing regulations was to (1) protect and conserve grazing land; (2) contribute to the well-being of the livestock industry; and (3) protect the interests of the settler against outside competition. Timber policy emphasized that the "prime object of the forest reserves is use." As long as safeguards for streams, soil, and the remaining forest were followed, timber could be sold and cut to meet "actual need." No upper limit was placed on timber sale size, other than the principle that monopoly would not be tolerated. The re-

26. USDA, Forest Service, *The Use of the National Forest Reserves: Regulations and Instructions* (hereafter cited as *Use Book*), July 1, 1905, p. 7.

27. *Use Book*, pp. 13–20. The Forest Homestead Act of 1906 was strongly advocated by Pinchot. The act was a tangible statement officially recognizing local primacy. The Forest Service welcomed permanent settlers but would not tolerate land speculators. To farm land better suited for agriculture was "to put lands within National Forests to their highest economic use." The Forest Homestead Act allowed selection of 160 acres of agricultural land. Property lines could conform to topographic features, being described by metes and bounds survey instead of the geometrically rigid public land system. The most fertile land frequently lay along valley floors, and the authorization to acquire a streamwide strip up to one and one-half miles in length enabled one settler to control access to a valley. For that reason and because the most valuable timber usually grew in valleys, there was some in-service opposition. For a good discussion see "Instructions and Discussions Affecting National Forest Lands," pt. 2, "Forest Homestead Act of June 11, 1906" (Oct. 1908), and E. T. Allen, "The Act of June 11, 1906 Purely from a Forester's Point of View," copies in drawer 381, Research Compilation File, 1897–1935, Record Group 95 (hereafter cited as RG 95–115), RFS.

mainder of the *Use Book* dealt with a miscellany of topics, including fire and reporting procedures. An appendix excerpted pertinent statutes and administrative decisions. Regulations were written in the positive language of authorization, rather than following the Land Office tradition of power to forbid. The *Use Book* was a model of brevity and instruction; to know its contents was to have a good grasp of duties and responsibilities. Pinchot was pleased and rightly so.[28]

The *Use Book* did not solve all administrative problems; it only served as a guideline. Reports to the next higher level were still a necessary part of the process. The Service Committee frequently reviewed the reporting procedure, in hopes of eliminating unnecessary paper work. Overton Price estimated in 1908 that probably one thousand periodic reports were required from the several administrative levels. Some claimed that paper work consumed one-third of the day; District Forester Olmsted thought he spent fully one-half of his time explaining or describing what he did with the other half. Arthur Ringland grumbled that reports in his southwestern district cost fifteen thousand dollars to prepare. Service-wide, the agency probably spent one hundred thousand dollars per year passing information from one administrative level to another.[29]

Many of the reports Ringland referred to were prepared by inspectors from the Washington, D.C., headquarters or district offices. Inspections, still common in the Forest Service, were to assist, advise, and encourage line officers in their duties. Inspectors examined field conditions and in their reports recommended any changes needed to correct deficiencies. Pinchot instructed his inspectors to make definite statements of field conditions, to make specific recommendations for action, and to be sure of all facts reported.[30] The system of inspections provided a constant test of the *Use Book*'s effectiveness and gave field men, distant from authority, an opportunity to find out how well they accomplished their assignments. Perhaps more important, inspections allowed desk-bound forest officers to get out into the field to observe firsthand the effectiveness of their directives and to learn for themselves the full scope of the Forest Service mission.

Pinchot wanted outside judgment as well. He commissioned the

28. *Use Book*, 1905, pp. 20–142; Pinchot, *Breaking New Ground*, pp. 267–68.

29. Minutes of the Service Committee, Sept. 8, Dec. 8, 1909, box 113, RG 95–8, RFS.

30. RG 95–115, RFS, Feb. 6, 1907. In his 1906 annual report (p. 5) Pinchot noted a "marked increase in cooperation between office and field men," showing the need for inspection. After he made a major reorganization to increase decentralization of administration, Pinchot reported that he had created six inspection districts (see p. 3 of 1907 report).

Model-T Ford adapted for railroad travel, Sierra National Forest, California, 1913. Courtesy of the U.S. Forest Service

Civilian Conservation Corps Fire Suppression Crew, Lassen National Forest, California, 1933. Courtesy of U.S. Forest Service

Forest Service fire fighter with pulaski, Kaniksu National Forest, Idaho, 1967. Courtesy of the U.S. Forest Service

Dispatcher locates forest fire from lookout reports. Courtesy of the U.S. Forest Service

Converted B-29 dropping fire-fighting chemicals, Wenatchee National Forest, Washington, 1970. Courtesy of the U.S. Forest Service

New York firm of Gunn and Richards to study the Forest Service and evaluate its organizational structure. The efficiency experts could not "praise too highly the personnel." The men were intelligent, loyal, and enthusiastic. Pinchot was cautioned, however, that rapid growth had in some instances caused the Forest Service to outgrow its organization. Noted, too, was a tendency to lose control because of decentralization. Since Pinchot's time, the Forest Service has juggled the need for centralized authority with the pragmatic advantages of decentralization. Its success at resisting centralizing forces has made it nearly unique among federal agencies and a source of much pride among its employees.[31]

Recruiting the Best

The men that Gunn and Richards had praised resulted from a hard-nosed personnel program. Pinchot's distaste for the spoils system has already been noted by describing his refusal to hire friends of the influential. Those already on the payroll had to perform. When an employee misstepped, the Forest Service was as ruthless in ridding itself of the miscreant as it was in trying to avoid hiring the unsuited in the first place. For example, when Forest Supervisor Everett B. Thomas was charged with falsifying his accounts prior to the 1905 reserve transfer, legal officer George Woodruff worked hard to polish his case for an assured conviction. To Woodruff, "the moral influence of such a conviction throughout the Service will be very great" by setting an example for other supervisors. Thomas' conviction would serve to prevent further fraud and cause all Forest Service officials "to consider the ethics of their actions. . . ." He was convicted, receiving a three-year sentence and seven-thousand-dollar fine. As a result of this sort of stern action, rather than a bureaucratic cover-up, and the recruitment of idealistic and enthusiastic men, scandal has rarely touched the Forest Service.[32]

Pinchot was proud of his men and set out to show them, by generous promotions, that he recognized their worth. Price brought him

31. "United States Department of Agriculture, Forest Service," Preliminary Report (New York: Gunn, Richards and Co., June 30, 1903). For a more recent analysis of Forest Service administrative technique, see Herbert Kaufman, *The Forest Ranger: A Study in Administrative Behavior* (Baltimore, Md.: Johns Hopkins Press for Resources for the Future, 1960).

32. George Woodruff to R. H. Charlton, Oct. 3, 1905, Selected Records Relating to the Administration of Gifford Pinchot, 1905–10, Record Group 95 (hereafter cited as RG 95–22), RFS; Forest Service, *Annual Report* (1906), p. 8. Scandal is a relative term. Some modern-day opponents of the Forest Service seem to judge much of its operation as scandalous.

down to earth with a warning from Secretary Wilson that he should slow down. Pinchot had recommended more advancements for his agency than those proposed for the rest of the Department of Agriculture. Wilson predicted that Congress would not tolerate mass promotions, and Price told Pinchot that he would cut the list back by 50 percent.[33]

For decades afterward, slow promotions and low salaries would plague the Forest Service. In 1905 Pinchot's rangers received $900 per year, the same amount that Fernow paid his typist in 1886, and were required to furnish their own horses and saddles. The rangers were not alone in their low salaries. District Forester Olmsted, who thought the basis of the whole salary system was "absurd," pointed out to Pinchot that the Forest Service paid its attorneys from $1,200 to $1,300 yearly and sent them to court against highly paid lawyers representing special-interest groups. Olmsted complained that his two best men were leaving over salary dissatisfaction. This news, on top of a presidential order declaring a moratorium on all promotions, must have been discouraging to all. Some predicted a rash of resignations, probably by the best men.[34]

To get the best men, Pinchot had insisted upon civil service examinations that specifically tested for the proficiencies required to get the job done. In fact, field men acquired from the Interior Department with the 1905 transfer suffered a loss of morale from Pinchot's published standards, as some believed that he was hinting they resign. A ranger had to be "thoroughly sound and able-bodied, capable of enduring hardships and of performing severe labor under trying conditions." He had to know woodscraft and horsemanship, as well as how to deal "tactfully with all classes of people." Pinchot sought "experience, not book education," but applicants were warned that they would be required to write "intelligent reports." If hired, the ranger could hold only one job; no outside work would be permitted. As an incentive, higher positions would be filled from the ranks, rather than bringing in outsiders at supervisory levels. Almost unnecessarily Pinchot added, "Invalids seeking light out-of-door employment need not apply."[35]

33. Price to Pinchot, June 16, 1905, drawer 49, RG 95–3, RFS.

34. F. E. Olmsted to Forester, Nov. 16, 1909, Office of Operation memorandum, July 16, 1909, drawer 153, RG 95–4, RFS.

35. Price to Pinchot, Oct. 26, 1905, drawer 49, RG 95–3, RFS; C. S. Chapman to Forest Supervisors, June 17, 1907, drawer 164, RG 95–4, RFS; *Use Book*, 1907, pp. 24–25.

The ranger examination was a tough test of proficiency. The men had to shoot, ride, use an ax, take a written test, and throw a diamond hitch, a knot of near-mythical difficulty used to lash freight on a mule or horse. Some forest supervisors complained that the test was too difficult and received permission to retest with more realistic standards those who had failed. Passing this examination was only part of the test. After a ranger was hired, he was subject to rigorous inspection. No housekeeping detail was too minor or too private to escape notice. Ranger stations were to be neat and sanitary, providing examples for campers. Rangers' privies had to be more than fifty yards from the house, with at least a six-foot vault.[36]

Other prospective Forest Service employees, the technical men straight out of college, did not escape the rigorous standards, either. To be forest assistants they had to be able to prepare working plans and function directly under the forest supervisor. At times they had to live down their college background, as locals usually valued practical knowledge over textbook theory. Yet, graduates of a certain college, some complained, started off with a plus. With the Forest Service resembling a Washington, D.C., chapter of the Yale Alumni Association, there was much disappointment among staff whenever Yale graduates faltered over the entrance requirements. In a few years, alleged preferential selection of Yale graduates would demand firm refutation by the chief.[37]

Pinchot broadcast the need for foresters by sending a stream of news releases to editors across the nation. He described a forester as a man who could "combine something of the naturalist with a good deal of the business man." The forester needed sound judgment and technical ability. Although his job brought modest financial rewards, it was indeed healthful and "national work of prime importance." Recruitment, examination, and inspection were key elements of the personnel program, and these efforts to get the best seemed to pay off. As a measure of esprit de corps, Pinchot once noted that his staff used

36. C. S. Chapman to Supervisors, Nov. 22, 1907; Clyde Leavitt to Supervisors, July 29, 1907, drawer 164, RG 95–4, RFS. The 1905 Transfer Act had required rangers and supervisors to be residents of the states in which they worked.

37. *Use Book*, 1907, pp. 22–23; Minutes of the Service Committee, June 2, 1909, box 113, RG 95–8, RFS. Much ill will was engendered whenever a member of the Robin Hood Society, a Yale forestry school club, was promoted over a nonmember. Professor J. W. Toumey of Yale agreed with Chief Henry S. Graves that the situation was serious, even though unfounded. See Toumey to Graves, Feb. 2, 1911, and Graves to Toumey, Feb. 4, 1911, Records of Forester Henry S. Graves, 1911, Record Group 95, RFS.

only 7 percent of earned sick leave and 66 percent of their annual leave. Considering the stereotype of a bureaucrat, such figures indicate a good measure of his success.[38]

Expanding the Reserves

In addition to rangers, supervisors, and other administrative jobs, there were specialized assignments. Pinchot gave boundary survey highest priority, pointing out that hastily drawn reserve boundaries required adjustment and proposed reserves had to be investigated. Those assigned to the work viewed themselves as members of an elite group. On horseback, working alone, these men accounted for perhaps twenty thousand acres each day. Some claimed 3 million acres per man during a field season. The men had a skilled eye: despite such cursory examination, only 2 percent of the land selected was later judged to be inappropriate for inclusion.[39]

Pinchot's chief of boundaries recounts an anecdote that gives a human touch to an otherwise dusty-dry job of locating new national forests. In 1907 Arthur Ringland was dispatched from Washington, D.C., to the Midwest to rendezvous with Pinchot, who was about to take an extended leave and wanted to get the forests designated without delay. Ringland carried the maps and descriptions in a golf bag, even sleeping with it to safeguard the valuable documents. Meeting in Lansing, where Pinchot was to receive an honorary degree from the Michigan State Agricultural College, they cleared Pinchot's hotel room of furniture and spread the maps on the floor. The two worked in their stocking feet, Pinchot's blue pencil outlining new forests as Ringland padded around reading off descriptions. The chore finished, Ringland hurried to Washington with seventeen proclamations for Roosevelt's signature. The shoeless pair in Lansing's Hotel Downey had created or enlarged national forests in Arizona, California, New Mexico, Nevada, and Utah.[40]

Responsible for land accumulating in great clumps, this bureau of young men—Pinchot was thirty-nine when the reserves were trans-

38. Forest Service news release, n.d. [1908], FHS clipping file; Forest Service, *Annual Report* (1901), p. 326.

39. Forest Service, *Annual Report* (1904), p. 171; Pinchot, *Breaking New Ground*, pp. 252–54; DuBois, *Trail Blazers*, pp. 27–46; Clyde Leavitt, "Methods in Determining Reserve Boundaries," *Forestry Quarterly* 4 (Dec. 1906): 274–81.

40. Arthur C. Ringland, *Conserving Human and Natural Resources*, Oral History Interview by Amelia R. Fry et al., 1970, Regional Oral History Office, Bancroft Library, pp. 56–57.

ferred in 1905—faced its challenges with zest. Lacking a pool of trained manpower to draw on, or even accurate maps of their domain, field men had to be resourceful. Supervisors had to recruit a force of rangers, explore and map the million or so acres of their forest, locate and build ranger stations, locate and build a road and trail system, and then set about the routine business of implementing programs as provided by the *Use Book*. The demands were exhilarating, remembered Elers Koch, supervisor of the Bitterroot National Forest in Montana: "The feeling of proprietorship . . . is so great that it is almost as though [the supervisor] owned the land himself, and he gets about the same pleasure and satisfaction out of it."[41]

In addition to general administrative programs, in the office and in the field, assignments were divided into specific functions. Public information, game management, range control, and timber management were represented in the early Forest Service. Many of the concepts advanced then are still in evidence.

Public information received much of Pinchot's energy. The Forest Service distributed free of charge, through forest supervisors, an abridged, popular edition of the *Use Book* entitled *The Use of the National Forests*. The forty-two-page book began with a notice "To the Public," explaining the importance of having a general understanding of the national forest system. Pinchot went on to assure the reader that the national forests were for use and did not lock up resources or prohibit settlement. The homeseeker, prospector, lumberman, stockman—all legitimate users of the forests would be appropriately accommodated. To counter arguments from those concerned about eroding the tax base with a large percentage of public ownership, he explained how 10 percent of all receipts was turned over to the counties in lieu of taxes. Counties received $153,000 as their share of 1907 Forest Service activities. The net result of these policies was to keep the land productive while utilizing it fully. "The man who skins the land and moves on does the country more harm than good. . . . National Forests are made, first of all, for the lasting benefit of the real home builder. They make it impossible for the land to be skinned."[42]

The Forest Service mailing list totaled two-thirds of a million names

41. Elers Koch, "Forty Years a Forester," unpublished memoirs, n.d., 230 pp.; permission to quote from Peter Koch, Pineville, La.

42. C. S. Chapman to Forest Supervisors, July 24, 1907, drawer 164, RG 95–4, RFS; Forest Service, *Annual Report* (1907), pp. 15–16; *Use of National Forests*, 1907, pp. 5, 7–10. In 1908 receipts to counties were increased to 25 percent. Roosevelt had used the same phrase about skinning the land in an earlier speech to the 1905 American Forest Congress.

in 1908, when a newly acquired addressing machine displaced thirty clerks. Herbert A. Smith, responsible for public relations, complained that with all of their efficiency, the list was still not used to its full capacity. He wanted to continue sending large volumes of popular publications but thought scientific information of permanent value should be published, too.

Press releases, as well as other publications, were favored weapons in battles against opponents of Pinchot's efforts. He instructed all forest supervisors to get acquainted with the local newspaper editor and to take pains to explain the various aspects of Forest Service policy. After all, Pinchot told his men, the users of national forests read those newspapers. "Use the press first, last, and all the time if you want to reach the public."[43]

Dealing with a grass roots constituency suited Pinchot well. A survey in Idaho and Colorado indicated his basic success: 80 percent of those questioned in Idaho were in favor of Forest Service programs, while Coloradans were evenly divided.[44] In the case of Colorado, to have even half of the citizens supporting the Forest Service at that time was a marked improvement.

With the effectiveness of Pinchot's public relations activities came protests from his opponents that he misused public funds to bypass congressional authority. Finally, Congress attached a rider to the 1909 Appropriations Act providing that no federal funds could be used to prepare or publish press releases, magazine articles, or other forms of information. Pinchot correctly read this act as a rebuke but authorized his men to answer official requests for information. No longer, however, could he report a mailing list of 670,000 names and news releases to papers with a combined circulation of nearly 10 million readers.[45] Slowly Congress was resuming command.

The Range—Wild and Domestic

His general constituency was far from Pinchot's only concern. Special interests demanded special attention. Pinchot decided that rangers

43. Pinchot to Forest Supervisors, Apr. 26, 1907, drawer 164, RG 95–4, RFS; Pinchot, "Foresters in Public Service," n.d., copy in Edward P. Cliff Papers, FHS.

44. Minutes of the Service Committee, Dec. 29, 1909, RG 95–8, RFS.

45. Forest Service, *Annual Report* (1908), p. 35; ibid. (1909), p. 40; USDA, *Annual Report* (1908), p. 442; Minutes of the Service Committee, Nov. 24, June 23, Sept. 29, 1909, Dec. 16, 1908, boxes 112, 113, RG 95–8, RFS. For fuller discussion see Harold T. Pinkett, *Gifford Pinchot: Private and Public Forester* (Urbana: University of Illinois Press, 1970), chap. 11.

would be permitted to serve as state game wardens. Since 1891 game animals had been under state control, but now rangers could cooperate with the states as long as such work did not interfere with official duties. Proposals to set aside game refuges within national forests, however, he met with caution: the livestock industry must not be interfered with, and agricultural land must be freely available to farmers.

The emphasis in the wildlife sector was on predator control, which concerned stockmen more than sportsmen. The Forest Service hired hunters and spread poisoned bait in an effort to substantially reduce wolf and coyote populations. In some areas, such as Wyoming and New Mexico, the Forest Service worked with livestock associations to exterminate, not just control, wolves. By 1909 Pinchot boasted that during the year 51 rangers spending a total of 107 working days had destroyed 108 bears, 96 mountain lions, 144 wolves, 62 wolf pups, 3,295 coyotes, 571 wildcats, and 81 lynx. Stockmen were not satisfied, Pinchot noted. "It was impossible to meet the demand for hunters."[46]

Stockmen, the main beneficiaries of Forest Service wildlife policies, were inherited along with the reserves. Binger Hermann had ruled that the Act of 1897 gave the government the right to regulate grazing on the reserves and to insist upon permit applications prior to use. Pinchot issued nearly eight thousand grazing permits the first year following transfer, but stockmen still contested federal authority to regulate. The Arizona Supreme Court upheld the Forest Service position, but the battle would continue for years.[47]

The Department of Agriculture issued grazing regulations that favored stockmen living on or near the reserves. A small fee, less than one-third of the value received, was levied to defray costs of management. President Roosevelt, undoubtedly at Pinchot's suggestion, wrote to Secretary Wilson offering strong support for the grazing policy. Pinchot collected over one-half million dollars in grazing fees in 1906, noting general agreement by stockmen, with only some objections, that the fees were right and proper. The following year saw

46. Joseph E. Kallenbach, *Federal Cooperation with the States under the Commerce Clause* (1942; reprint ed., Westport, Conn.: Greenwood Publishers, 1962), p. 115; Minutes of the Service Committee, March 4, 1908, box 112, RG 95–8, RFS; Minutes of the Service Committee, Feb. 3, 1909, box 113, RG 95–8, RFS; Forest Service news release, Apr. 1909, copy in FHS clipping file; Forest Service, *Annual Report* (1906), p. 18; *Annual Report* (1909), p. 27.

47. Hermann to WTS, May, July 19, 1900, RG 95–22, RFS; Forest Service, *Annual Report* (1905), pp. 206–7.

more than a 50 percent increase in grazing receipts. By the end of Pinchot's administration, grazing contributed over one million dollars annually to the federal treasury.[48]

Range administration was closely tied to the Forest Service legal program. Although Congress had enacted laws applicable to grazing, the laws had to be interpreted. From Pinchot's point of view, Interior Department attorneys had a tradition of showing why a goal was not achievable under the law. Now that the reserves were in Agriculture, he wanted legal support.

Pinchot first selected a friend from his Yale days, George W. Woodruff, to be his law officer. When Woodruff transferred to the Department of the Interior in 1907, he was succeeded by Philip P. Wells. Both Woodruff and Wells shared Pinchot's philosophy. Utilizing cumbersome injunction procedures as an interim measure, they sought judges known to be supportive of Forest Service policies, filed charges against selected violators, and won favorable decisions. Avoiding vigorous prosecution when judges might side with the defendant, Wells accumulated a series of precedents in support of policy.[49] Two spectacular successes were not finally resolved until a year after Pinchot and Wells had left office.

The cases began in 1906. Stockmen defied grazing regulations to gain a court test of Forest Service jurisdiction. Fred Light challenged Forest Service permit requirements by turning 500 head of cattle loose to graze on the Holy Cross Forest Reserve in Colorado. Light ignored a U.S. District Court injunction against this open trespass, appealing to the U.S. Supreme Court for relief. The Colorado state legislature, frequently an opponent of the Forest Service, passed a special appropriation bill to defray Light's legal expenses. To no avail, however; on May 1, 1911, the higher court upheld the lower court injunction.[50]

Concurrently with Light's test case, Pierre Grimaud ran his sheep on the Sierra South National Forest in California, also without a

48. T. Roosevelt to James Wilson, Dec. 26, 1905, box 7, AFA Records, FHS, Santa Cruz, Calif.; Forest Service, *Annual Report* (1906), p. 16; *Annual Report* (1907), pp. 30–33; *Annual Report* (1909), p. 11.

49. Philip P. Wells, "Philip P. Wells in the Forest Service Law Office," *Forest History* 16 (Apr. 1972): 22–29 (excerpted from a lengthy memorandum to Pinchot, dated March 27, 1913, copy in Pinchot Papers); Forest Service, *Annual Report* (1907), p. 5.

50. Dana, *Forest and Range Policy*, p. 146; Forest Service news release, May 22, 1911, copy in FHS clipping file.

permit. Grimaud wanted to test the validity of making a violation of Forest Service regulations a penal offense. On May 3, 1911, the Supreme Court reversed a district court decision and substantiated Forest Service contentions. The case had been complicated because it was not clear whether the 1897 Sundry Civil Act applied to California. The two cases upheld the laws of 1891 and 1897, which gave to the secretary of agriculture the right to make rules and regulations for the national forests and to the government the right to levy grazing fees, and decided that the granting of a grazing permit did not grant a vested right to use.[51]

The Light and Grimaud cases, favorable to the Forest Service, reflected a sophisticated legal program. Superficially, the Forest Service during the Pinchot administration appears to have been a small band of zealots working long hours for low pay to rough out the national forest system and administrative machinery. To a certain extent this view is indeed accurate. Yet programs were well thought out and patiently executed. Wells continued Woodruff's campaign to carefully establish legal precedents for the conservation movement. The Forest Service strategists understood well what many refuse to accept, that courts by their decisions do make the law. By picking and choosing cases—and judges who were supporters of conservation—Wells was able to build up an impressive corpus of common law to give substance to hard-won legislative battles. *U.S.* v. *Grimaud* and *U.S.* v. *Light* were outstanding achievements of a dogged strategy.

Timber and the Timber Famine

Grazing was and is important on national forest lands. As with other uses, grazing had to be administered in conjunction with other activities. We have already seen how wildlife management directly served livestock needs; so, too, did grazing have to yield to larger priorities. District Forester E. T. Allen, soon to go to work for an industrial association, wrote from Oregon that "grazing is a matter of secondary importance on lands that should properly be given up to timber raising. . . ." Allen was particularly concerned about reseeding programs. On burned-over areas, careful consideration should be

51. Dana, *Forest and Range Policy*, pp. 146–47; Minutes of the Service Committee, Jan. 26, 1910, box 113, RG 95–8, RFS.

given to planting trees rather than sowing forage grasses. Chief and staff supported his position.[52] Timber on national forests was to receive higher and higher priority.

Timber policies evolved rather awkwardly during the late nineteenth century. The 1897 law provided authority necessary to cut and sell mature or dead timber on the forest reserves. Circular 21 provided cooperative machinery to produce management plans for private and public owners. The Department of the Interior had sought assistance from the Bureau of Forestry in the Department of Agriculture in administering the forest reserves. Since 1905 the Forest Service had handled the massive timber supplies on the national forests.

Foresters and lumbermen quickly recognized the influence the Forest Service could exert on American timber policies. Its ability to affect the price of logs and lumber by selling or not selling timber from its vast reservoirs was a powerful lever. Forest Service policy, however, held the public forests to be supplemental to, not competitive with, private holdings, and national forest timber was to be made available only to meet local needs.[53] Statistics show that this policy was followed with care.

Pinchot in 1905 had only twenty-six technical foresters assigned to make cutting plans for 63 million acres of forest lands. Despite the sparsity of men, he saw this handful as the definite beginning of the practice of forestry. Timber sales were minuscule, however, compared to national production: less than 100 million board feet came from Forest Service lands, out of more than 40 billion cut on all lands. Forty percent of this 100 million was dead and damaged—salvage sales only. Sales tripled the next year, and in 1907 tripled again, reaching 950 million board feet valued at 2.5 million dollars. But 1907 was the peak year of lumber production, and the Forest Service's sales represented only 2 percent of the 44 billion total.[54] It would be nearly four decades before Forest Service timber sales constituted an important part of the national cut.

Pinchot, committed philosophically to running the Forest Service as a business venture, outlined for his staff a plan to ask Congress for a

52. Minutes of the Service Committee, June 23, 1909, box 113, RG 95–8, RFS.

53. Forest Service, *Annual Report* (1906), pp. 51–52, 267; William L. Hall, "Some Comments on the Government Timber Sale Policy," 1906, ms. in box 1, WFCA Records, Oregon Historical Society, Portland.

54. Forest Service, *Annual Report* (1905), pp. 209, 206; ibid. (1906), p. 19; ibid. (1902), p. 26.

5-million-dollar loan, using the forest reserves as collateral. The cash would be working capital to get desired programs underway. After ten years the Forest Service would start paying off the loan at a rate of 1 million dollars per year. He asked Hall, Benedict, and Sherrard to prepare a list of expenses to be charged to this account. The staff discussed the proposal at subsequent meetings and decided to proceed. They learned from Price that a House committee had listened to Pinchot's proposal with favor but had thought a 5-million-dollar lump sum was too large. Better to loan in installments, the congressmen believed, beginning with 2 million dollars. Pinchot had told Price that the committee seemed to favor direct appropriations, rather than a loan.[55] This approach, like others, failed to bear fruit.

The goal of self-sufficiency eluded Pinchot. Statistics published in 1906 show the trend.

Year	Revenue	Expenditures	Deficit
1901-2	$ 24,431.87	$325,000.00	$300,568.13
1902-3	45,838.08	300,013.50	254,175.42
1903-4	58,436.19	379,150.40	320,714.21
1904-5	73,276.15	508,886.00	435,609.85
1905-6	767,219.00	979,519.00	212,300.00

The following year was little better: 1.53 million dollars in receipts failed to offset 1.83 million dollars spent. The gap remained. More encouraging were the increases in national forest acreage, up 11 percent to 168 million acres, and in timber sales, up 236 percent. Even then, receipts failed to pass relentlessly growing expenditures.[56] Pinchot finally accepted the inevitable.

When Pinchot gave up trying to show a cash profit, he rationalized that "the National Forests exist not for the sake of revenue to the Government, but for the sake of the welfare of the public." He pledged the Forest Service to work toward stabilizing local industry and encouraging development where none existed. He promised that the Forest Service would not take advantage of local needs to extract monopoly prices. Sales would be offered only where there was a local demand, thus avoiding competition with private supplies. Pinchot clearly stated, however, that scientific forestry could only "be brought

55. Minutes of the Service Committee, Oct. 25, Nov. 14, 28, 1906, Jan. 10, 1907, box 112, RG 95–8, RFS.

56. USDA, *Yearbook of Agriculture* (1906), p. 525; USDA, *Annual Report* (1907), p. 62.

about by the axe." Public timber must be cut in order to transform national forests "from a wild to a cultivated forest. . . ." He reminded the readers of his 1908 annual report that based upon current inventories, the total timber supply on national forests could meet demands for only ten years, if full national dependence fell to these stands.[57] Obviously, adequate future supplies relied heavily on private holdings, and official attention to nonfederal forestry affairs remained substantial.

Following the reserve transfer, industrial applications for assistance under Circular 21 continued, numbering 167 in 1905. Since the beginning of the cooperative program, working plans had been prepared for nearly 11 million acres at an average cost to the owner of two cents per acre. From a questionnaire sent to all cooperators Pinchot learned that 75 percent had adopted the plans "and are now lumbering conservatively or in some other way apply practical forestry." Within three years of the transfer, however, he reported that the greatly increased administrative burden was forcing a reduction of emphasis on cooperation. Even so, he formally recognized an obligation to provide assistance and in 1908 established a division of State and Private Forestry. One of the division's first assignments was to aid states in legislative studies in forest taxation, a topic of great and continuing concern to private owners.[58]

The Forest Service supported industrial forestry in ways other than working plans and tax studies. Since most of the wood needed in the future would come from private holdings, Pinchot felt fully justified in promoting what he called conservative logging. Beginning on July 1, 1908, the Forest Service published a monthly wholesale lumber price list for the twenty principal market areas. Industrial trade associations had been circulating price lists until the Justice Department cracked down on the practice as a violation of antitrust legislation. Pinchot offered his services as a substitute, because it seemed unlikely that the attorney general would intervene with another federal agency.[59]

Pinchot considered the price lists to be in the public interest.

57. Forest Service, *Annual Report* (1908), pp. 18, 19, 15.

58. Ibid. (1905), pp. 210–11; ibid. (1907), p. 29; ibid. (1908), pp. 29, 34–35; ibid. (1909), pp. 30–33.

59. The whole issue of lumber trade associations and price lists has yet to be explored fully. For a brief summary, see Harold K. Steen, "Forestry in Washington to 1925," Ph.D. diss., University of Washington, 1969, pp. 188–95, 280–82.

Lumber was a necessity and susceptible to monopoly profit as supply decreased. Additionally, "cutthroat" competition would bring wasteful use and earlier exhaustion. To him, the only acceptable alternative to public ownership was to strike a balance, a balance perhaps requiring public regulation. Pinchot rationalized that the price lists, of great potential value to the lumbermen, also provided a means to judge whether the public interest was being served.[60]

The Forest Service supported the lumber industry, too, in its efforts to retain a tariff on lumber. Tariffs to protect fledgling American industries from competing British and European enterprise and to produce revenue for government operation were enacted by the first Congress and signed by George Washington. Lumber had received protection along with most manufactured items. After more than a century of shelter, lumbermen faced with apprehension the possibility of a free international market. Realizing the futility of asking to have profits protected, lumbermen, as did the Forest Service, presented tariffs as a conservation measure. A trade journal portrayed two scenes following logging, one neatly cut when lumber prices were high, the other with waste piled high during a market slump. The contrast was striking and the relation between high prices and fuller utilization was graphic.[61] Tariffs obviously were in the public interest.

To Pinchot's staff, reducing or eliminating the tariff seemed to serve no useful purpose. If any profits were to be made, Canadian lumbermen, not the American public, would realize them. Shying away from open involvement in national politics, Pinchot and his staff maintained that their sole interest in the lumber tariff was as a conservation measure. "It is foolish and unjust to denounce the lumberman as a vandal, or to accuse him of willfully destroying our forests," wrote one of Pinchot's staff. After all, the lumbermen could sell only what the public would buy; it was not his fault if no one wanted low-grade lumber. The lumberman was a businessman, not a philanthropist, and in general was as "conservative" (that is, practiced conservation) as he could afford to be.[62]

Tariff was a conservation measure in the following sense. Every tree contains both low- and high-grade lumber; and when a tree is cut,

60. Forest Service, *Annual Report* (1909), p. 35.

61. *Timberman*, Dec. 1908, p. 40G.

62. Minutes of the Service Committee, Feb. 27, 1909, box 113, RG 95–8, RFS; "Forest Conservation and the Tariff upon Lumber," 1909, ms. copy in drawer 239, RG 95–115, RFS.

only those parts that can hold their own in the marketplace are used. Most western lumber had to be shipped to a distant market, as local markets were saturated. Since it cost as much to ship a low-quality as a high-quality board, only the better materials could justify shipment. This situation left large quantities of low-quality lumber to glut western markets, giving the lumbermen incentive to leave much material at the scene of logging rather than haul it to a mill. If tariffs would keep prices up, then less material would be left to rot in the woods. Thus, tariffs contributed to conservation, or at least to a reduction in waste.

Pinchot testified in this vein to the House Ways and Means Committee in 1909, supporting tariff retention. Trade journals lauded his position, and one lumberman admired his "moral courage" to support the industrial position. Industrial testimony was carefully groomed to support the contention that conditions were grave. The industrial campaign with Pinchot's support was partially successful: Congress retained the tariff on lumber, but at a reduced rate.[63]

Pinchot thought that his support of the lumber industry position on tariff, as well as holding national forest timber off the market in order to be noncompetitive, obligated the industry to operate in the public welfare. If the industry failed to redeem this responsibility, then the government could justifiably intervene. At that time, cooperation was still the watchword for Forest Service–industrial relations. The only serious attempt at regulation had been initiated by the lumbermen themselves, who were willing to exchange some of their independence in return for federal permission to form a trust, or private utility. They worked out the details with Philip Wells of the Forest Service Law Office, but at the last minute some members of the proposed association bolted and the regulation issue simmered for a decade.[64]

Despite Pinchot's open support for the industry, he is best remembered for launching a high-powered conservation campaign. The precise causal relationship is difficult to establish, but in 1907, the same year a glutted lumber market and depressed prices distracted industrial attention away from cooperation, Pinchot reported the need for

63. U.S. Congress, House, Ways and Means Committee, *Tariff Hearings*, 60 Cong. 1 and 2, 9 vols., 1909, 8: 8079; *American Lumberman*, March 13, 1909, p. 35; T. B. Walker to Pinchot, Apr. 3, 1909, drawer 380, RG 95–115, RFS; Frank H. Lamb, "We Pioneers," unpublished typescript, University of Washington library, 1950, p. 157; F. W. Taussig, *The Tariff History of the United States*, 8th rev. ed. (New York: Capricorn Books, 1964), pp. 380–83. The tariff issue must also be examined in context with President Taft's policies and the concurrent congressional reaction.

64. *Tariff Hearings*, p. 8079; Wells, "Philip P. Wells," pp. 26–27.

"concerted action to avert the calamity of an exhausted timber supply."[65] The "timber famine," predicted by Secretary of the Interior Carl Schurz in 1879, became a near motto for Pinchot during the latter part of his administration. It is incongruous that Pinchot could discuss the problems caused by glutted lumber markets and problems of scarcity both at the same time, but he did. No one seemed bothered by the apparent inconsistency.

Three years later Pinchot wrote *The Fight for Conservation*, a short book that brought his fears into sharp focus. Acknowledging that estimates of timber supply were subject to error, he believed them reliable enough to be "certain that the United States has already crossed the verge of a timber famine so severe that its blighting effects will be felt in every household in the land." Cost of houses and buildings would rise, mining would become difficult on account of shortages of mine props, railroads would run out of ties, and food prices would increase as water supplies diminished. All this havoc would occur, according to Pinchot, because of the "suicidal policy of forest destruction" which Americans had allowed. Using the analogy that trusts spread their power like an oil slick on water, Pinchot predicted that little could be done once such market coverage was complete. Conservation would have to save the little man; conservation would result in national efficiency. The Forest Service was one of the chief agencies of the conservation movement. Public support for the bureau and its goals would stem the tide.[66]

The whole doom and gloom notion of a timber famine has remained a vivid picture in the public mind. Ironically, chronic surplus, as measured by the marketplace, seemed to be a major contributing factor. Using the same logic that made tariffs a conservation measure, high prices in general would cause the lumberman to reject his wasteful ways. It was supply in excess of demand that drove lumber prices down. Surplus and scarcity were directly related, but explaining the apparent contradiction to a busy public was a problem. It seemed more prudent to illustrate predictions of famine with photographs of waste. To bring up glut would only weaken the argument that something immediate had to be done. At least in one sense the Forest Service campaign was successful; the visage of a woodless world be-

65. Forest Service, *Annual Report* (1907), p. 4. Roosevelt had used the phrase "timber famine" in a speech to the 1905 American Forest Congress.

66. Gifford Pinchot, *The Fight for Conservation* (1910; reprint ed., Seattle: University of Washington Press, 1967), pp. 15, 16, 17, 29, 50.

came burned into the public eye. Sixty years after Pinchot left office, the Forest Service would be trying to convince a skeptical—even cynical—public that logging in itself did not necessarily contribute to scarcity. But the specter of a timber famine would be difficult to eradicate.

Conservation Movement

Pinchot's activities ranged far wider than the Forest Service. He occasionally served as presidential emissary, but usually his attention focused on some aspect of conservation. As an old man, he liked to remember how the whole concept of conservation had suddenly flashed through this mind while riding horseback in Rock Creek Park of Washington, D.C. He saw that all resources were to some degree related and had to be dealt with by a unified approach. Comparing the value of his revelation to that of the Declaration of Independence, and with support from Overton Price and W J McGee, he took it to Roosevelt. In Pinchot's words, the president "understood, accepted, and adopted it without the smallest hesitation. It was directly in line with everything he had been thinking and doing. It became the heart of his administration." They christened their newly found philosophy "conservation."[67]

The impact upon American consciousness was dramatic. Telling measures of this impact were political platforms. In 1904 Democrats and Republicans held up the 1902 Reclamation Act as the salvation of arid lands but made no other allusions to conservation. No third-party demands for conservation were in evidence. Four years later the Democrats came out four-square for the Republican president's conservation crusade and taunted the GOP for not supporting Roosevelt. Not to be outdone, the Republicans accepted credit for progress in the field of conservation and approved "all measures to prevent the waste of timber."[68] From 1908 onward, conservation was a common element of political rhetoric, reportedly causing President William H. Taft to lament that whatever conservation was, everyone was in favor of it.

Water—either too much or not enough—was a vital element of the

67. Pinchot, *Breaking New Ground*, pp. 319–26. It can be appropriately argued that others before Pinchot had grasped the same conservation principles. What is important is the effectiveness of Pinchot's use of the concept.

68. Kirk H. Porter and Donald B. Johnson, comp., *National Party Platforms, 1840–1968* (Urbana: University of Illinois Press, 1970), pp. 133, 137, 149, 160.

conservation movement. Controlling floods and assuring water supplies had been a forestry issue from the beginning. Western support for reclamation and irrigation in the latter part of the nineteenth century fit well with growing concern in the East for the nation's forests. A reading of Pinchot's autobiography suggests that he frequently viewed water problems as more than just strictly forestry problems. His key role in merging reclamation with forestry was a major contribution to the success of the conservation movement. Throughout his tenure as chief, he saw the Forest Service and his responsibilities in the broadest of contexts. Never again would the agency be so aggressively cosmopolitan.[69]

As part of this broad perspective, Roosevelt appointed the Inland Waterways Commission in 1907 to design multiple-purpose development of river basins. Irrigation, navigation, flood control, and hydropower production were to be coordinated. No longer would single-use projects dominate a region. The commission's greatest success was in advertising conservation. Roosevelt was pleased and invited the governor of each state to attend a White House conference to examine the many ramifications of conservation.[70]

The governors assembled at the White House in May of 1908. The president told them that "the conservation of our natural resources is the most weighty question now before the people of the United States." Governors or their representatives spoke in turn, each expressing belief in conservation. Industrialists, such as Andrew Carnegie and James Hill, gave similar speeches. Oddly, only one lumberman, R. A. Long from Kansas, was given the floor. Other missing faces—John Muir, Edward Bowers, Charles Sargent, Bernhard Fernow—are even more difficult to explain. Pinchot dismissed these omissions by claiming that it was not a forestry conference, having grown out of the Inland Waterways Commission.[71] This was a rather feeble excuse, considering Pinchot's broad-gauged definition of conservation. It is apparent that he chose not to share the considerable limelight with others or risk challenges from the floor.

69. For an excellent analysis of that broad view of conservation, see Samuel P. Hays, *Conservation*.

70. Ibid., chap. 6; Pinchot, *Breaking New Ground*, pp. 326–33. Two decades later the Inland Waterways Commission would serve as a model for the Tennessee Valley Authority.

71. Pinchot, *Breaking New Ground*, pp. 344–55; U.S. Congress, *Proceedings of a Conference of Governors in the White House, Washington, D.C., May 13–15, 1908*, 1909, p. 3 and *passim*; Robert U. Johnson to Pinchot, Sept. 4, 1908, and Pinchot to Johnson, Sept. 15, 1908, box 576, Pinchot Papers.

A useful product of the governors conference, besides its immediate publicity and creation of an annual tradition of similar conferences, was the recommendation to establish the National Conservation Commission. Roosevelt promptly appointed such a commission "to inquire into and advise me as to the condition of our natural resources." As with the Public Lands Commission, Inland Waterways Commission, and the governors conference, Pinchot received a key assignment and was named chairman. The three-volume report of the National Conservation Commission is a compendium of thought covering the broad aspects of conservation, which by now had been expanded to include health of children, civic beauty, and waste in war. From the national conference developed the North American Conservation Conference, then the World Conservation Conference.[72]

The Roosevelt-Pinchot duet cannot be faulted for lack of zeal, and the series of conferences they organized obviously had merit. Perhaps they attempted too much, too fast. They failed to secure their flanks before plunging ahead. In retrospect, the turning point came in 1907. Foes of the conservation movement, largely in Congress, had been knocked off balance by Roosevelt's aggressive assertion of executive authority to implement Pinchot's proposals. The regaining of governmental equilibrium was nearly imperceptible, but effective.

Much of the opposition to the conservation movement came from westerners angered by what they viewed as eastern-based interference with their prerogatives. They charged that easterners wanted to prevent use of western resources, which would cripple economic growth. Where, they demanded, was the concern for conservation when the eastern portion of the nation was developing? It seemed hypocritical to express interest in conserving resources only when they belonged to someone else. This attitude partially explains the western explosion prompted by President Cleveland's Washington's Birthday reserves in 1897. Creation of forest reserves, or national forests, seemed to many to be a loss of local control over valuable resources.

Actually, the loss was only potential, since the forests were simply transferred from one form of public ownership to another. The difference, however, was that the vast bulk of the public domain was originally destined for private ownership under state and local juris-

72. U.S. Congress, Senate, *Report of the National Conservation Commission*, 3 vols., Sen. Doc. 676, 60 Cong. 2, 1909, *passim*; Pinchot, *Breaking New Ground*, pp. 355–67; Hays, *Conservation*, p. 176.

diction, while the national forests would forever be controlled by the distant Washington, D.C. As a point of fact, Pinchot's insistence upon decentralization was largely designed to overcome this sort of opposition.

In 1907 Senator Charles W. Fulton of Oregon, who a few years earlier had been implicated in the land scandal that had caused Binger Hermann to resign as commissioner of the General Land Office, introduced an amendment to the agricultural appropriation bill. The amendment forbade presidential creation of national forests in the six western states of Washington, Oregon, Idaho, Montana, Wyoming, and Colorado, authorizing Congress alone to establish reserves.

Arthur Ringland recounted one of the events that triggered Fulton's action. Ringland, as one of Pinchot's elite "boundary boys," working throughout the West, in late 1906 was ordered to examine public lands near the Canadian border of Washington State. He had to hurry, as the General Land Office had scheduled opening the lands for entry on February 6, 1907. Ringland wired his report to Washington, D.C., and on January 7, 1907, thousands of acres of prime Douglas fir were placed under Forest Service control by presidential proclamation.[73]

"Then the storm broke." To Ringland, the uproar seemed unparalleled. The local press, chamber of commerce, and Washington's congressional delegation protested the reservation, charging that it created a great hardship on local residents. The Seattle *Times* belittled Ringland's efforts as "making black spots on the map of Washington" and demanded drastic measures to curtail future activities of the "alleged expert." Senator Samuel H. Piles threatened legislation to restore the area to the public domain. Pressure became so intense that Pinchot yielded, and Roosevelt wired Washington's Governor Albert E. Mead his regrets for the "clerical error" that had caused the whole incident. Revealing the full extent of the pressure the president faced, he added: "It is not to be charged to our forest policy."[74]

The act of March 4, 1907, renamed the forest reserves as national forests, permitted export of national forest timber from the state in which it was cut (except for South Dakota, as a sop tò Senator Kittredge), and allowed only Congress to create or enlarge national

73. Arthur Ringland, *Conserving Human and Natural Resources*, p. 43 H.

74. Seattle *Times*, Feb. 11, 1907; Ringland, *Conserving Human and Natural Resources*, pp. 45–50. Nearly fifty years later the area Ringland had recommended be set aside was owned by private companies and managed as commercial tree farms. Harold C. Chriswell to Ringland, Nov. 9, 1965 (letter in Ringland's possession).

forests in the six states. The law further abolished the special fund whereby Forest Service receipts could be spent directly on national forest projects, increased the agency's appropriations by $1 million, and raised Pinchot's salary by over 40 percent to $5,000 per year.[75] Obviously, reassertion of authority, not punishment, was Congress' goal.

Roosevelt signed the bill into law. Needing appropriations, he had no real choice. Prior to signing, however, he proclaimed 16 million acres of new reserves in those same six states. The boundary survey work had provided a backlog of data, which by grueling, round-the-clock sessions was quickly translated into presidential proclamations. Roosevelt defended his action by claiming that he had saved vast tracts of timber from falling into the hands of the "lumber syndicate." Remembering the congressional uproar that ensued, he later gleefully recounted how "the opponents of the Forest Service turned handsprings in their wrath." He saw their threats as a "tribute to the efficiency of our action."[76] Maybe so, but the setting aside of the so-called midnight reserves was really the last flamboyant act of the conservation movement. Perhaps realization of this fact caused Pinchot, a couple of years later, to take a truly drastic step.

Insubordinate and Bounced

Pinchot's relationship with the Department of the Interior had been congenial since 1897, when he had worked as a confidential agent. The one important exception had been Binger Hermann, commissioner of the General Land Office, who resigned in 1903 under a cloud during the Oregon land scandals. Secretary Hitchcock had been amenable, if unimaginative, toward Pinchot's views. Since Roosevelt's re-election in 1904, James Garfield had been secretary, and Garfield and Pinchot got along famously. With Secretary of Agriculture Wilson providing free rein and with vigorous support from Roosevelt, Garfield's cooperation rounded out whatever executive needs Pinchot

75. 34 Stat. 1256, 1269.

76. Roosevelt, *Autobiography*, pp. 404–5; Pinchot, *Breaking New Ground*, p. 300; memorandum dated March 2, 1907, in Theodore Roosevelt, *The Letters of Theodore Roosevelt*, ed. Elting E. Morrison, 8 vols. (Cambridge, Mass.: Harvard University Press, 1951), vol. 5: *The Big Stick, 1905–1907*. It is interesting to note that the local press portrayed the new reserves as playing into the hands of the lumbermen. The proclamation made private holdings even more valuable by reducing the supply of public domain, which would eventually be in private ownership. Seattle *Times*, March 5, 1907; Seattle *Post-Intelligencer*, March 7, 8, 1907.

may have had. Then in 1908 the American electorate selected William H. Taft to succeed Roosevelt.

Actually, Roosevelt had handpicked Taft, believing that this choice would guarantee continuation of his policies. From Pinchot's perspective, Taft's first breach of confidence came early when he replaced Secretary Garfield with Richard A. Ballinger.

Ballinger had served briefly as land office commissioner under Garfield and then as mayor of Seattle. He quickly put a stop to Pinchot's easy access to Interior personnel. The chief forester would have to go through prescribed channels, meaning Pinchot to Wilson to Ballinger to whomever in the department he wanted to see. Ballinger also retrieved administration of forestry affairs on Indian reservations from the Forest Service, a responsibility that Pinchot had been particularly fond of.[77] A clash was inevitable.

Alaskan coal claims prompted the final battle. Pinchot viewed the processing of these claims by the General Land Office as contrary to the public interest, and probably illegal. Blatantly ignoring bureaucratic protocol, he publicly criticized Ballinger's handling of the situation. The instant he did so, Pinchot forced his own firing.[78]

Taft, in fact, did support the conservation-oriented policies inherited from Roosevelt. He had even promised to keep Pinchot on as "a kind of conscience." But Taft was a stickler for proper administrative procedures. He saw Pinchot's battle with Ballinger as "a row over nothing very serious, except a kind of jealousy and friction between Departments that ought to be avoided." The deeper problem, to Taft, stemmed from Roosevelt's having given Pinchot powers that were better held at the secretarial level. The president was determined to prevent any of his department heads from being undermined by anyone, even Pinchot. Nor would he overlook an open attack by the chief of a bureau in one department on the secretary of another. To do so would be "very bad for government discipline." Taft saw Pinchot as "a good deal of a radical and a good deal of a crank" and suspected that he was looking for martyrdom. If so, the president would oblige, though reluctantly. For Taft to seek Ballinger's resignation when un-

77. J. P Kinney, *The Office of Indian Affairs: A Career in Forestry*, Oral History Interview by Elwood R. Maunder, 1969, FHS, pp. 9–11.

78. For a detailed account see James Penick, Jr., *Progressive Politics and Conservation: The Ballinger–Pinchot Affair* (Chicago: University of Chicago Press, 1968); Pinkett's *Gifford Pinchot*, pp. 114–29, contains an excellent summary. See also Pinchot, *Breaking New Ground*, pp. 413–58.

fairly attacked would make him a "white livered skunk."[79] It was Pinchot who must go, and Taft asked for his resignation on January 7, 1910.

Pinchot was leaving his home to have dinner with Philip Wells when a presidential messenger handed him two envelopes. He opened one, then hurried to join his friends. He arrived a little late, cheerfully showing his dinner companions the two envelopes and announcing, "This tells me I've been bounced. Excuse me if I try to find the reason." Then he read Taft's letter of dismissal aloud, saying, "Possibly you may care to hear what it says." The other letter was from Secretary Wilson, naming Albert Potter as acting chief. Pinchot, in high spirits, telephoned the news to friends and associates. Other guests were stunned. When Wells's twelve-year-old son burst into tears, Pinchot tried to stem the flow: "I'm all right. I'm just as happy as a clam."[80]

Pinchot remembered his firing as a moment of triumph. In one sense, he was right. The lengthy congressional investigation into the affair received widespread publicity and gave Pinchot full opportunity to take his case to the public. Congress evaluated the contest along strictly political lines, and Ballinger was formally exonerated. Pinchot was satisfied, however, for he was now unencumbered by government regulations, and he believed the Forest Service he had shaped could pretty well run itself. Hindsight shows that he misjudged.

79. Donald F. Anderson, *William Howard Taft: A Conservative's Conception of the Presidency* (Ithaca, N.Y.: Cornell University Press, 1965), pp. 72–76.

80. Thornton T. Munger, "Recollections of My Thirty-eight Years in the Forest Service, 1908–1946," *Timberlines*, supplement to vol. 16 (Dec. 1962): 13–14. Munger was Wells's brother-in-law and present at the dinner. He quotes his diary entry for Jan. 7, 1910; Taft's letter of Jan. 7 is reprinted in Pinchot, *Breaking New Ground*, pp. 451–53.

CHAPTER V

Administration, Cooperation, Research

PINCHOT WAS STILL jubilant over being fired when he bade farewell to his staff the next morning. "Stick to the Forest Service," he advised in an inspirational mood. Associate Forester Overton Price, also fired, was too upset to attend. Philip Wells, Pinchot's law officer, would soon resign in protest. The assembled crew, stunned by the news that the chief had been fired, was ready to "fight to the death for Pinchot and his policies." Loyalty was at such a fever pitch that Albert Potter, the acting chief forester, felt it necessary at the next staff meeting to caution that Pinchot's address was not a call for greater loyalty to him than to the administration. Potter warned the men that although they were free to discuss forestry matters with Pinchot, they "must not seek his advice regarding the business of the Forest Service." The staff assured Potter that they had not misinterpreted Pinchot's remarks and would act properly.[1]

Rumors that Taft would name A. P. Davis of the Bureau of Reclamation to be chief prompted a meeting between Herbert Smith and Pinchot. They quickly agreed that the best candidate would be Henry Graves, Pinchot's close associate from the 1890s and now dean of forestry at Yale. Here might be a figure acceptable to all—an eminent professional forester, yet an outsider not formally committed to Pin-

1. Thornton T. Munger, "Recollections of My Thirty-eight Years in the Forest Service, 1908–1946," *Timberlines*, supplement to vol. 16 (Dec. 1962): 14; Samuel T. Dana, *The Development of Forestry in Government and Education*, Oral History Interview by Amelia R. Fry, 1967, Regional Oral History Office, Bancroft Library, p. 62; S. T. Dana, personal communication, Jan. 29, 1974; Minutes of the Service Committee, Jan. 12, 1910, box 113, RG 95–8, RFS, National Archives.

chot's policies. Smith wired Graves to tip him off and at the same time asked United States Treasurer Lee McLung to persuade another Yale man, Anson P. Stokes, to bring Graves's name to the president's attention. Taft had considered Potter for the post, but the range specialist withdrew in favor of Graves. Taft accepted the recommendation and on January 12, 1910, only five days following Pinchot's dismissal, named Graves chief, effective February 1.[2]

Graves, reluctant to leave Yale, discussed the situation with Pinchot and finally took a one-year leave of absence. At the end of that year Taft, satisfied with his chief forester, asked President Hadley of Yale to extend Graves's leave. But Graves, as have many others, gave in to a direct presidential call to service and resigned from the deanship, knowing that now he would remain in Washington. He waited nearly three months before telling Pinchot of his decision, but to an associate he explained that he could not return to Yale "as long as I succeed in doing my part effectively." He would stay because of "very hard fights ahead" until the "Forest Service and the Service policies are safe. . . ."[3]

Administration of the National Forests

Graves was qualified by education and experience to administer the Forest Service. His background was assurance that professional foresters, not unqualified political appointees, would run the agency. It was probably good that he lacked Pinchot's flamboyance, for Congress seemed in no mood for another maverick. He did have a strong personality in his own right. Men who worked under Graves almost uniformly remember his piercing black eyes. He could be arbitrary: he frowned on smoking in the Forest Service headquarters and forbade whistling. Official reports lacked humor or zest. Yet, for all the strongly puritanical, no-nonsense administration suggested by contemporary accounts, Graves commanded respect and even affection from his staff.[4]

2. Herbert Smith, memo dated Sept. 11, 1941, box 576, Gifford Pinchot Papers, Library of Congress; Graves to Pinchot, Jan. 19, 1910, ibid.; Hadley to Graves, March 11, 1911, Records of Forester Henry S. Graves, 1911, Record Group 95 (hereafter cited as RG 95–29), RFS; Graves to Stokes, May 20, 1911, RG 95–29, RFS.

3. Graves to Pinchot, May 29, 1911, RG 95–29, RFS; Graves to F. E. Olmsted, March 4, 1911, ibid.; Henry Clepper, *Professional Forestry in the United States* (Baltimore, Md.: Johns Hopkins Press for Resources for the Future, 1971), pp. 55–56. Graves also had industrial support to remain as chief. E. T. Allen to Graves, Dec. 29, 1910, box 68, Henry Graves Papers, Yale University.

4. Dana, *Development of Forestry*, p. 64; George L. Drake, *A Forester's Log: Fifty Years in the*

Facing the huge task of rebuilding shattered morale, Graves concentrated on internal operation. Pinchot's conservation juggernaut had been permanently deflected by presidential support for Ballinger; Forest Service opponents who had fumed for nearly a decade now moved to see to it that the agency remained in check. The new chief first had to mend the relations between his office and Secretary Wilson. Wilson believed in giving his bureau chiefs a relatively free hand but had felt that Pinchot spent too much time on nonforestry matters. For a time, Wilson required forest officers to ask for his permission before making speeches. After all, Pinchot had stood at the center of a major public controversy and had been fired for insubordination. The secretary wanted to restore discipline and to re-establish a work routine. Graves was appalled when he discovered that the secretary was conducting a private investigation of the Forest Service and corresponding directly with field men without routinely routing the letters through him. Graves protested to Taft and approached Wilson directly, promising loyalty in return for trust. Their relationship improved immediately and remained satisfactory.[5] Graves turned to other matters.

The weakened condition of the Forest Service rekindled interest in state control of the national forests, harking back to 1897 and the furor over Cleveland's Washington's Birthday reserves. Graves believed that waterpower and mining interests were prompting western congressmen to advocate removal of the forests from federal supervision. This transfer would result in a loss of "at least two billion dollars worth of property" and would jeopardize the headwaters of navigable streams. Even worse, in Graves's view, would be the loss of the Forest Service's professional supervision to political appointees in rather ragtag state agencies.[6]

Graves worked strenuously to retain the national forests. He cautioned his staff, however, not to use scandals in the Department of

Pacific Northwest, Oral History Interview by Elwood R. Maunder, 1975, FHS, Santa Cruz, Calif., p. 132; Ralph S. Hosmer, "Early Days in Forest School and Forest Service," *Forest History* 16 (Oct. 1972): 8; Minutes of the Service Committee, Jan. 5, 1910, box 113, RG 95–8, RFS.

5. Dana, *Development of Forestry*, p. 64; Willard L. Hoing, "James Wilson as Secretary of Agriculture, 1897–1913," Ph.D. diss., University of Wisconsin, 1964, pp. 241–52; Overton Price to Jessie B. Gerard, Apr. 29, 1910, box 576, Pinchot Papers; Henry Graves, "Personal Recollections," cited in Clepper, *Professional Forestry*, p. 56.

6. Graves to Charles Lathrop Pack, Dec. 19, 1912, drawer 173, Records of the Office of the Chief, Record Group 95 (hereafter cited as RG 95–4), RFS; Graves, "Personal Recollections," cited in Clepper, *Professional Forestry*, pp. 57–58.

the Interior, which was enmeshed in local politics, as means for blocking transfer. He believed that many in Congress, even supporters like Senator Francis Newlands, the reclamationist so supportive of Pinchot's conservation efforts, favored the eventual return of the forests to Interior. The stronger strategy would be to show that the Forest Service was the best-qualified agency to manage the national forests. It worked. A series of bills to transfer the forests back to Interior came to naught. For the time being, at least, the national forests were safe. But more significant to Forest Service history is the constant drain of administrative energy that these maneuvers cost.[7]

"Forces of reaction," Graves thought in 1916, threatened the integrity of the national forest system. Even if the forests stayed under Forest Service control, it seemed as though his prerogatives were eroding. Often the Forest Service would contest land claims within national forests, claiming fraud. In Graves's view, Department of the Interior investigators approved the contested claims without adequate study. He blamed the situation on lack of direct presidential opposition to the constant "nibbling away" at the forests. Also at fault was a lack of teamwork between the Forest Service and Interior. Graves was convinced that he was charged with protecting the integrity of the national forests, and he saw that the two departments must work together.[8]

Shortly after Pinchot's departure in 1910, Acting Chief Potter attended a meeting between Ballinger and Wilson. The two secretaries had a "very satisfactory" discussion, sending a letter to Taft outlining the policy agreement. The communiqué stressed cooperation. Watershed lands would remain in national forests unless they were of greater value for agriculture; only land more valuable for timber production would be retained in national forests, and all other lands would be eliminated. Potter had further meetings with Ballinger and reported that the Interior secretary was willing to meet the Forest Service "more than half way" on matters of mutual interest to the General Land Office.[9]

Despite these official pronouncements of good will, Graves was

7. Graves to Sherman, June 18, 1912, drawer 173, RG 95–4, RFS; Clepper, *Professional Forestry*, pp. 59–60; John Ise, *The United States Forest Policy* (New Haven, Conn.: Yale University Press, 1920) (chapters 8 and 9 discuss transfer attempts and additional sources of contention between the Forest Service and its western opponents).

8. Graves to Secretary of Agriculture, June 19, 1916, box 13, Graves Papers.

9. Minutes of the Service Committee, Feb. 9, Apr. 13, June 16, 1910, box 113, RG 95–8, RFS.

nervous. Taft had ordered an investigation of the Forest Service accounting system and conferred with Potter on the matter. Potter tried to reassure Graves that there was no reason for alarm: the president only wanted to show congressional Democrats that the Forest Service operated efficiently. But then Taft suggested, perhaps benignly, that Graves reduce his budget request in order to prove it could be done.[10] Anyone would be skeptical.

The congressional Democrats to whom Taft was referring had bothered Potter, too. He warned his colleagues to be aware of the situation in the Senate, where a pending amendment to the appropriations bill required a report on all Forest Service expenditures from 1900 to 1910. Potter explained that many senators and congressmen thought "the Forest Service must settle down and stop the continual agitation." He added that Secretary Wilson had agreed and insisted "that the most important thing required of the Service now is strict economy and that it must guard against expenditures which may be classified as extravagant."[11]

Timber Management. Charges of extravagance and perhaps even insolvency were leveled at the Forest Service. Receipts from timber sales, grazing permits, and recreation leases perennially totaled less than expenditures. Critics claimed that Graves spent too little money directly on the national forests and too much on his clerical force, travel, and public relations.[12] Part of the dispute arose from different accounting philosophies, questions such as whether research and other expenditures not directly related to national forest administration should be defrayed by national forest receipts. Such issues can never be definitely resolved. Although the purpose of government is service, not revenue production, critics can and do seize upon dollar return versus expenditures, and the administrator must have an answer.

Fiscal pressure continued as Congress hacked at Graves's budget requests. A reduction of more than 1 million dollars in 1911 included almost a quarter million less for roads, trails, and bridges and a whopping $800,000 off the emergency fire-fighting fund.[13]

Graves knew the importance of getting congressional favor, and he counseled his field men accordingly. They worked for a federal service

10. Potter to Graves, Apr. 5, 1911, drawer 202, RG 95–4, RFS.

11. Minutes of the Service Committee, March 8, 1911, box 113, RG 95–8, RFS.

12. Ise, *United States Forest Policy*, pp. 284–88.

13. Minutes of the Service Committee, Dec. 20, 1911, box 114, RG 95–8, RFS; Pinchot to Garfield, March 21, 1912, box 117, James F. Garfield Papers, Library of Congress.

and were responsible to Washington, D.C. Now Congress wanted to know precisely how the Forest Service spent its money. The audit of expenditures would help. Potter was delighted with the results of the congressional inquiry, because it proved the Forest Service had managed its funds with care. For example, in 1900 travel accounted for 22 percent of the appropriation, but by 1910 this figure was reduced to less than 7 percent. Decentralization deserved a great deal of the credit for this saving, but in the main the Forest Service had simply been frugal.[14]

Even this vindication of financial responsibility did not satisfy Graves. He felt that to report a gross income of only 2.25 million dollars from millions upon millions of national forest acres could be interpreted as poor business management. First, he wanted to show that much of the forests was of value only for watersheds, not timber production, and that other areas would be used for future timber supplies. But most important, and reminiscent of Pinchot's similar aspirations, Graves wanted to show Congress how the Forest Service could be made self-sufficient.[15]

Listing the forests individually, Graves cited forty-four as already self-supporting. Eighteen forests he classed as primarily protective, destined always to operate in the red. Graves predicted twenty-two others would be in the black within one year, twelve more in two years, and the others during the next twenty-five years. In sum, within three years the Forest Service would return more to the treasury than it received, except for research costs and the maintenance of administrative offices. Everything would be paid for, Graves predicted, within twelve years.[16]

To the public, Graves confessed that national forest receipts were down but claimed the decline was due to a slump in the economy. He felt that it was legitimate to question why the service allowed costs to increase and receipts to diminish. If the Forest Service stopped all non-timber-related activities, Graves suggested, then the "forests could easily be made to show a net profit." He quickly added that such a decision would be a mistake, because it was more important to view the national forests as investments and not as immediate income.

14. Proceedings of the First Annual Supervisors' Meeting, district 3, Nov. 9–14, 1911, drawer 37, Research Compilation File, 1897–1935, Record Group 95 (hereafter cited as RG 95–115), RFS.

15. Minutes of the Service Committee, Apr. 3, 1912, box 115, RG 95–8, RFS.

16. Minutes of the Service Committee, Nov. 19, 1914, Dec. 10, 1913, ibid.

Anyway, to increase receipts by additional timber sales would injure private owners by further glutting the lumber market.[17]

Shortly thereafter, economist John Ise defended Graves's timber sale policy. To critics who said the Forest Service should charge more for its timber because lumbermen were making large profits on federal sales, Ise replied that elementary economic theory showed a commodity priced higher than market value would not sell. To critics who demanded lower prices, Ise invoked the same equation to show that competition would only drive nonfederal prices lower. In that case, private enterprise would be injured and the federal treasury would reap no increase in returns because of the lower prices. National forest timber should be sold at or above its appraised value. Besides, Ise reminded, that was the law; the Forest Service had no choice. To criticize the agency for congressional requirements was unfair.[18]

To assist him in developing a timber policy, Graves in 1910 had summoned William Greeley from Montana, where he had been district forester. Greeley would be head of timber management in the Washington office. He was confronted by editorials in the *Saturday Evening Post* which were critical of Forest Service timber policies. His response explained that over half of the national forest timber was "ripe for the axe and deteriorating in value." Timber could not be stored indefinitely, unlike coal. When timber rotted on the stump, it was a waste of public property, and waste, said Greeley, was the "very antithesis of conservation." He estimated that the Forest Service could increase timber sales twelvefold without exceeding annual growth. Since sales should take place only "under conditions practicable for the lumber operator," lack of markets or the "presence of privately-owned timber now being cut" would cause the government to hold back. Under no circumstances would the Forest Service "sacrifice the intrinsic value" of public property by a "bargain-day policy" of dumping timber on a glutted market. Additional safeguards to the public interest were competitive bids, price adjustment clauses in sales contracts, and the requirement that every railroad carrying national forest timber become a common carrier. This latter requirement prevented an operator from tying up adjacent timber by monopolizing the most desirable access.[19]

17. Forest Service, *Annual Report* (1912), pp. 13–15.

18. Ise, *United States Forest Policy*, pp. 289–92.

19. William B. Greeley, "Timber Sales in the National Forests," 2 drafts dated Aug. 17, 1912, and Jan. 20, 1913, copies in RG 95–115, RFS; *Saturday Evening Post*, Jan. 18, 1913.

Policies had to be flexible enough to accommodate the unexpected. In 1910 a 3-million-acre fire in the mountains of Idaho left an enormous amount of dead timber to dispose of before it decayed beyond use. The timber staff was concerned that private owners, already financially hurt by the same fire, would further suffer by influx of federal timber into the market. They decided, however, that the Forest Service was "morally obliged" to make use of all marketable timber. The staff concluded that in this case the impact on local markets would probably be minimal. A year later, Greeley reported that half a billion board feet of fire-killed timber from the Idaho burn had been sold, equaling the amount cut from all other national forest lands and nearly 14 percent of all timber cut in the United States from all sources in 1911.[20]

Even the most rudimentary developments in timber policy challenged the imagination and initiative of Forest Service staff. Nothing is more fundamental to forest conservation, for example, than reforestation following logging. Reforestation, like everything else, required expenditures of public funds, and expenditures needed authorization. When first approached, the solicitor for the Department of Agriculture would not allow the cost of planting trees following logging to be included in the sale contract. Such inclusion, he claimed, would be tantamount to increasing administratively the reforestation appropriation, which was a congressional prerogative. The solicitor reconsidered after learning from Greeley that the market value of seed trees left by the logger ranged from $25 to $40 per acre but planting following logging cost only $8 to $10. The straight economic benefits gained from planting seedlings instead of leaving seed trees were muted, however, by the fact that without seed trees a logged area would be a clearcut. Silvicultural knowledge at the time suggested that clearcutting be held to a minimum, along the lines of an experimental program, until more could be learned about the process.[21]

On the whole, however, timber management problems on the national forests were minimal, as the agency during its early years performed largely custodial functions. Of more interest were privately owned forest resources. In late 1906, South Dakota's Senator Kit-

20. Minutes of the Service Committee, Sept. 30, Dec. 8, 1910, box 112, Oct. 18, 1911, box 114, RG 95–8, RFS. Timber sold in one year does not mean that it was cut in the same year. Logging of the Idaho fire-killed timber probably spread over several years.

21. Minutes of the Service Committee, June 5, 1912, and Feb. 19, 1914, boxes 114 and 115, ibid.

tredge introduced a resolution calling for the secretaries of labor and commerce to investigate and report on high lumber prices. Specifically, Kittredge wanted to know of "any combination, conspiracy, trust, agreement, or contract intended to operate in restraint of lawful trade." A month later, apparently having decided for himself that a lumber trust existed, he introduced another version. Lumber was of great importance to the farming community, Kittredge told the Senate, and "the lumber trust is the king of combinations in the restraint of trade." To substantiate his charges, he claimed to have obtained uniform lumber price lists, supposedly from different regions but printed with the same type face on the same press.[22]

Kittredge found support, and by resolution Congress directed the Bureau of Corporations to study the lumber industry to see if its practices were in the public interest. Six years in preparation, the report indeed described monopolistic conditions, that is, concentration of timberland ownership. Additionally, the report accused the industry of protesting the national forest system as restrictive, while at the same time lumbermen were deliberately tying up access to public land for private gain.[23]

Advance reports of the Bureau of Corporations' findings disheartened industrial supporters. At Greeley's suggestion, the Forest Service countered by proposing a study of its own. The secretary of agriculture strongly endorsed a "constructive" review of overproduction and wasteful exploitation. Taxes, market conditions, fire, and other hazards were probable causes for much destructive logging. The Forest Service would work cooperatively with the Bureau of Corporations, the Forest Service studying trees from the stump to the sawmill, and the bureau concentrating on the mill to the market. Lumbermen professed full support, but Greeley reported that they still were a little suspicious.[24]

Whatever doubts may have existed, outwardly the Forest Service was hopeful. Claiming that the lumber industry recognized the need for and even welcomed additional studies, and noting rapidly diminishing timber supplies, the Forest Service stressed the urgency of the investigation. Top men were assigned. Greeley himself pieced

22. *Congressional Record*, 59 Cong. 2, Dec. 6, 1906, Jan. 18, 1907.

23. U.S. Department of Commerce, Bureau of Corporations, *Report on the Lumber Industry*, 4 pts., 1913–14, p. xxii. During the investigation, the Forest Service issued price lists, as industrial trade associations feared to do so. The Forest Service produced lists from May 1908 to the third quarter of 1912. Lists may be found in drawer 220, RG 95–115, RFS.

24. Minutes of the Service Committee, July 2, 15, 1914, box 115, RG 95–8, RFS.

together specific national aspects of the situation. He kept his associates informed on progress, related in optimistic terms.[25]

Although the industry generally supported the Forest Service–Federal Trade Commission (successor to the Bureau of Corporations in 1915) studies, some industrialists felt a still broader approach was needed. In the Far West, E. T. Allen suggested to Greeley that the Federal Trade Commission, the Forest Service, Department of Commerce, and the lumber industry should study the situation jointly. The secretary of agriculture was favorable to Allen's proposal and wrote to the secretary of commerce. We should "consider forest problems of common concern," Secretary David F. Houston wrote, adding that the cooperative efforts that the Bureau of Corporations report had stimulated ought to be kept alive.[26]

In an influential monograph, Wilson Compton, soon to become secretary-manager of the National Lumber Manufacturers Association, analyzed the situation from the industrial point of view. Lumbermen had acquired vast tracts of low-cost timberland as a speculative venture, believing that prices would certainly rise. Normally an entrepreneur would choose to hold on to assets until prices rose to an acceptable level, but many lumbermen could not meet carrying charges out of pocket. They were compelled to manufacture and market lumber even when prices were low, having no other means to pay off their creditors. This situation led to glutted markets, driving prices even lower. It was a vicious circle, and lumbermen seemed at a loss to find their way out.[27]

Greeley concurred with the industrial view. His study in response to the Bureau of Corporations report saw overcapitalization in forest land as the worst debilitating factor lumbermen faced. Interest rates simply were too high to allow holding land until prices were adequate. Also, since railroad freight rates were based on weight, not value, there was no export market for the lower grades of lumber, further glutting local yards with low-quality lumber. In desperation, western lumber was shipped eastward—unsold—in search of markets, making

25. Forest Service news release, July 27, 1914, FHS clipping file; Minutes of the Service Committee, May 27, Dec. 16, 30, 1915, box 116, RG 95-8, RFS.

26. E. T. Allen, "A Suggestion for Continued Co-operation between Agencies Responsible for Improving the Country's Forest and Lumbering Policies," 1916, box 1, WFCA Records, Oregon Historical Society, Portland; Greeley to Allen, March 23, 1917, ibid.; Houston to Secretary of Commerce, March 22, 1917, ibid.

27. Wilson Compton, *The Organization of the Lumber Industry* (Chicago: American Lumberman, 1916), pp. 5–6.

Chief Forester Henry S. Graves examining logging debris on private land within the White Mountain National Forest, New Hampshire, 1919. Courtesy of the National Archives

Alpine range, watershed, and recreation area, Gifford Pinchot National Forest, Washington, 1949. Courtesy of the U.S. Forest Service

The automobile begins its impact on outdoor recreation, Pike National Forest, Colorado, 1915. Courtesy of the U.S. Forest Service

Chief Forester Henry S. Graves and party inspect high range country of the Gallatin National Forest, Montana, 1917. Courtesy of the U.S. Forest Service

Herding sheep, Beaverhead National Forest, Montana, 1945. Courtesy of the U.S. Forest Service

the seller extremely vulnerable to hard-bargaining retailers. Greeley insisted that the industry needed help, not criticism; the Forest Service should use the national forests in a fashion most beneficial to the industry. Policy would continue unchanged: avoid competing with private enterprise by withholding federal timber until private supplies were exhausted; sell only to meet purely local shortages; protect the national forests from fire and other disasters. Even though Pinchot had strongly disagreed, calling the report a "whitewash of destructive lumbering," Greeley's moderate views held sway.[28]

Recreation Policy. If timber policies remained steady, others did not. Shifts in Forest Service recreation policies over the decades reflected changes in public attitudes. Since public agencies are supposed to respond to public preferences, shifts were appropriate. At times critics pounce on agencies for changing—construing change as tacit proof that administrative resolve is lacking. At other times agencies are chastised for not changing, static policy being viewed as tacit proof that changing societal values are being ignored. The administrative hide must toughen to the frustrations of public service.

To some, timber management and recreation use were incompatible. From the Forest Service point of view, all legitimate use could be accommodated without serious dislocation. Staff had to develop policies for a number of uses and deal with ever-constant pressures from special-interest groups, mainly representing timber, water, grazing, or mining. Mapping and bringing under even rudimentary management the over 180 million acres of the national forest system was a great burden. The agency dealt first with what it saw as the great problems, leaving lesser issues to take care of themselves as best they could. To the Forest Service, recreation was a lesser use.

Soon after the Antiquities Act became law in 1906, Pinchot notified his men of their additional responsibilities. The act was needed, he said, and they should report objects or areas eligible within the national forests for designation in order to safeguard these pieces of national heritage.[29] This directive and general references to hotels and cabins in the *Use Book* mark the beginning of Forest Service policies on recreation. Game management (that is, predator control programs) was of interest to hunters and also an aspect of this policy. By the time

28. William B. Greeley, *Some Public and Economic Aspects of the Lumber Industry*, USDA Report No. 14, 1917, pp. 13–100 *passim*; Greeley, *Forests and Men* (Garden City, N.Y.: Doubleday, 1951), p. 118.

29. Pinchot to Forest Officers in Charge, Nov. 21, 1906, drawer 164, RG 95–4, RFS.

Henry Graves became chief in 1910, the constant creation of national parks out of national forests and the need for overall park administration had become the main recreational concern in the Forest Service.

As early as 1904 and before he even had the forest reserves, Pinchot had moved to have the national parks transferred to the Department of Agriculture. Preservationist spokesmen in Congress managed a delay until a counterproposal to create a park bureau in the Department of the Interior had gained momentum. At this time the parks were administered by a variety of agencies. To Pinchot, a special park bureau was unnecessary. In his words, it was "no more needed than two tails to a cat."[30]

A bitter episode in park history occurred at about this time, and its outcome caused Pinchot and the Forest Service to be branded as being anti–national parks. The Sierra Club and its leader-founder John Muir had successfully petitioned Congress to add the Hetch Hetchy Valley to Yosemite National Park in California. San Francisco wanted to develop the natural reservoir site as a municipal water supply. Following the disastrous 1906 earthquake and fire, the city gained enough sympathy and support to wrest the valley away and construct a dam, which Pinchot had outspokenly favored. Muir, once his friend, became a bitter rival who predicted, "We may lose this particular fight, but truth and right must prevail at last."[31]

As long as he remained in office, Pinchot continued to pursue the parks. His staff reviewed a proposal to place parks under Forest Service control and decided to recommend enabling legislation. In the interim they would approach the secretary of the interior to arrange a cooperative agreement "to adopt such regulations for the administration of National Parks, as will not conflict with the interests of the adjoining National Forests."[32]

Graves picked up the thread in an exchange with the secretary of the interior. Ballinger asked the Forest Service to comment on draft legislation creating a national parks bureau. Graves had felt since 1907 that the national parks belonged in Agriculture because of the need to salvage the dead and down timber. He therefore suggested to Bal-

30. Samuel P. Hays, *Conservation and the Gospel of Efficiency: The Progressive Conservation Movement, 1890–1920* (Cambridge, Mass.: Harvard University Press, 1960), pp. 196–97.

31. Ibid., pp. 192–94; see also Elmo R. Richardson, *The Politics of Conservation: Crusades and Controversies, 1897–1913* (Berkeley: University of California Press, 1962), pp. 43–44, and Holway R. Jones, *John Muir and the Sierra Club: The Battle for Yosemite* (San Francisco, Calif.: Sierra Club, 1965), pp. 82–169.

32. Minutes of the Service Committee, Nov. 13, 1908, box 112, RG 95-8, RFS.

linger that the bill be amended to allow Forest Service experts to help the proposed park bureau with its timber management problems.

On other points of the proposal, Graves stiffened. The chief forester, always distraught by low federal pay scales, bristled over a proposed $6,000 annual salary for the commissioner of parks while, he pointed out to Ballinger, the commissioner of the General Land Office was paid only $5,000. He then reiterated his skepticism about the $75,000 added to the Department of the Interior budget, when Interior already was tending parks. Presumably it would not cost more to administer the parks; only a staff reorganization was involved.[33]

Pinchot's sometimes intemperate battle with John Muir over Hetch Hetchy had given many the impression that the Forest Service was hostile to the concept of national parks themselves. Pinchot's reference in 1911 to the park bill as a "frame-up" and branding Interior employees as crooks did little to smooth troubled waters. An increasing segment of the public believed that parks were important in their own right. J. Horace McFarland, a staunch park supporter, insisted that parks deserved to be administered "as a primary and not a secondary matter." Graves denied the existence of hostility with Interior, describing instead a spirit of "close cooperation." He then repeated his views about not needing a park bureau. Parks could better be administered as a part of the national forest system.[34]

When Graves described his view of park management, however, it is obvious why there was growing opposition to Forest Service control. He saw the parks as only "slightly different" from national forests. The difference would be one of purpose, not administration. Under Forest Service guidelines, there would "not be in the Parks as extensive a cutting of timber for commercial purposes" as in the national forests, but both types of reserves required technical foresters.[35]

On a California trip, Graves referred to the raging battle over Yosemite National Park and Hetch Hetchy and acknowledged that Forest Service commitment to a policy of "full utilization" had caused many to fear that the natural beauty of the Sierra was threatened. By and large, however, Graves was puzzled by adverse public reaction to his proposal for logging under controlled conditions in national parks.

33. Graves to Ballinger, Jan. 5, 1911, RG 95-29, RFS.

34. J. Horace McFarland to Graves, Feb. 14, 1911, ibid.; Graves to McFarland, Feb. 18, 1911, ibid. See also Hays, *Conservation*, pp. 194–95.

35. Graves to Herbert Parsons, Jan. 20, 1911, RG 95-29, RFS. Pinchot had long shared this view on logging in parks; see Hays, *Conservation*, p. 195.

To him, "the parks should comprise only areas which are not forested or areas covered only with protective forest which would not ordinarily be cut." He believed that Congress should be told if there were valid reasons why areas proposed for parks were better left undesignated. In this way, there would be little conflict of purpose and the existing system of park management would be adequate. That Graves fully understood the significance of opposition to his own views was evident in the caution to his men to avoid any action that could be interpreted as being anti–national parks.[36]

Graves was not really hostile toward parks, but he was not ready to assign them much of a priority. It took time for him to rethink the situation. Certainly it was not clear at that time which way the parks would ultimately be treated. When, for example, a national park was proposed for the region around Mount Lassen in California, the first Forest Service reaction was to oppose inclusion of commercially valuable timber. Associate Chief Albert Potter pointed out the inadvisability of an inflexible position. Asked for a general policy statement on parks, Potter replied that he saw the parks operating on a nonprofit basis, with "timber cutting . . . done as would be an improvement to the stand, and not with a view to its commercial possibilities." The general staff consensus, however, continued to advocate including in parks only timber having aesthetic value.[37]

Still, Graves asked the district foresters to describe scenic areas in national forests that would be better suited as parks. To clear up some confusion, he stated: "The idea was that areas of unusual value from a scenic point of view and for recreative purposes should be protected from any form of use which would destroy their value or would permanently embarrass their future use as National Parks." Graves explained that "foreseeing the possibilities of national park development" would meet with a great deal of approval and would be of ultimate benefit to the Forest Service.[38]

During the Graves years, it is impossible to separate Forest Service recreational policy from its attitude toward national parks and creation of a national park bureau. The beleaguered Forest Service was constantly confronted with myriad issues. It seems fair to say that Forest Service leaders at least accepted the need for parks and recreation but

36. Minutes of the Service Committee, July 26, 1911, box 112, RG 95-8, RFS.
37. Minutes of the Service Committee, Jan. 31, Feb. 21, July 31, 1912, box 114, ibid.
38. Minutes of the Service Committee, Jan. 29, 1913, box 115, ibid.

were unwilling to alter their own priorities. Fire, timber, grazing, and evolution of the basic organization received most of their attention. After all, a public agency is obliged to respond to all of the public, and at that time there definitely was not massive pressure for increased recreational facilities in the nation's forested areas.

That said, Forest Service officials firmly believed that recreational responsibilities had been given their due. One district forester saw that the 1905 *Use Book* had begun "a very definite recreation policy." The earlier policies had focused primarily on fish and game but also included marking roads and trails for convenient public use. In fact, many roads and trails were designed to accommodate recreation. To the district forester, there was no doubting Forest Service intent and action. The agency deserved public support for its recreational program.[39]

Instead, public support for a park bureau grew. Although the Forest Service still did not acknowledge the need for a separate park agency, more and more it recognized the changing public attitudes. When logging was planned for the White Mountain National Forest in New England, Potter predicted public disapproval. Studies had suggested that clearcutting was needed because selective logging would result in wind damage, but clearcutting areas of scenic beauty, he assured staff members, would be controversial even though the proposal was technically sound. In the Far West, Graves cautiously approved a large timber sale in Oregon, despite "a good deal of attention . . . being paid to the scenic value." Since the district forester was leaving a screen along public roads, he pointed out, the scenery would not be harmed. In both cases, the Forest Service heavily publicized "the fact that the sale will not detract from the scenic and recreational value of the Forest."[40]

Publicity would help. Herbert Smith, responsible for public relations, saw recreation as a great asset for Forest Service popularity. Graves agreed and instructed him to give special emphasis to the recreational uses of national forests in his press releases. He predicted that increased use of autos would substantially accelerate recreational demand. As news stories of the time reported, "getting 'back to nature' " was becoming a national habit because of urban life with its

39. Smith Riley to M. B. Tomblin, Dec. 8, 1915, box 103, RG 95-115, RFS.
40. Minutes of the Service Committee, Apr. 29, 1914, July 2, July 13, 1916, boxes 114 and 115, RG 95-8, RFS.

"industrial pace." Americans were resorting to a "primitive way of life . . . as a remedy for worn nerves."[41]

Meanwhile, pressure and support for a park bureau continued, including for transfer of national monuments within national forests to the new agency. Stephen Mather, soon to be head of the Park Service, created in 1916, favored inviting Forest Service representatives as well as others to a planning session in Berkeley, California, to discuss future responsibilities for national monuments. Graves saw the import of the meeting but was unable to attend, so dispatched District Forester Coert DuBois in his stead.[42]

DuBois met again with Mather and William E. Colby of the Sierra Club, this time to discuss extending Sequoia National Park into the neighboring national forest. DuBois argued for the "highest public good" in the region as a whole. He acknowledged that recreation was of high value, but so were timber, mining, and grazing. Anyway, he said, recreation was not incompatible with these commodity uses. Mather apologized to DuBois for what must have seemed like anti–Forest Service propaganda; antagonism was not intended.[43]

Potter, too, met with Mather, but this time about the Grand Canyon. Still a national monument under Forest Service jurisdiction, the canyon was naturally of interest to Mather as an obvious choice for a national park. Mather told Potter that he would want more land than was currently included in the monument. Much of this land was nonscenic but necessary in order to control access to the proposed park. Potter felt that the Forest Service should resist giving up that much range and timberland, and that offering right-of-way for park access should be sufficient. He admitted that congressional attitudes on such matters were unknown and thought the Forest Service should make an effort to explain its case. Contrasts should be drawn, he insisted, between benefits to local communities from having nearby forests as opposed to parks. Parks tied up development, but national forests favored settlement, mining, timbering, grazing, and summer homes. Extensive propaganda was not advisable, according to Potter; but "whenever Forest officers were asked for information," they

41. Minutes of the Service Committee, Sept. 23, 1915, box 115, ibid.; Washington, D.C., *Evening Star*, Jan. 11, 1915, FHS clipping file.

42. Mather to Secretary of the Interior, Feb. 12, 1915, , drawer 163, RG 95-4, RFS; Graves to Potter, Feb. 24, 1915, ibid.

43. R. Headly to Forester, Oct. 6, 1915, drawer 155, ibid.

should advise the inquirers "as to the exact status of the lands under both systems of management."[44]

Graves was distraught. "There seems to be continuous trouble over the National Parks." He believed that the American Civic Association, inspired by Enos Mills, author and parks enthusiast, was spreading reports that Graves was opposed to national parks. He also saw park supporters stepping up the tempo of their campaign after becoming more and more exasperated by Forest Service opposition to new parks.[45]

Forest Service efforts to stave off creation of a separate park bureau ended in failure on August 25, 1916, when President Woodrow Wilson signed enabling legislation. Secretary of the Interior Franklin K. Lane named Stephen Mather to head the new agency. National parks and most national monuments were administered by Interior; national forests were under Agriculture. The logic of having two distinct agencies in different departments with overlapping functions was questioned again by the Forest Service at the time of enactment. Regardless, park supporters favored separate agencies, mainly because of mistrust of the Forest Service. Pinchot, after all, had been a leader in the Hetch Hetchy Dam project, and Graves's views were at best benign.

Both agencies sought to avoid abrasion, but some sensitivities could have been anticipated. Assistant Chief Forester Edward A. Sherman felt that lack of animosity toward the Forest Service at a national park conference merited specific mention at a staff meeting: "nothing inimical to the interest of the Service had been said in any of the discussions or papers." Yet apprehension remained, and Potter reported to Graves that Secretary Houston had approved a carefully honed policy statement to be sent to Secretary Lane. If the two secretaries could reach an accord, "it will place us in a sound position to head off National Park projects which are desired to meet local demands rather than national needs."[46]

The agreement was reached. Forest Service staff received assurance that Mather and the Park Service would more freely consult with

44. Minutes of the Service Committee, Jan. 6, 1916, RG 95-8, RFS. By executive order in 1933, President Franklin D. Roosevelt placed all national monuments under the Department of the Interior.

45. Graves, Diary, Apr. 5, 14, 1916, Graves Papers.

46. Minutes of the Service Committee, Jan. 5, 1917, RG 95-8, RFS; Potter to Graves, Dec. 18, 1917, drawer 202, RG 95-4, RFS.

them on the merits of withdrawing proposed parks from national forests. Even so, uneasiness remained. Graves heard rumors that Mather had been critical of the Forest Service and sharply demanded supporting evidence. To those who carried rumor, he snapped, "I propose to have this sort of thing stopped at once and finally." He wrote to Mather, explaining that there had been reports claiming "fundamental difficulties." He told Mather that certainly they could work out some broad areas of policy agreement. Mather replied quickly and amiably that Graves's idea was a good one and suggested lunch sometime at Washington's Cosmos Club to talk over the situation.[47]

It is not clear whether the Forest Service increased its recreational programs in response to this new competitor or not. It seems more than coincidental, however, that the year following creation of the National Park Service, Frank A. Waugh completed an intensive study of Forest Service recreation facilities. The study was "to determine as clearly as possible what policies should govern in the development and administration of these recreation uses, in what ways and to what extent the Forest Service might direct this development, and what methods of organization and administration would give maximum efficiency in the field."[48]

Waugh toured the nation, inventorying recreational resources in national forests. It would be difficult, he admitted, to place an economic value on recreation, but he recommended somehow recognizing the full worth of these resources. Maybe recreation cost the Forest Service two cents per user-hour to administer, Waugh pondered; in any case, its worth was "at least as valuable as the cheapest wholesale commercialized entertainment." He calculated that national forest recreation was worth a minimum of $3 million annually. It was clear to Waugh that recreational values substantially exceeded costs: "The National Forests are certainly a paying investment for the American people."[49]

47. Minutes of the Service Committee, March 20, 1919, RG 95-8, RFS; Graves to George H. Cecil, Sept. 13, 1918, drawer 192, RG 95-4, RFS; Graves to Mather, Sept. 30, 1918, drawer 192, RG 95-4, RFS; Mather to Graves, Oct. 1, 1918, drawer 192, RG 95-4, RFS.

48. Frank A. Waugh, "Recreation Uses on the National Forests—A Report to Henry S. Graves," 1917, bound typescript, copy in FHS library, pp. 1–2. Donald C. Swain credits the Park Service with "goading" the Forest Service into recognizing the value of preserving natural beauty; see his *Wilderness Defender: Horace M. Albright and Conservation* (Chicago: University of Chicago Press, 1970), p. 98.

49. Waugh, "Recreation Uses," pp. 121, 124; Forest Service news release, 1918, FHS clipping files.

To clarify his own views, Graves issued a sixteen-page policy statement on national parks. He began by emphasizing the need for stability. Permanent national forest boundaries were necessary to good administration—"every boundary change is unsettling." He saw a "new form of reservation" as a disruption of the integrity of the national forest system. A "proper" park policy would alleviate fears within the Forest Service, Graves explained, as he elaborated his plan for settling differences with the Park Service.[50]

Strongly supporting the "National Park principle," Graves rejected the idea that foresters were inherently disinterested in parks. It seemed unfortunate to Graves that during earlier conservation campaigns foresters had found it necessary to advocate *use* so strongly in order to counteract western opposition to forest reserves. This earlier strategy had precluded the opportunity to support parks and recreation. Even more unfortunate, he felt, was the public outcry against Pinchot's stand on Hetch Hetchy. "Very unjust conclusions" were drawn in spite of Pinchot's repeated statements "that he was absolutely opposed to timber sales and similar commercial uses of the Yosemite and other Parks." Graves supported Pinchot's contention that supplying water to San Francisco was not equivalent to commercial timber operations.[51]

Graves again raised the issue of there being no administrative need for a separate park bureau. He insisted that the Forest Service was fully capable of administering the national parks, but he acknowledged that the lack of confidence in his agency was not limited to enemies. Since even "the most vigorous champions of the National Forests" had doubts about Forest Service recreational policies, it was "essential that our real position, our real purposes, our real methods be made crystal clear." To Graves "the fact that economic forestry and scenic preservation are not inconsistent does not mean that they should always be combined." He agreed that some areas should remain untouched but then added that "it will be a long time before the National Parks will have to be drawn upon for raw materials." Unless, of course, too much was set aside. A few national parks contained resources of great commodity value, and Graves felt that the United States had enough wealth to afford leaving something alone. He

50. "Policy Letters from the Forester, No. 5—National Parks," Jan. 6, 1919, copy in box 209, RG 95-4, RFS.
51. Ibid.

voiced his opposition to "any action to commercialize timber, grazing, or other resources in the Parks."[52]

Graves came back to the bothersome prospect of too many parks. Even though parks must be areas of "supreme beauty and grandeur," it did not logically follow that all beautiful and grand places must be parks. Graves advocated full funding and support for existing parks but cautioned against their "multiplication," a lowering of standards, and "depreciating the value of all." Mount Rainier National Park in Washington State he agreed was an obvious asset, but the proposed Mount Evans park near Denver was not so obvious and should be set aside only if it was part of a grand plan.[53]

Graves demanded a "concrete policy" and a "farsighted program" for parks. He pointed out how the Forest Service without consultation had been placed in an awkward position by Department of the Interior decisions on parks. Promising "every effort" to work cooperatively with the Park Service, Graves hoped that "we will not again find ourselves facing such an embarrassment."[54]

All in all, Graves's policy letter on parks is a remarkable document. It shows clearly his genuine concern and support for parks. It shows equally clearly that to men like Graves, recreation was a valuable resource but not as valuable as other forest resources. Some day, Graves predicted, the parks would be a supply of raw materials, but until that time they should be fully protected. It is no wonder that Mather and his supporters wanted their own agency. The differences between a Graves and a Mather may have been largely philosophical, but they were basic. These fundamental differences would remain a serious obstacle to a unified approach to public recreation management—a goal Graves earnestly sought, on his own terms, but never achieved.

The Weeks Law of 1911. Until 1891, public land policy had been committed to transferring the public domain into private ownership. With passage of the Forest Reserve Act, some lands would remain public. By this time, however, the only land eligible for reservation was in the West, prior land policy having successfully placed the

52. Ibid.

53. Ibid. The Forest Service prevented the Park Service from acquiring the Mount Evans area and developed its own recreation area there. An outspoken park supporter charged that Horace Albright had "sold out" and "abandoned the principles of the national parks" by letting Mount Evans get away. See Swain, *Wilderness Defender*, p. 94.

54. "Policy Letters from the Forester, No. 5."

eastern portion of the nation in private hands. If there were to be forest reserves in the East, then purchase was in order. The concept of purchase was a departure from the century-long tradition of disposal. The shift was dramatically abrupt; the same decade Congress decided to retain portions of the public domain, advocates appeared for buying back still more.

As it turned out, the question of forests and floods was the key. In 1864 George Perkins Marsh had claimed that forests diminished the frequency and violence of floods. Reading widely in European scientific literature, Marsh also concluded that forests affected weather and climate. Hough's reports from 1877 to 1882 contained similar contentions. The consensus of the late nineteenth century supported their views. In 1873 Congress passed the Timber Culture Act, which allowed planting a certain number of trees to serve as a substitute for the residence requirements of the Homestead Act. In addition to providing a supply of wood on the treeless prairies, it was thought that the new forests would increase rainfall. After all, wherever there were forests there was rain. Logic of the time saw forests bringing rain rather than the reverse.[55]

Three years later, the year Congress established the embryonic Forest Service in the Department of Agriculture, there was an effort to secure protection of forested watersheds. On February 14, 1876, Congressman Greensborough Fort of Illinois introduced legislation to provide that "all public timber lands adjacent to the sources of affluents of all the rivers be withdrawn from public sale and entry."[56] Fort's efforts came to naught, but the Forest Reserve Act of 1891 in general terms and the Sundry Civil Appropriations Act of 1897 specifically made watershed protection the law of the land. Meanwhile, efforts to secure eastern reservations focused on recreational values.

As early as 1892, Charles Sargent had suggested setting aside a portion of the southern Appalachians to serve recreational needs. By 1899 the Appalachian National Park Association had formed and was petitioning Congress for a park. Concurrently the Society for the Protection of New Hampshire Forests joined its southern colleague

55. George P. Marsh, *Man and Nature* (Cambridge, Mass.: Harvard University Press, 1965 [reprint of 1864 edition]), Chap. 3; see also David M. Emmons, "Theories of Increased Rainfall and the Timber Culture Act of 1873," *Forest History* 15 (Oct. 1971):6–13; Charles R. Kutzleb, "Can Forests Bring Rain to the Plains?" *Forest History* 15 (Oct. 1972):14–21.

56. H. R. 2075, 44 Cong. 1, Feb. 14, 1876.

and supported the recreational value of New England forests as well.[57]

Both the House and Senate saw bills introduced in support of eastern reservations. In 1900 Congress appropriated $5,000 for a study, which the Division of Forestry and Geological Survey made jointly. President Roosevelt told Congress the investigation showed "unmistakably" that a forest reserve was needed in the South. The president justified the purchase on grounds of bolstering the southern economy and improving flood control. The following year, the Appalachian National Park Association mailed over one million pieces of literature to gather support. Then in 1903 the group changed its name to the Appalachian National Forest Reserve Association to reflect changing sentiment away from an Appalachian park to a forest reserve. Supporters became demoralized when their efforts north and south seemed to be of little avail; but persistence would pay off.[58]

The year 1907 marked the turning point in the campaign to create federal reservations in the eastern United States. At its congress in 1905, the American Forestry Association passed a resolution supporting establishment of reserves in the East and began an active role in the venture. The national group helped weld efforts by the two regional associations into a stronger, better-organized program. Overstatement of fact by partisans had given House Speaker Joseph Cannon and other congressional skeptics an excuse to view the proposals as poorly conceived. Further, lack of a broad popular support in the South had allowed legislative attention to wander.[59]

Cannon, a powerful opponent, declared "not one cent for scenery." He stopped a bill in its tracks when he judged the proposed eastern reserves not to be a proper federal function. He took this action in spite of over seventy resolutions sent to Congress from chambers of commerce, conservation groups, universities, garden clubs, and industrial associations in support of reserves in the White Mountain–Appalachian region. Some dejected supporters blamed the American Forestry Association for "showing the white feather in the eleventh hour [after it had] waved a brave flag and sounded a splendid battle cry." Pinchot agreed that the situation called for "a little rioting,"

57. Frank B. Vinson, "Conservation and the South, 1890–1920," Ph.D. diss., University of Georgia, 1971, pp. 42–44; Charles D. Smith, "The Movement for Eastern National Forests, 1899–1911," Ph.D. diss., Harvard University, 1956, pp. 6–64 *passim*.

58. Smith, "Movement for Eastern National Forests," pp. 72–73; Vinson, "Conservation and the South," pp. 45, 49, 59–60.

59. AFC, *Proceedings* (1905), p. 450; Vinson, "Conservation and the South," pp. 62–65.

because Congress could not long ignore public opinion. He did not blame the AFA: there had been no letup in its efforts; the organization simply was not powerful enough to carry the load alone.[60]

The AFA did play a major role in the eventual enactment of legislation to establish forest reserves in the East. Its monthly journal, then known as *Forestry and Irrigation* (later *Conservation*, and now *American Forests*), printed detailed accounts of events in and out of Congress relevant to the struggle. Suggestions that these reserves would be unconstitutional, that Congress had no authority to acquire land, prompted the editor of *Forestry and Irrigation* to cry "obstructionism." All other arguments having failed, opponents were using the constitution as a last resort. With the great need for reserves, the editor found it incredible "that intelligent people ever intended thus to bind themselves, hand and foot, with a bit of writing." The whole notion of unconstitutionality was, to the AFA editorial writer, unacceptable.[61]

The American Forestry Association notwithstanding, the constitutional question was central. William L. Hall, whom Pinchot had assigned to the Appalachian–White Mountain project, explained to chief and staff the significance of the House Judiciary Committee's entry into the issue: the committee had sweepingly ruled that the federal government lacked authority to purchase land. The whole national forest question might be affected by this judgment. Meanwhile, the bill for eastern reserves was accordingly amended to limit purchase to land protecting watersheds of navigable rivers. With this step, proponents of the reserves were asserting that because the commerce clause of the constitution gave Congress authority over navigation, the federal government surely could purchase land to protect these headwaters. And if it could be shown that forests had a beneficial effect on flooding, then logically the purchase of certain forested lands was constitutional.[62] The House Judiciary Committee agreed, but the relationship between forests and floods still had to be proved. Proving it turned out to be a lively affair.

"The statement shows so complete and thorough a misunderstand-

60. Benjamin A. Spence, "The National Career of John Wingate Weeks," Ph.D. diss., University of Wisconsin, 1971, p. 40; Vinson, "Conservation and the South," p. 71; copies of resolutions in box 61, AFA Records, FHS; Allen Chamberlain to Pinchot, Jan. 28, 1907, box 575, Pinchot Papers.

61. *Forestry and Irrigation* 14 (March 1908):121–22.

62. Minutes of the Service Committee, March 20, 25, 1908, box 112, RG 95-8, RFS; Gordon B. Dodds, *Hiram Martin Chittenden: His Public Career* (Lexington: University Press of Kentucky, 1973), p. 159; U.S. Constitution, Section 8, Article 3: "[Congress has the power] To regulate commerce with foreign nations, and among the several states, and with the Indian tribes."

ing of the most elementary foundations of forestry as to invalidate at once in the mind of any man acquainted with the forest, the author's observations or conclusions." Thus argued Gifford Pinchot against Hiram M. Chittenden of the Army Corps of Engineers, who, at a 1908 meeting of the American Society of Civil Engineers, had presented a lengthy paper on the relation of forests to stream flow. He claimed to support the conservation movement in the main but believed that the effects of forests on flood prevention had been grossly exaggerated and could not allow these unsupported theories to go unchallenged. In a counter-rebuttal to Pinchot, Chittenden saw that distortion of the importance of forests was due to the conservation cause. According to Chittenden, a long campaign to achieve congressional favor had forced conservationists into making unsupported claims.[63]

Not surprisingly, Pinchot and the Forest Service saw forests as specifically and directly preventing destructive floods. Pinchot earlier had been less positive about the effect of forests on floods. In his *Primer of Forestry* he wrote that forests only tended to prevent floods. When the conflict with Chittenden broke, however, Pinchot became dogmatic, specifying that forests did prevent floods. Testifying to congressional committees, he showed the legislators a photograph of a denuded area, held it at an angle, poured water on the inclined illustration. The lawmakers could see how quickly the water ran off. Pinchot then would pour water on a sloping blotter, calling attention to how the water was absorbed and how little ran off the bottom of the sheet. The analogy, he insisted, was appropriate to forested and deforested areas. Forests prevented floods.[64]

In contrast to Pinchot's analogies, Chittenden used facts. He had encountered serious flooding in Yellowstone National Park, an area that obviously was fully forested and had suffered no logging. Certainly, he claimed, this sort of specific evidence made it difficult to accept the view that forests automatically prevented flooding. Chittenden argued well and at length.

Part of Chittenden's concern was to aid the Corps of Engineers, recently rocked by scandal and scorned for having been so unimagina-

63. Hiram M. Chittenden, "Forests and Reservoirs in Their Relation to Stream Flow, with Particular Reference to Navigable Rivers," in American Society of Civil Engineers, *Transactions* 62 (March 1909):465, 245–46, 510.

64. Gifford Pinchot, *A Primer of Forestry*, 2 vols., Bulletin 24, Bureau of Forestry, 1905, 2:68; Chittenden, "Forests and Reservoirs," pp. 245–46.

tive as to have viewed the Panama Canal as a poor project. The Inland Waterways Commission, with its emphasis on multiple-purpose development, also seemed to be a threat to the corps, which stoutly maintained that levees and dams were the only way to handle floods. Chittenden himself was not doctrinaire, but he did side with the corps. He was pleased that he had dislodged some of his adversaries. When the time came, he supported the Weeks bill but still insisted that it should not be based on the ability of forests to prevent floods.[65]

Congressman John Weeks first showed interest in purchasing forest reserves in 1906 when he supported a petition to acquire two reserves to protect industrial water supplies. Speaker Cannon appointed Weeks to the House Committee on Agriculture in 1908 in hopes that he would put businessman's sense into pending forestry legislation. Cannon's judgment was sustained when the House passed a forest reserve bill in 1908, calling for states to protect their own forests and establishing a National Forest Commission to study the relation of forests to floods. The following year the House narrowly passed another reserve measure, which would have used timber and grazing receipts to purchase additional reserves. States, of course, opposed losing their share of the receipts, and this bill was killed in the Senate by threat of filibuster by Henry M. Teller, ex-secretary of the interior.[66]

The House Judiciary Committee in 1910 reiterated its view that the federal government had "no power to purchase forest lands at the headwaters of navigable rivers for the purpose of preserving the forests;" it only had power to purchase land to maintain navigation. Worse news for Pinchot, Chittenden's view that forests did not prevent floods was supported by William S. Bixby, past president of the Mississippi River Commission. The Weather Bureau concurred. But engineers in the U.S. Geological Survey supported Pinchot and thus contradicted their colleagues in the Corps of Engineers. The House Committee on Agriculture overwhelmingly supported the view that forests prevented floods.

Weeks, no longer on the Agricultural Committee, reintroduced the bill but dropped the clause about using receipts for purchase. This version passed by a nineteen-vote margin in the House. After the bill

65. Dodds, *Hiram Chittenden*, pp. 159–61; Chittenden, "Forests and Reservoirs," pp. 466–545.

66. Spence, "National Career of John Wingate Weeks," pp. 40–44, 63–64; Vinson, "Conservation and the South," pp. 83–89.

was amended to include a commission to certify that each acre purchased contributed to the protection of navigation, the way was clear for enactment. Western opposition in the Senate to federal purchase of eastern forests had gone up in the smoke of a 1910 holocaust in Idaho. President Taft signed the Weeks Bill into law on March 1, 1911.[67]

The Weeks Law authorized creation of the National Forest Reservation Commission, consisting of the secretaries of agriculture, interior, and war, plus two senators and two congressmen. The Forest Service was to search for land to purchase. The Geological Survey would examine the tracts to determine whether they fulfilled the constitutional requirements for protection of navigation. The Forest Service would keep its cost of land-searching distinctly separate from the rest of its operation, and field work would include a 10 percent sample of the timber on each tract under study. Graves assigned thirty-five men to the examination task. His agency had been gathering data for ten years. At its first meeting on March 7, 1911, the commission elected Secretary of War Henry L. Stimson as its presiding officer. Since then, the secretary of war, more recently the secretary of the army, has served as president of the commission. A Forest Service officer serves as secretary.[68]

Commission responsibilities were so grave and growing pains so extreme that many questioned congressional wisdom for enacting the law at all. Sums of money for acquisition had to be dispensed within fiscal years, and procedures were ill-defined. Some landowners saw the confusion as an opportunity to ask exorbitant prices. Two years following enactment, however, William Hall, who had been assigned the job of overseeing Forest Service participation, could report an improved situation. Despite the problems of the first two years, over seven hundred thousand acres had been added to the national forest system in the East, costing less than five dollars per acre. By the end of Graves's administration in 1920, more than 2 million acres had been purchased; in 1961, fifty years after enactment, the Forest Service

67. *Congressional Record* 61 Cong. 2, 1910, pp. 8742–45, 8974–77, 8984, 8997; U.S. Congress, House, Hearings, *Acquisition of Forest Lands*, n.d., pp. 119–42. Later, Fernow commented to Chittenden about Pinchot's performance during the congressional hearings: "It is to be understood that at such meetings considerable buncombe needs to be performed, if you want to handle the half-informed legislators" (Sept. 1, 1916, quoted in Dodds, *Hiram Chittenden*, p. 183).

68. Minutes of the Service Committee, March 18, 1911, box 114, RG 95-8, RFS; Forest Service, *Annual Report* (1911), pp. 61–62; USDA, Forest Service, *The National Forest Reservation Commission*, 1961, pp. 6–7.

reported nearly 20 million acres acquired, almost entirely in the East.[69]

The purchase program created two types of national forests. Those in the West had been carved out of the public domain, but eastern national forests have been purchased under Weeks Law authority, bolstered by subsequent legislation. Until 1956, congressional interior committees heard matters concerning national forests reserved from the public domain, but agricultural committees passed on those purchased in the East. Recently, national forest appropriations are reviewed by an Interior committee alone. As far as the administration of the national forests is concerned, however, there are no basic policy differences for East and West. Forest Service policies are uniform.[70]

Cooperation with State and Private Forestry Organizations

Section 2 of the Weeks Law had been scarcely discussed, let alone debated, during lengthy congressional deliberations. Yet in retrospect it could be argued that this section was the most important. The whole concept of federal matching money to boost local incentive began with Section 2's authorization of $200,000 to be used as federal matching funds for states having a forest protection agency which met government standards. These funds could support protection of nonfederal lands in the watersheds of navigable streams. Protection agencies, underfinanced and frequently staffed with political appointees, had appeared in many states by this time. In fact, state programs often preceded federal. Weeks Law support came at a crucial period in the development of state agencies and provided minimum standards of operation.

Cooperative programs of course were not new. Pinchot's Circular 21 had been popular throughout the century's first decade. But the idea of direct federal funding of nonfederal programs marked the

69. Hall to Pinchot, Dec. 4, 1913; "The New Eastern National Forests," speech presented Jan. 8, 1913, to the AFA, RG 95-84, RFS; Forest Service, *National Forest Reservation Commission*, p. 26.

70. The Weeks Law not only authorized federal purchase of lands that would contribute to the protection of the watersheds of navigable streams but also approved condemnation under powers of eminent domain, allowing acquisition for public use with just compensation. The courts upheld this aspect of the law and the concept of purchase for navigation protection. In 1926 Malcom Griffin of Virginia argued that the Weeks Act was unconstitutional; the court found in favor of the government, thus upholding the right of federal purchase. *Griffin* v. *US*, 58 F.2d 674.

beginning of a new era. Later legislation would greatly expand the scope and effectiveness of Section 2.

The first question the Forest Service faced in administering the cooperative program was whether funds could be spent only in states where purchases had been made under the Weeks Law. The USDA solicitor's office decided informally that any state that had a forestry agency complying with federal standards could qualify. Chief Henry Graves placed a ten-thousand-dollar limit for any one state. New Hampshire was the first to apply, and the secretary of agriculture earmarked $7,200. The Forest Service used the agreement with the New England state as a model. Soon agreements were made with Minnesota for $10,000; New Jersey, $1,000; Wisconsin, $5,000; Maine, $10,000; and Vermont, $2,000. In each case, federal monies were to be used only for patrolmen's salaries, and the state had to spend at least as much on fire protection. By the end of his administration, Graves listed twenty-three states as Weeks Law cooperators and complained that the appropriations fell short of meeting the need.[71] In eight short years, federally assisted state agencies had assumed a major role in forest fire protection.

Meanwhile, federal support also was rapidly urging the states into the conservation movement. Early state attention had focused on two points: (1) improving commerce by regulating log brands, incorporating companies, and developing uniform measuring practices; (2) protecting private forest land from fire. By the turn of the century, fire was of greatest interest at the state level, although more and more attention was being given to forms of taxation that would encourage desired logging practices. State forestry agencies were most of all fire departments, requiring spark arrestors on logging equipment and other safeguards, and engaging in both fire patrol and suppression. After the threat of white pine blister rust in 1920 prompted eastern states to band together, the group soon expanded into the National Association of State Foresters. More and more the Forest Service would view state agencies as complementary to federal efforts. Usually sensitive to state fears of federal domination, the Forest Service continued to expand its cooperative programs in fire protection and control of insects and disease.[72]

71. Minutes of the Service Committee, Apr. 14, May 17, 1911, box 114, RG 95-8, RFS; Forest Service, *Annual Report* (1911), pp. 62–63; *Annual Report* (1919), p. 27.

72. For a useful summary of the history of state forestry activities, see Ralph R. Widner, ed., *Forests and Forestry in the American States: A Reference Anthology* (Washington, D.C.: National

There is no doubt that uniform federal standards and encouragement to parsimonious state legislatures in the form of matching funds have been major factors in the growth of state forestry. Doubtless, too, grass-roots political support from cooperative forestry has been of great value to the Forest Service in Congress. Section 2 of the Weeks Law soon dominated the programs of State and Private Forestry, a unit that had first appeared on Forest Service organizational charts in 1908. Joining with National Forest Administration, State and Private Forestry became a major Forest Service subdivision. Within four years of the Weeks Law, the agency as organized to date would be rounded out with the advent of the Branch of Research.

Research Policy and Facilities

The first thing needed in trying to stop floods by tree growth was to "drop all exaggerations about it." So testified the chief of the Forest Service in 1927, referring to the decade-long controversy over the relation of forests to floods. Pinchot had won a political victory over Chittenden on the question as part of getting the Weeks Law through Congress, but still needed was scientific support. In 1910 the Forest Service began a long-term cooperative experiment with the Weather Bureau at Wagon Wheel Gap in Colorado to measure quantitatively and qualitatively forest–flood relations.[73] Research was coming into its own.

When Pinchot took over from Fernow in 1898, he was highly critical of his predecessor for emphasizing technical and theoretical aspects of forestry rather than dealing with "practical" problems. Pinchot's aggressive program of cooperative forestry outlined by Circular 21 epitomized his approach. Yet, despite his enthusiasm for practical forestry, the first year he was chief he found it necessary to establish the Section of Special Investigations, a research arm. By 1902 the section had achieved division status and boasted fifty-five employees. Together the Division of Forest Investigation and the Division of Forest Management conducted all technical studies for the Bureau of

Association of State Foresters, 1968). J. Girvin Peters was the Forest Service officer assigned to organizing cooperative programs with states. Ten of his notebooks are in drawer 398, RG 95-115, RFS.

73. Quoted in Ashley L. Schiff, *Fire and Water: Scientific Heresy in the Forest Service* (Cambridge, Mass.: Harvard University Press, 1962), p. 135. Many judged the experiment to be inconclusive, in any regard. Over the years the Forest Service has advocated a broader and less doctrinaire approach to watershed management.

Forestry. When all such work was merged in 1902, Forest Investigation accounted for one-third of the agency's $185,000 budget.[74]

Obviously, much of the effort by the Division of Forest Investigation could not be classified as research by modern standards, because it offered empirical conclusions rather than tested hypotheses. The authorized studies reflect, however, a realization that programs of practical forestry could succeed only if supported by substantial efforts to obtain information. Even though his administrative burden soared following the 1905 reserves transfer, Pinchot announced that "the research side [of the Forest Service] will receive more attention during the coming year." He was afraid that the agency was losing its scientific point of view.[75]

By 1908 the Forest Service had an experiment station in Arizona and investigations were being conducted throughout the West. The Office of Silvics now administered research needs. A series of experimental areas similar to permanent sample plots yielded practical information, such as Thornton Munger's work in the Pacific Northwest on Douglas fir growth rates. In California the Forest Service established a nursery on the Shasta National Forest. Experiments there would show "exactly what class of trees furnish the most and best seed." Both studies would yield practical, but still fundamental, information to an agency engaged in seed collecting and marking timber for cutting.[76]

Forest Products Laboratory. Even as the Forest Service was accumulating rudimentary technical information on forests, plans were underway to make comprehensive studies of wood itself. Obviously, if wood made available by logging could be utilized more fully and with greater durability, then in effect supply would be increased.

Fernow's earlier extensive studies on the physical properties of wood—"timber physics"—had been praised by the wood-using industries for their usefulness. The Forest Service under Pinchot continued to study wood, frequently contracting this specialized research out to forestry schools. As the need for a permanent, centralized testing

74. George B. Sudworth, "Scope of Work of the Division of Forest Investigation, with Notable Investigations," Apr. 4, 1902, memorandum in box 356, RG 95-115, RFS.

75. Minutes of the Service Committee, June 27, 1906, box 112, RG 95-8, RFS; T. T. Munger, "Fifty Years of Forest Research in the Pacific Northwest," *Oregon Historical Quarterly* 56 (Sept. 1955):227.

76. Emanuel Fritz, "Recollections of Fort Valley, 1916–1917," *Forest History* 8 (Fall 1964); William T. Cox to F. E. Olmsted, Apr. 10, 1908, box 359, RG 95-115, RFS; T. D. Woodbury to W. B. Rider, Oct. 6, 1909, ibid.

facility became apparent, McGarvey Cline, chief of the Office of Wood Utilization, set to work. After studying several alternatives, Cline convinced Pinchot and Price that a cooperative agreement with a university was the best solution. On October 28, 1908, Cline queried seven schools: Purdue University, Carnegie Institute of Technology, University of Illinois, Yale University, Cornell University, University of Michigan, and the University of Wisconsin. Yale and Carnegie disqualified themselves immediately. After visiting the campuses, Cline decided that Cornell, Purdue, and Illinois were inadequate, leaving Michigan and Wisconsin as serious contenders.[77]

Cline's proposal required the cooperating institution to provide approximately thirteen thousand square feet of floor space, special foundations for the testing machines, heat and utilities, and administrative offices. The Forest Service would provide a staff of fourteen technical men and six assistants, with a payroll of twenty-eight thousand dollars per year, all testing equipment, and test materials. Facilities would be available to graduate students, and laboratory staff could deliver lectures on their specialties. Finally, all publications would give credit to the university as cooperator.

The ensuing battle of the universities grew heated, each state calling on its congressional delegation to help sway Pinchot's decision. Michigan exerted extreme pressure through the House Committee on Agriculture, and Pinchot reversed an earlier decision to award the contract to Wisconsin. We need not concern ourselves here with the details of the furor, complicated by the late entrance of the University of Minnesota, at Pinchot's invitation, as a serious contender. After four more months of negotiation and repeated staff preferences for Wisconsin, Pinchot selected Madison as the site for the Forest Products Laboratory.

Although the doors had opened months earlier, the lab was dedicated by a host of officials in June 1910. Henry Graves, as new chief forester, spoke of the lab's role. Citing the appalling waste between the stump and the consumer—25 percent of the tree was left on the ground, half of the wood reaching the mill was discarded, and still more whittled, shaved, sanded, or sawn in final manufacture—he

77. For an excellent analysis of increasing supply through improved utilization see Sherry Olson, *The Depletion Myth: A History of Railroad Use of Timber* (Cambridge, Mass.: Harvard University Press, 1971). In the following account the author has drawn freely on Charles A. Nelson, *History of the U.S. Forest Products Laboratory, 1910–1963* (Madison, Wis.: Forest Products Laboratory, 1971).

predicted that new by-products developed at the lab would substantially reduce this waste. Lab Director Cline, speaking more specifically about how the impressive new facility would be a "laboratory of practical research," offered a list of purposes: (1) to study mechanical and physical properties of commercial woods; (2) to study and develop principles of wood preservation treatment, fiber products, and chemical derivatives; (3) to develop practical uses for wood now wasted; (4) to disseminate information; and (5) to cooperate with wood-using industries.

The wood industry offered enthusiastic support, petitioning Congress for increased appropriations. The lab in turn responded with vastly improved methods of distillation for the naval stores industry, less destructive techniques for collecting oleoresin, better ways to employ chemical preservatives, and more efficient designs for beams, timbers, and containers. Research results increased the forest economy of the South in particular, and national defense capabilities were boosted greatly. New chemical treatment substantially extended the durability of railroad ties, bridge piers, and piling.

The lab proved to be both effective and popular, but as with many scientific endeavors, there were other factors to consider. After the lab had been in operation for several years, cooperative efforts with several universities seemed less necessary, so Assistant Chief William Greeley called for centralizing all products research in Madison. As the regional laboratories were popular locally, repercussions were immediate. For example, the Forest Service had been conducting timber tests on the University of Washington campus since 1906. When Hugo Winkenwerder, dean of forestry in Seattle, heard of the centralization plan, he immediately asked E. T. Allen to intervene with Graves to have the decision reversed. Allen wrote Graves the next day and said that closing the laboratory on the University of Washington campus would lose support in the big Pacific Northwest lumber industry for Forest Service programs. Winkenwerder also got Senator Wesley Jones of Washington to pin Graves down on the status of the Seattle lab. Graves acceded to pressure, and the lab remained at the University of Washington, although with reduced funding.[78] These satellite wood-research labs were gradually phased out, but the

78. Minutes of the Service Committee, Jan. 2, 1913, box 114, RG 95-8, RFS; Hugo Winkenwerder to E. T. Allen, March 27, 1913, Allen to Graves, March 28, 1913, Wesley Jones to Graves, March 31, 1913, and Charles H. Flory to Winkenwerder, Nov. 7, 1913, all in Records of the College of Forest Resources, University of Washington, Seattle.

Forest Service continued to fund on-campus research by awarding contracts to faculty and promising graduate students. It would not be until the 1960s that Forest Service research facilities would again appear on university campuses.

Field Investigations. Research with well-defined parameters and under controlled conditions is possible in laboratories. In the field, however, research could be a different matter, particularly before the Forest Service completed its network of experiment stations. Studies of grazing impacts on alpine ranges, the best methods to achieve reforestation, and efficiency of logging techniques are representative of field research. Topics of practical interest received the most rigorous investigation.

Fire, of course, was a prime concern to all, including those in research. During his first year as chief, Pinchot had authorized studying the history of forest fires to better understand damage. His staff cataloged five thousand fires occurring since 1754 as a basis for the study.[79] Information gained from this and similar studies was invaluable in swaying departmental fiscal officers during budget sessions. Fire research evolved from this humble, practical beginning into today's highly sophisticated analyses of fuel, combustion, and weather. Proving that fires are destructive is no longer the mainstay of fire research.

Proposals to use fire beneficially have recurred from time to time. Supposedly Indians had used fire as a tool for eons. Graves was appalled by such thoughts, labeling these early fires as "enormously destructive." Burning the forests periodically to reduce fuel and thus hazard was "utterly out of the question," and it was "inconceivable" that anyone would seriously advocate such a policy. Then he learned of California lumberman T. B. Walker's interest in fire use and approached him to underwrite a chair of fire protection at the Yale School of Forestry. He asked for $100,000, saying that the National Lumber Manufacturers Association and Frederick Weyerhaeuser had each donated the same amount for another chair, Andrew Carnegie had pledged $100,000, and Mr. E. H. Harriman and Mrs. Morris K. Jessup had given $100,000 each. The Pinchot family gave $200,000.[80]

With a fire patron for Yale in tow, Graves held his outspoken opposition in check. He authorized S. B. Show, forest examiner for

79. Forest Service, *Annual Report* (1899), p. 98.
80. Ibid. (1910), pp. 17–18; Graves to T. B. Walker, May 19, 1911, RG 95-29, RFS.

reforestation, to make a fire study on Walker's property, the forests of the Red River Lumber Company of Westwood, California. Walker was intrigued with the notion of protecting his forest lands by the controlled burning of fuel during nonhazardous seasons. Show found little damage to reproduction on the burned sites, but the process proved more expensive than conventional protection techniques. Another fire study, also in California, received little official enthusiasm: Coert DuBois was denied permission to publish the results of his brush-burning studies but was at least allowed to continue his investigation.[81]

Forest Service opposition stemmed mainly from anticipating public confusion over good and bad fires. The sort of popularization that Graves and his staff feared materialized in the February 1920 issue of *Sunset*. Novelist Stewart Edward White prepared an article that advocated burning forests to reduce hazard. When Graves learned of the forthcoming article, he was upset and ordered a campaign to counter it. Through interviews, letters to trade journals, and articles, the Forest Service sustained a steady flow of information showing that burning was disastrous. *Sunset* promoted White's article by placing posters in city buses proclaiming, "Your forests are in danger. The Forest Service won't save them but fire will, says Stewart Edward White in a smashing article." When the Forest Service protested to the mayor of Seattle, in at least that city the posters came down. *Sunset* itself allowed a retreat from the posters because of a belief that the Forest Service was about to bring suit. The episode quickly subsided, but the Forest Service remained uneasy that the public would not understand technical issues involved and would either not support Forest Service fire prevention policies or advocate indiscriminate use of fire to solve all protection problems.[82]

Not only singular topics such as beneficial uses of fire could engender controversy; in some quarters the whole idea of research as an appropriate Forest Service function was still challenged. Scientific investigations could and would yield information vital to routine administration of the national forests, but Graves and his staff still feared

81. S. B. Show, "Results of Light Burning on Red River Lumber Co. Holdings, near Westwood, California," Progress Report, Jan. 25, 1912, drawer 455, RG 95-115, RFS; Minutes of the Service Committee, Dec. 30, 1915, RG 95-8, RFS.

82. Minutes of the Service Committee, Jan. 2, 1920, March 18, 1925, RG 95-8, RFS. Light burning for fire protection in California should not be confused with prescribed burning in forests of the southeast, where forest lands are burned to encourage regeneration. For a thorough discussion see Schiff, *Fire and Water*.

that criticism would arise from those who did not understand the importance of research. So that no one could accuse the Forest Service of spending money on research at a time when national forests were not fully protected from fire, Graves ordered that no general expense money would be used to support experiment stations. Instead the Forest Service would ask Congress directly for research funds.[83]

As another step in developing a research program, Chief Graves in 1912 set up a Central Investigative Committee. Raphael Zon was in charge of silvicultural research; James T. Jardine, grazing; and Carlisle Winslow, products. Each district named similar committees, its members nominated by the district forester and approved by Graves. One of the committees' tasks was to distinguish between investigative and administrative projects. Zon scratched Graves's opinion on the bottom of a memo: investigations should involve fundamental principles, while administrative studies would seek additional information about known factors.[84]

Independence for Research. In June 1915 Graves established the Branch of Research, replacing the investigative committees, which had been strongly tied to administrative discretion at the district level. Placing Earle H. Clapp in charge, Graves directed him to "bring together . . . the various lines of research on investigative work conducted by the Forest Service." Graves believed that the new branch gave Research and its personnel "the fullest possible recognition." This point was important because for years researchers had been complaining that they were denied recognition and independence comparable to that of the administrator. The chief forester himself would settle disputes between the chief of research and the chief of administrative branches. District foresters would be instructed through administrative channels how to implement research findings. At the field level, research and administration were separated, a vast improvement over the previous operation under advisory committees. Graves intended eventually to place field researchers again under the district forester.[85]

83. Dana to Graves, Sept. 21, 1911, box 359, RG 95-115, RFS.

84. Service Order no. 21, "Organization of Investigative Work," Jan. 2, 1912, copy in drawer 164, RG 95-4, RFS; Central Investigative Committee to Forester, Feb. 19, 1913, drawer 220, ibid.

85. Service Order no. 45, "Establishment of the Branch of Research," May 14, 1915, copy in drawer 164, ibid. On May 6, 1908, Zon had sent a memo to Pinchot strongly recommending increased emphasis on research conducted independently from administration. Pinchot penciled his reply on the memo: "I am for this with some changes." It remained until 1915, however, before such independence was achieved. Zon memo reprinted in *Service Bulletin* 17 (June 19, 1933):2.

More and more, research grew into an important, officially recognized function of the Forest Service. In fact, to Potter, research was the reason for the Forest Service being in the Department of Agriculture rather than in the Department of the Interior: "Were it not for research work conducted by the Forest Service, it would be merely an administrative organization."[86]

If the importance of research was generally accepted, its relation and responsibility to the administrative arm of the Forest Service was not. District foresters asserted that they would take more interest in the research conducted in their areas if they had control of what was done. Thornton Munger feared that the district office in Portland would become merely "a whistling post for investigative projects," since apparently all final action on research was handled in Washington. Greeley shared Munger's concern and insisted that the experiment stations should work to solve "local problems of immediate importance, and that the District Offices should have full power to call upon these men for such work."[87] Debate continued on whether the district office or the Washington office would establish research priorities. The decision drifted toward the chief's office.

Earle Clapp was relieved that his Branch of Research was responsible only to Washington. In his view, direct district participation had stifled research, for it was impossible to develop "real research" if the investigator had to cater to local whims. According to Clapp, before 1915 researchers had a "very subordinate and uncertain status," compared to administrative personnel. Districts transferred unsuccessful administrators to Research: "It is significant that an activity fundamental to good forest practice should have been used to unload Incompetents." Clapp observed that the policy of dumping unwanted men onto Research had so discredited the program that qualified men were reluctant to enter the field. Moreover, with research subordinate to administration, programs were disrupted at the slightest provocation.[88]

The separation of research from administration accomplished two things that Clapp failed to mention. First, not only could researchers operate independently from the daily pressures of administering the

86. Proceedings of the Meeting of Forest Investigators, Feb. 26–March 2, 1917, p. 1, drawer 359, RG 95-115, RFS.

87. Ibid., pp. 2–5.

88. Clapp to F. Silcox, March 19, 1934, box 1, Earle H. Clapp Papers, Record Group 200, National Archives (hereafter cited as RG 200).

national forests, but their efforts also attained the appearance of objectivity to onlookers. Creditability is vital, and separation from administration added a sense of reliability to research results. The other advantage of separation, probably not noted at the time, was to reduce the researcher's involvement in direct, technical advice to the administrator. Now the district forester would have to develop a staff of specialists to translate the increasing flow of technical information into field applications. The dual role of staff and research had not been satisfactory. Since 1915, the Forest Service has developed an effective combination—line, staff, research.[89]

Were Clapp's views a damning indictment that first Pinchot and then Graves had paid only lip service to research? Or were they trenchant observations that the scientists' demands for free action had conflicted with the administrators' focus on fiscal years and daily pressures? Graves understood some of the sources of abrasion, and three years after setting up the independent Branch of Research he tried to set the issues in perspective.

He told Forest Service personnel that too often they thought of research as something separate rather than as the means of acquiring "knowledge that is necessary to do our work." He acknowledged that the administrator often became impatient with the lengthy periods the researcher took to come up with answers, adding that sometimes these answers were expressed in highly technical terms which held little meaning to the nonspecialist. Good forest management was the objective of everyone in the Forest Service; everyone's job held importance. Research results should be presented in terms useful to the practicing forester, Graves insisted. Administrator and researcher should acknowledge each other.[90]

While its place in the Forest Service continued to evolve, the Branch of Research expanded its programs. In addition to the Forest Products Laboratory at Madison, experiment stations appeared in the West. The first was a station near Flagstaff, Arizona, in 1908; in 1921 there were seven more. By the end of the 1920s, a basic network of twelve regional stations was in place and all major forest regions were represented. One new station, headquartered at Amherst, Massachusetts, served seven northeastern states. In the Far West, an ex-

89. The author is grateful to Verne L. Harper, retired deputy chief for Research, for making these observations.

90. "Policy letters from the Forester, No. 4—Concerning Research," Dec. 5, 1918, copy in box 209, RG 95-4, RFS.

periment station at Berkeley served California; the station in Portland investigated the problems of Washington and Oregon. Six stations were in the East and emphasized problems of private forest management. The six western stations had the additional responsibility of providing for the research needs of the massive national forest system. To Clapp the stations constituted "the most significant development in Forest Service research."[91]

Congressional Recognition. In annual reports, chiefs continued to speak of the need for research. Foresters were confronted by "myriad technical puzzles" during the "revolution from timber mining to timber growing." Needed most was both technical and economic research, for it was "impossible to deal with production independently of utilization." Yet, with all that, the Forest Service research program lacked focus and recognition of its importance. More and more it had the manpower and the facilities to cope with demands for information, but it needed unifying legislation and specific congressional support.[92]

Responding to a speech in 1924 by the chief of the Forest Service on the importance of research, the Washington, D.C., section of the Society of American Foresters established a Special Committee on Forest Research, naming Clapp chairman. Charles L. Pack's American Tree Association published the 232-page committee report—a blueprint for forestry research. Explaining why research was needed and thoroughly describing all aspects of forestry investigations then conducted, Clapp outlined what he called an organic act for Forest Service research.[93]

The industry-based National Forestry Program Committee proved a valuable ally. Clapp fed them information for the forthcoming campaign: he wanted a million-dollar budget for the Forest Products Laboratory, a million for the experiment stations, a half-million to inventory forest resources, and more than a quarter of a million dollars to study grazing. Confidentially, he told Zon, now director of an experiment station in Minnesota, it was up to the pair of them to do the hard work because the committee did not have the same "vital inter-

91. Susan R. Schrepfer et al., *A History of the Northeastern Forest Experiment Station, 1923 to 1973*, Forest Service Technical Report NE-7, 1973, p. 4; Earle H. Clapp to Clyde Leavitt, July 18, 1923, RG 95–115, RFS.

92. Forest Service, *Annual Report* (1922), p. 44; ibid. (1924), p. 34.

93. Earle H. Clapp, *A National Plan of Forest Research* (Washington, D.C.: American Tree Association, 1926).

est." They would have to be discreet, however, in acquiring support from the various special-interest groups involved.[94]

The American Forestry Association threw its support behind Clapp's proposal, which nearly cleared congressional hurdles in 1927. Ohio Congressman John R. McSweeney had introduced the bill in March 1927, but Congress adjourned before its supporters in the House could overcome delaying tactics. Senator Charles L. McNary of Oregon joined with McSweeney late that same year, and on May 22, 1928, the McSweeney–McNary Act had sailed through Congress. Clapp got what he wanted. The authorized budget ceiling was 3 million dollars, with internal breakdown much the same as he had outlined two years earlier. Clapp was deservedly elated. The law recognized the importance of research to forestry "in a way that it has never been recognized before." Research now stood "on a par" with other forestry activities. Clapp measured his success in part by noting that even though the appropriation ratio of research to other purposes was still only one to twenty in the Forest Service, the same comparison for the Department of Agriculture as a whole was one to eighty-five.[95]

Achievements of Leadership. By creation of the Branch of Research in 1915, the basic structure of the modern-day Forest Service was completed. Decentralized administration of the national forest system carried over from Pinchot's time, State and Private Forestry had special funding from Section 2 of the Weeks Law, and Research with its impressive laboratory in Madison and administrative autonomy in the districts rounded out the program. Graves could be proud of his achievement. He had left his deanship, for what he thought would be one year, to fill the void caused by Pinchot's dismissal. Instead, he led the Forest Service for ten.

Graves knew failure, too. From his point of view, a major disappointment had been his inability to obtain salary increases. Pinchot had complained of the same problem. The situation continued to deteriorate through Graves's decade. During the fourteen months be-

94. Clapp to E. T. Allen, Dec. 28, 1926, WFCA Records; Clapp to Zon, March 31, 1927, Clapp Papers, RG 200; Ringland to Kellogg, Jan. 13, 1927, NFPC Records, Cornell University.

95. AFA, *Annual Report* (1927), copy in AFA Records; 45 Stat. 699–702; Earle H. Clapp, "Forest Research and the McSweeney–McNary Act," Jan. 26, 1929, unpublished ms. in box 360, RG 95-115, RFS. Almost simultaneously with the McSweeney–McNary Act, Congress passed the McNary–Woodruff Act (45 Stat. 468) as an amendment to the Weeks Act. Congress authorized 8 million dollars for land purchase during a three-year period.

tween June 30, 1918, and August 31, 1919, 460 technical forest officers resigned. Some men who left the Forest Service took jobs paying nearly 170 percent more than they had been receiving. In 1920 rangers still received $1,100 per year, the same as in 1910. Forest supervisors responsible for a million acres were paid between $2,000 and $2,500 per year.[96] Graves struggled in vain to increase the monetary rewards for Forest Service employees.

World War I further proved a major source of disruption in terms of men and material and taxed Graves's constitution to the limit. While on military duty in France in the fall of 1917, he noted in his diary, "If I keep on losing flesh I shall soon look like a skeleton." Chain-smoking caused a serious abscess in his mouth, and he resolved to cut down.[97]

Returning home did little to ease the burden. After an inspection tour of the western national forests, he came back to a desk piled high with paper. Catching up required "day and night work and no relief in sight." For two months Graves neglected his diary, then: "I have been chiefly preoccupied by the heart-breaking work of running the Forest Service." To maintain control of the agency, sapped by loss of men into the Army, Graves insisted that all decisions cross his desk for review.[98]

His health failed. He suffered from Menier's Syndrome, a disorder of the inner ear which caused dizziness and nausea. A thorough medical examination revealed very low blood pressure and sluggish reflexes. His physician ordered him to take at least one month's vacation; two would be better. He was not to read correspondence or in any way be involved with work. Graves cooperated only partially and took a short holiday, returning to an unusually difficult fire season and "a disagreeable experience with Stephen Mather." He broke down again. This time he took a long vacation and seemed refreshed when he returned to his desk.[99]

96. "Reasons for Readjustment of Salaries . . .," n.d., drawer 2, RG 95-115, RFS.

97. Graves, Diary, Sept. 15, Oct. 7, 1917, Graves Papers. In mid-1917, the Tenth Engineers of the U.S. Army were in France to assist the Allies in meeting their demands for wood. Later that same year the Twentieth Engineers formed a second forestry regiment. In 1918 the two groups were combined into the Forestry Division in the Service of Supply. Graves had been in France to make preliminary arrangements. Many other Forest Service employees and other American foresters saw military duty as part of the Forestry Division. It is a colorful episode but adds little to the history of the Forest Service.

98. Graves, Diary, May 1, July 20, 1918.

99. G. A. Waterman, M.D., to Graves, Dec. 27, 1918, Graves Papers; Graves, Diary, Nov. 25, 1919, box 82, ibid.

Graves's recovery was only an illusion, as his diary reveals. By late 1919 he saw the Forest Service as a "crippled organization . . . with troubles piling up of every description." He dabbled with a new program to improve industrial forestry but now saw Pinchot, a friend and colleague for three decades, as an enemy. He expected at any moment to be denounced by the prestigious ex-chief for holding too-moderate views about the lumber industry.[100]

It was too much. Within three months, he announced his resignation. The afternoon of March 9, 1920, Graves called his staff together to tell them of his decision before they read it in the newspapers. It was obvious to him that his health was too unstable to withstand the great burdens of the office, and he wanted to leave before he impaired the agency's operation. Graves recounted his aspirations: to rebuild the Forest Service to its prewar condition, and to start a whole new movement in American forestry. His health would endanger realization of these goals. Another goal had been to improve salaries, from the top down. The secretary of agriculture, Edwin T. Meredith, was "at this minute" telling the Senate Committee on Agriculture and Forestry that Graves had resigned in protest over pay scales. They hoped that his action would help increase the "pitifully low" financial rewards of working for the Forest Service. "If I remain here in acceptance of the situation, certainly we cannot expect others to realize its urgency." His resignation, he hoped, would successfully emphasize the fact that the "Government cannot expect men to remain indefinitely in positions of such great responsibility as that of the Chief Forester at the present standards of remuneration."[101]

Secretary Meredith accepted Graves's resignation with a formal statement of regret, effective April 15. He was glad, however, that a man of William Greeley's caliber was available to take over. Later in the year, Graves sent out an announcement: "Henry S. Graves, formerly Chief Forester of the Forest Service, United States Department of Agriculture, announces that he has opened offices at 1731 H Street, N.W., Washington, D.C., for the general practice of forestry."[102]

100. Graves, Diary, Nov. 25, 1919; Minutes of the Service Committee, Feb. 13, May 1, 1919.

101. Henry Graves, March 8, 1920, box 354, RG 95-115, RFS; "Remarks of Col. Graves to members of the Forest Service, on the afternoon of March 8, 1920," box 68, Graves Papers.

102. Meredith to Forest Service personnel, March 16, 1920; copy of announcement in box 202, RG 95-4, RFS.

Two years later he resumed his forestry deanship at Yale, where he carried on a distinguished program in forestry education, retiring in 1939. He remained active in forestry affairs, and his personal papers show that he kept well informed on Forest Service matters. Henry Graves died March 7, 1951, at the age of eighty.

CHAPTER VI

Reorganization, Recreation, Range, and Routine

"GREELEY HAD the highest mark of any of our recent graduates. He is a special star and I recommend him for almost any work which may come along."[1] So wrote Dean Henry Graves to Overton Price when Greeley applied to the Bureau of Forestry in 1904. The young forester got the job.

William Buckhout Greeley was born in 1879, the son of a New York State minister. He graduated from the University of California in 1901 and received his master's degree in forestry from Yale three years later. Greeley quickly showed himself worthy of Graves's praise. He was forest supervisor on the Sierra South National Forest (since renamed the Sequoia National Forest) when sheepman Pierre Grimaud, with support from large landowners, decided to test penalties levied for violating Forest Service grazing regulations. Greeley had anticipated the move and had handled the situation in the field with care. The case moved up through the courts and in 1910 reached the Supreme Court, where the challenge was defeated.[2]

Pinchot had moved Greeley to Montana in 1908 as district forester. Two years later he faced "The Big Blowup"—3 million acres of Idaho and Montana forests burned in late August of 1910. Seventy-eight fire fighters lost their lives, and the fire destroyed about one-third of Wallace, Idaho. Heroism was routine. Ranger Edward Pulaski saved his crew by holding them in a mine shaft at gunpoint as the fire roared

1. Graves to Price, July 13, 1904, General Correspondence of the Forest Service, 1898–1908, Record Group 95, RFS, National Archives.

2. William B. Greeley, *Forests and Men* (Garden City, N.Y.: Doubleday, 1951), pp. 79–80.

past. Great administration skills were needed for the efforts to suppress the massive fire, and Greeley passed the test. Significantly, Greeley, who had superintended a catastrophe, received reward, not punishment, as at times happens when a bureaucracy seeks a scapegoat.[3]

At thirty-one Greeley was promoted to assistant chief in charge of silviculture and transferred to Washington, D.C., in the fall of 1910. He was welcomed by Henry Graves, who was now chief. Greeley seems to have quickly assumed leadership: minutes of weekly staff meetings show that Greeley and Associate Chief Forester Edward A. Sherman dominated the proceedings. Graves apparently preferred to delegate large blocks of authority, as had Pinchot. Greeley's assignments and reports dealt with the tough, fundamental policies of timber management. It was Greeley who led the Forest Service response to the Bureau of Corporations' study of the lumber industry, which was so critical of the private sector. Greeley was also in charge of research, which, under Earle Clapp's direction, rapidly developed into one of the three main Forest Service programs.

World War I brought added responsibilities to the Forest Service, which was still regaining its balance following the Pinchot–Ballinger uproar. The need to provide wood for barracks, trench timbers, aircraft, and myriad military uses to the Allied forces in France prompted a shift in Forest Service priorities at home. After American entry in 1917, many personnel, Graves and Greeley included, went to France to coordinate lumber production nearer the front. Graves soon returned to his duties in Washington. Greeley remained in France until after the armistice, supervising production of vital forest products. In France, Greeley first saw effective cooperation between the government and lumbermen.[4] Commissioned as a major, Greeley was promoted to the rank of lieutenant colonel at war's end. As did many

3. For a more complete view of Greeley, see George T. Morgan, Jr., *William B. Greeley: A Practical Forester* (Saint Paul, Minn.: Forest History Society, 1961). For a detailed description of the fire see Betty G. Spencer, *The Big Blowup: The Northwest's Great Fire* (Caldwell, Idaho: Caxton Printers, 1956). For a view of Greeley's administrative ability, see Samuel T. Dana, *The Development of Forestry in Government and Education*, Oral History Interview by Amelia R. Fry, 1967, Regional Oral History Office, Bancroft Library, pp. 68–69. For Pulaski's account of the fire, see Ovid Butler, ed., *Rangers of the Shield: A Collection of Stories Written by Men of the National Forests of the West* (Washington, D.C.: AFA, 1934), pp. 77–82. Pulaski later developed a popular, if awkward and unbalanced, fire-fighting tool and saw it named in his honor.

4. Morgan, *William B. Greeley*, pp. 37–38. There are many accounts of the forestry/military effort which culminated in the 20th Engineers, the largest regiment in the Army. Greeley's account is in *Forests and Men*, pp. 87–95.

men of that generation, he preferred to be addressed by his military title: to the end of his life, he was Colonel Greeley.

After finishing pesky details in France and having been awarded the British Distinguished Service Order, Greeley reported back to the Forest Service in the fall of 1919. Graves was suffering from overwork. Perhaps the two discussed his pending resignation and Greeley's willingness to succeed him. On April 16, 1920, a still-youthful forty-one years of age, Greeley became chief forester. He promptly wrote to Secretary of Agriculture Meredith, expressing his gratitude for the promotion from within, feeling that such advancement to the top through the ranks would add to the agency's esprit de corps.[5]

In his first annual report, Greeley broached the subject of consolidating national forest ownership. At a later staff meeting, Greeley proposed that land acquisition through purchase and exchange be a major Forest Service program. A combination of haste to withdraw national forests from the public domain and the checkerboard ownership patterns resulting from nineteenth-century railroad land grants had resulted in the enclosure of 24 million privately owned acres within national forest boundaries. This nonfederal ownership posed obvious administrative headaches for the Forest Service and at times the private owners, because private owners were governed by a variety of state laws, and the Forest Service had jurisdiction only over federally owned land. Greeley wanted to "block up" units in order to have solid federal ownership.[6] Since Weeks Law purchases were limited to land in the watersheds of navigable streams, other means were needed to implement the proposal.

As part of the legislative program for forestry in the early 1920s, Congress in March 1922 enacted the General Exchange Act. Probably because its logic spoke for itself, the act encountered no serious opposition in its move through Congress. Under the new law, nonmineral lands or timberlands on the public domain could be exchanged for private or state-owned land within national forest boundaries. No money would change hands. The traded properties had to be of equal value and within the same state. When President Harding signed the exchange act on March 20, Greeley was pleased and predicted that

5. News release, March 15, 1920, FHS clipping file; Greeley to Secretary of Agriculture, March 26, 1920, William B. Greeley Papers, University of Oregon.

6. Forest Service, *Annual Report* (1920), p. 7; Minutes of the Service Committee, Feb. 18, 1921, March 3, 1921, box 115, RG 95-8, RFS.

within fifty years "this law would probably be regarded as one of the half-dozen most important laws affecting the National Forests."[7]

Reorganization

Pinchot had finessed the forest reserve transfer to Agriculture in 1905, after abortive efforts by the Department of the Interior to set up its own forestry division. Pinchot had also gained from Interior brief control of forestry practices on Indian reservations, a contributing factor to the Pinchot–Ballinger controversy. Under Graves came the fuss over a new park bureau. Relations between Interior and Agriculture remained strained but proper until President Harding in 1921 chose Albert B. Fall to be his secretary of the interior.[8] Alaska, central to the earlier Pinchot–Ballinger affair, again became an interdepartmental issue.

The Forest Service viewed Alaska as a unique, second-chance opportunity. With its vast resources and still underdeveloped society and economy, the area could benefit from mistakes made in the lower forty-eight. An enthusiastic Associate Chief Sherman told Greeley of the opportunity to build a permanent lumber industry in southeastern Alaska. The region could become a "second Norway industrially," with coordinated use of all natural resources. The two had frequently exchanged memos on the northern territory. Also, Greeley was proud that receipts from Alaskan national forests exceeded expenditures: $549,000 spent and $674,000 paid in between 1911 and 1920, with 25 percent of the receipts refunded to Alaska. For the Forest Service as a whole during the same period, figures were less heartening: appropriations approached $64 million but receipts were $38 million, little more than half that much. Twenty-five percent or $8.5 million had been returned to the states, leaving less than $30 million for federal coffers.

A 1921 timber sale prospectus on the Tongass National Forest advertised 335 million cubic feet of pulpwood to be logged over a fifty-year period. The purchaser would be obligated to construct a

7. 42 Stat. 507, 520; Minutes of the Service Committee, March 23, 1922, box 115, RG 95–8, RFS.

8. Pinchot was suspicious of Fall's intentions from the beginning. See Burl Noggle, *Teapot Dome: Oil and Politics in the 1920s* (New York: W. W. Norton, 1965), pp. 22–28. For a view of forest management on Indian reservations, see Jay P Kinney, *The Office of Indian Affairs: A Career in Forestry*, Oral History Interview by Elwood R. Maunder, 1969, FHS, Santa Cruz, Calif.

100-ton capacity pulp mill in Alaska as part of the transaction. Despite these efforts, the Forest Service was accused of discouraging the paper industry when it rejected earlier proposals for industrial development. In truth, the agency, mindful of the financial plight of paper plants in neighboring British Columbia, had rejected what it considered bad proposals. High transportation costs and failure to consider the long-term market had disabled many Canadian enterprises; the Forest Service insisted that in Alaska a stable economy be built.[9]

A proposal by Fall placed the whole program in jeopardy. In office less than two months, Fall drafted for Harding's signature an executive order that would have transferred the Forest Service to Interior. The transfer seemed logical to the new president, as it would place all of the public lands in one department. Wishing to avoid controversy so early in his administration, however, Harding took no action. Fall then turned to Congress.[10]

He began by asking for the national forests in Alaska. Forest Service supporters mounted a major publicity campaign to defeat Fall's proposal, blanketing the nation's newspaper editorial offices with anti-transfer releases. Their zeal nearly backfired when the American Forestry Association mailed a propaganda sheet of reprinted articles against transfer, including excerpts from an article by Greeley that had been critical of General Land Office operations in Alaska. Fall used this open attack by a bureau chief on another department as an excuse for outrage. He carried his views to a sympathetic Harding. Secretary of Agriculture Henry C. Wallace was concerned enough about the course of events to make public a letter he had written to Senator Arthur Capper, who had asked about the apparent conflict between the two departments. Wallace dismissed Fall's charge as baseless, lamely telling Capper that since Greeley had written the article under the preceding Wilson administration, it did not represent current views. He assured the senator that his dealings with Fall had "at

9. Sherman memo to Greeley, [1922], drawer 196, Records of the Office of the Chief, 1908–47, Record Group 95 (hereafter cited as RG 95–4), RFS; Greeley to Dan A. Sullivan [Sutherland], Aug. 12, 1921, drawer 189, ibid.; Sale Prospectus, [1921], copy in drawer 189, ibid.; "Administration of Timber on National Forests in Alaska," memo to file [1923], copy in drawer 189, ibid.

10. The author is heavily indebted to Donald L. Winters, *Henry Cantwell Wallace as Secretary of Agriculture, 1921–1924* (Urbana: University of Illinois Press, 1970), chap. 8, for background on the transfer attempt.

all times been friendly." Privately, Greeley told Wallace that he had directed all Forest Service employees to avoid any discussion of Alaska until Fall calmed down.[11]

Pinchot, long since ousted as chief, had retained a vigorous interest in the Forest Service. Disgusted with what he interpreted to be Greeley's acquiescence, he confronted Fall directly. Fall assured the skeptical Pinchot that he did not really want the Forest Service. The former chief remained dubious, but there was little more he could do. He had misjudged Greeley, however, who had not given up but was following Wallace's earthy advice to not get into "a pissin' contest with a skunk." Let Fall fulminate, counseled the secretary; there were quiet ways to handle Interior. Greeley accepted the advice, withdrew a letter of resignation that he had tendered to allow him freedom to attack Fall openly, and waited.[12]

The waiting had its bad moments. The Washington *Post* supported government reorganization. In an editorial entitled "The People's Patrimony," the *Post* held up the Forest Service as a good example of why reorganization was desirable. The newspaper claimed that the agency was inefficient and refused to cooperate with the Department of the Interior. Lamenting the fact that the Forest Service had changed from a research to an administrative bureau, the *Post* charged that it "halts farmer's cows in search of a drink of water" and kept wood from the farmer while selling abroad. Greeley denied all charges and issued a news release which described the *Post* as "seriously misinformed."[13]

Resolutions favoring the Forest Service from the Society of American Foresters and National Association of State Foresters, and open support from the American Forestry Association and other conservation groups, had discouraged Harding from moving the bureau to Interior. Yet, the issue remained alive. As an added safety factor thereafter, Greeley moved to strengthen his agency's ties with Agriculture, pointing out that previously the Forest Service had been a part of Agriculture mainly in the terms of appropriations. Greeley now prompted his staff to overcome the "separatist tendency" and identify the work of the Forest Service more closely with the "func-

11. USDA news release, March 11, 1922, FHS clipping file; William B. Greeley, "What Is Wrong with Alaska," *American Forestry* 27 (Apr. 1921):198–201; Greeley memo to Wallace, March 6, 1922, drawer 192, RG 95-4, RFS.

12. Winters, *Henry Wallace*, pp. 173–74; Greeley, *Forests and Men*, pp. 95–101.

13. USDA news release, March 1922, FHS clipping files.

tions and scope" of the department. They must learn to relate forestry to agriculture.[14]

Alaska still required attention, and a presidential trip seemed in order. If Harding could see firsthand what was happening, he might settle the lingering dispute himself. Accompanied by a large official party, including Greeley and Secretary of Agriculture Wallace, the president toured southeastern Alaska during July 1923. Whenever possible, Greeley and Wallace slipped away to chat with local Forest Service personnel or to visit an agricultural experiment station. After talking to many Alaskans about forestry matters, Greeley decided that most criticism of his agency had originated in Seattle from Ballinger's supporters and not from Alaskan residents. Preparing to return home, Greeley confided to his wife that the trip had gone well for the "Forest Service and conservation" and predicted that "the President's Alaska policy will be very close to what we would like to have it."[15]

In Seattle a surprised audience heard Harding reverse himself on his previous Alaskan resource policies. The trip had made him wonder, he said, if opponents of the Forest Service were more interested in exploitation rather than development. One of the few unsurprised was Greeley, who knew of Harding's change of heart in advance, having helped draft the speech before leaving Alaska. The president's speechwriter had warned that there would be a "helluva row" if Harding tried to bring together secretaries Wallace, Herbert Hoover of Commerce, and Hubert Work, Fall's replacement in Interior. It would be better to present them with a *fait accompli*. Harding agreed, and Wallace, Greeley, and Judson C. Welliver, Harding's literary secretary, drafted the speech. The president made no changes of substance.[16]

Upon his return to Washington, Greeley reported to his staff that Harding's speech was a "very satisfactory resume" of the Alaskan situation. Greeley, showing his unconcealed disdain for what he saw as an overcentralized, bureaucratic General Land Office, said he was gratified that the Forest Service had won out over Interior mainly by good men doing good work at the local level. An inspection report of

14. SAF, Resolution, Dec. 28, 1921, copy in FHS clipping file; Association of State Foresters, Resolution, Sept. 20, 21, 1921, copy in FHS clipping file; AFA, Resolution, Jan. 16, 1922, box B-6, AFA Records, FHS; Winters, *Henry Wallace*, p. 179; Minutes of the Service Committee, Apr. 11, 1923, RG 95-8, RFS.

15. Greeley to Gertrude Greeley, July 11, 22, 1923, Greeley Papers.

16. Judson C. Welliver to John E. Ballaine, Nov. 13, 1940, ibid.

Alaska that year affirmed Greeley's judgment, calling Harding's speech "a turning point in the administration of the Alaskan forests." Quashing the transfer ended the uncertainties that had impeded long-range planning.[17]

Harding's change of heart could have resulted from firsthand observation of Alaskan conditions or from sudden revelations of the resource plundering involved in the Teapot Dome scandal, which soon was to rock the country. Six days after his Seattle speech, still on the homeward leg of the Alaskan trip, Harding died in San Francisco. With Fall to be indicted and ultimately convicted for his part in Teapot Dome, and Harding's strongly pro–Forest Service speech, the agency seemed safe in Agriculture for the time being. For how long was not clear, however, and E. T. Allen predicted that if a strong conservationist were to be named secretary of the interior, it would be difficult in the future to argue against transfer.[18] He would prove to be prophetic.

Recreation

With the Forest Servive secure in Agriculture, Greeley looked to other matters. Recreational or noncommodity use of national forests was legitimate, but still ill-defined. At the insistence of Navy Undersecretary Theodore Roosevelt, Jr., and his fellow sportsmen from the Boone and Crockett Club, President Calvin Coolidge in April 1924 publicized the need for a definite national policy on outdoor recreation. A month later 309 delegates from 128 organizations formed the National Conference on Outdoor Recreation, meeting in Washington, D.C. Forester Arthur Ringland was soon named executive secretary. The first act of the conference was to inaugurate an inventory of outdoor recreation resources to serve as a basis for national planning. The conference remained in existence until mid-1929.

As part of the overall conference program, the Forest Service was asked to furnish information for a joint study of recreation areas on federal lands by the American Forestry Association and the National Parks Association. Greeley hesitated because it would be difficult to supply maps showing areas of unsurpassed scenic quality and all stop-

17. Minutes of the Service Committee, Aug. 2, 1923, box 115, RG 95–8, RFS; E. E. Carter, "Tongass National Forest Inspection," Sept. 6, 1923, ms. copy in drawer 206, RG 95–4, RFS.

18. Allen to Ovid Butler, Feb. 25, 1924, WFCA Records, Oregon Historical Society, Portland.

ping places for tourists; but since the data would be of value to the Forest Service, too, he complied. The AFA–NPA report to the commission offers insights into the recreational thinking of the time.[19]

Explaining that national forests represented conservation and national parks preservation, the report pointed out that earlier Forest Service recreation policies had looked to revenue from special uses — resorts and summer homes. At first the Forest Service had been criticized for not developing its wilderness areas; now, to some, it had overdeveloped them. The Park Service, too, was overdeveloping its wildlands, the report charged. In defense of twenty-one proposed wilderness areas containing 12.5 million acres, the report maintained that these areas represented the least accessible and least commercially valuable tracts. If each of the anticipated 25,000 visits was assigned a value of $250, less 4 cents per acre for protection, the wilderness would yield $5,750,000, per year. As the areas themselves were appraised at $100 million, this return would be a 5 3/4 percent dividend. Continuing the comparison, the report held that since the total worth of the United States was approximately $320 billion, "surely no one could object to setting aside a mere pittance of 100 million dollars worth of real estate for a purpose that would pay an indirect dividend of nearly six percent."[20] These calculations are similar to those that Waugh made for Graves in 1917 about national forest recreational values. In any event, attempts to place a dollar value on recreation and wilderness resources continue to this day.

The joint American Forestry Association and National Parks Association committee found that not all Forest Service personnel favored wilderness preservation. The committee predicted, however, that public opinion would eventually decide the issue in favor of nondevelopment. The Park Service, too, fell short of its public responsibilities. Reminiscent of Graves's earlier concern, the committee found that "people's playground" had replaced natural wonders as

19. Arthur Ringland, *Conserving Human and Natural Resources*, Oral History Interview by Amelia Fry et al., 1970, Regional Oral History Office, Bancroft Library, pp. 139–57; Minutes of the Service Committee, Nov. 6, 1924, RG 95–8, RFS; "Outdoor Recreation Resources of the United States Federal Lands," copy in box 71, SAF Records, FHS; "Preliminary Inventory of the Records of the National Conference on Outdoor Recreation, 1924–29," Record Group 220, National Archives, (hereafter cited as RG 220), 1962, compiled by Kenneth W. Munder and Richard Bartlett, processed, pp. 1–8. For a full discussion of Forest Service recreation policy see James P. Gilligan, "The Development of Policy and Administration of Forest Service Primitive and Wilderness Areas in the Western United States," Ph.D. diss., University of Michigan, 1954.

20. "Outdoor Recreation Resources," pp. 145–46, 154–63.

justification for park reservation, causing inclusion of attractive but not spectacular areas. Since the automobile was going to be a major factor in all outdoor recreation, the committee challenged the National Conference on Outdoor Recreation to take a stand on tame versus wilderness recreation.[21]

Wilderness preservation was a tough problem. In April 1922 Will C. Barnes, assistant chief for grazing, met with George H. Lorimer, editor of the *Saturday Evening Post*. Lorimer suggested that the Forest Service reserve portions of the national forests in their undeveloped condition. Hearing Barnes's report, Greeley explained that the Forest Service had authority to designate inaccessible, noncommercial areas as recreation units. To Greeley, the knotty problem was predicting changing future demands for resources, but he agreed that for the time being wilderness use could be a major aspect of recreation policy.[22]

Lorimer responded with an editorial, "Cutting the [Grand] Canyon Forests," which expressed despair over Forest Service attitudes toward wilderness. Robert Sterling Yard, executive secretary of the National Parks Association, sympathized with Lorimer but counseled patience, as the Forest Service was "intensely human" and would respond to public concern. To force the agency would be viewed as an attempt to restrict its authority. The Forest Service, Yard reminded Lorimer, "is very strong in Congress" when defending its authority. Within ten years, he predicted, the Forest Serive would lead the way in wilderness preservation.[23]

Arthur H. Carhart, a landscape architect, resigned from the Forest Service in December 1922, frustrated by what he saw as too little support for his proposals. Three years earlier he had studied the area around Trapper's Lake in Colorado. Carhart advocated leaving the area roadless and denying applications for summer homes at lake's edge. His innovative ideas intrigued his superiors, but Carhart eventually resigned, despairing over the improbability of seeing his proposal adopted.[24]

During his study Carhart had exchanged views on wilderness reserves with a forester in the Southwest, Aldo Leopold. Leopold had advocated designating portions of the Gila National Forest in New

21. Ibid., pp. 163–217.

22. Minutes of the Service Committee, Apr. 6, 1922, box 115, RG 95–8, RFS.

23. *Saturday Evening Post*, May 27, 1922, p. 22; Yard to Lorimer, June 3, 1922, John Merriam Papers, Library of Congress.

24. Donald N. Baldwin, *The Quiet Revolution: The Grass Roots of Today's Wilderness Preservation Movement* (Boulder, Colo.: Pruett Publishing Co., 1972), pp. 30–42.

Mexico as a wilderness area or national hunting ground since at least 1921. Although he, too, would leave the Forest Service, Leopold was more patient and more successful than Carhart. In 1924 Greeley, using his administrative discretion, converted portions of the Gila National Forest into the nation's first wilderness area.[25]

Leopold was quickly recognized as a wilderness advocate, although his interests and expertise lay in game management. Franklin W. Reed, an industrial spokesman, believed that Leopold's attitude toward wilderness was advanced over others in the Forest Service. Reed thought that summer homes and similar forms of recreation tended to crowd out wilderness considerations. It seemed that public opinion needed to "lead & direct the thought of the official organization." Leopold agreed: "It is undoubtedly a fact that Foresters as a whole, and especially the Forest Service, tend to be unfavorable to the wilderness idea." He thought that foresters too often considered economic returns, but he felt confident that "they will come around later."[26]

Greeley acknowledged increasing public sentiment in favor of wilderness. In his 1926 annual report, he stated that the Forest Service recognized the validity of wilderness reserves and was sympathetic with the public. Yet, he asserted, wilderness must be correlated with the "other obligations and requirements of national forest administration." Wilderness preservation would be unfair, for instance, to communities dependent upon the forest for their livelihood. There would be no general Forest Service policy; each area would be considered on its own merit. Greeley added that in some cases grazing or water storage might be compatible with wilderness.[27]

After an inspection trip through the West in 1927, Greeley again wrestled with wilderness policy. For the second time, and with strong support from conservation groups, he had rejected plans for a commercially operated tramway on Oregon's Mount Hood. He saw a definite need to marshal public opinion against such incompatible uses to assure "retention of some of the best of our western mountain areas in their wilderness condition." Roads, too, were unacceptable in wilderness areas, and he proposed to prohibit both roads and summer

25. Susan L. Flader, *Thinking Like a Mountain: Aldo Leopold and the Evolution of an Ecological Attitude toward Deer, Wolves, and Forests* (Columbia: University of Missouri Press, 1974), pp. 79–81.

26. Reed to Leopold, Feb. 17, 1926, box 38, RG 220; Leopold to Reed, Feb. 25, 1926, ibid.

27. Forest Service, *Annual Report* (1926), p. 34.

homes. Access to wilderness must be by foot, horse, or canoe. Still, he wavered from committing the Forest Service to a "future course in dealing with urgent economic demands for all time to come."[28]

Leon F. Kneipp, assistant chief for the Division of Lands, asked Ringland to comment on Greeley's wilderness policy. Offering his support, Ringland believed that recreation had to be coordinated with other uses and that wilderness was a special type of recreation requiring restricted use. Ringland reiterated Greeley's statement that no one could foresee the future and therefore policies must be flexible. Ringland later recalled that the Forest Service and the Park Service had the same goal of serving the public, and "they met differences in a broad-minded manner, as you would expect of public servants of their stature."[29]

Much has been made of their clash over wilderness, but correspondence in the records of the National Conference on Outdoor Recreation reveals that neither the Forest Service nor the National Park Service had clear notions about completely unimproved areas. The Park Service was under increasing pressure from the automobile and mass recreation; the Forest Service was commodity oriented and reluctant to make irreversible decisions that would prevent future generations from drawing upon needed resources. The Forest Service did, however, administratively reserve wilderness, starting with the Gila in 1924. In 1929 the agency announced two new designations: research reserves and primitive areas. The research reserve would be preserved for scientific and educational purposes, while primitive areas would provide the "nature lover and student of history a representation of conditions typical of the pioneer period."[30] Until enactment of the 1964 Wilderness Act, the Forest Service set aside 9 million acres of wilderness on its administrative authority alone.

Recreational policy encompassed far more than wilderness preservation. By the 1920s, summer homes, resorts, and family camping occupied most of the Forest Service attention, as well as how to deal with the ever-increasing automotive onslaught. In terms of broad policy, however, how to coordinate recreational use on national forests

28. Greeley to Caspar S. Hodgson, June 13, 1927, drawer 233, RG 95–4, RFS.

29. Kneipp to Ringland, Oct. 19, 1926, box 28, RG 220; Ringland to Kneipp, Nov. 1, 1926, ibid.; Ringland, *Conserving Human and Natural Resources*, p. 153.

30. Donald C. Swain, *Federal Conservation Policy, 1921–1933* (Berkeley: University of California Press, 1963), pp. 137–38; F. W. Reed to Philip W. Ayres, Feb. 19, 1926, box 28, RG 220; USDA news release, Apr. 1, 1929, copy in FHS clipping file.

with that on the national parks continued to plague Greeley, as it had Henry Graves before him.

The official relationship between the two agencies was cooperative, and Greeley repeatedly ordered his men to be cordial. Yet, sources of abrasion persisted, and suspicious minds read sinister significance into every act. When the Forest Service published a map of the national forests in California, the national parks were left as blank spaces. An offended Mather complained. On the Forest Service side, Allen S. Peck, district forester in Colorado, complained to Greeley that "in accordance with your expressed wishes every effort is being made to cooperate with the offices of the National Park Service," but the Park Service did not respond in kind. Peck added that although he got along well with Horace Albright, superintendent of Yellowstone National Park, Mather ought to be told of his agency's shortcomings. Earle Clapp answered for Greeley, encouraging Peck to maintain friendly relations.[31]

Greeley described his recent meetings with Mather to the district foresters. They had agreed that the two agencies would jointly examine those portions of national forests under consideration for national parks. Only national forest land "where the dominant resource consists of scenic features of such a character as to have national importance" would be transferred to Parks. The two bureau chiefs further agreed that "areas whose dominant resources are economic or whose scenic and recreation features are not of outstanding importance should remain national forests." The Forest Service would protect entrances to national parks that were within national forests and take special care with logging and grazing near parks. Mather received assurance that Forest Service personnel would give the Park Service credit for its worthwhile program when speaking about recreation.[32]

Even so, incidents similar to the blank park acres on Forest Service maps plagued the relationship. A since-abandoned Sierra Club tradition was to elect the chief of the Forest Service as an honorary vice-president. When Sierra Club President William F. Bade sent Greeley congratulations on his election in 1921, he mentioned that Arthur Carhart had disrupted the relations between the Forest Service and the Park Service. Bade said that Carhart had jumped to erroneous

31. A. S. Peck to Greeley, Nov. 23, 1920, drawer 175, RG 95-4, RFS; Clapp to Peck, Nov. 26, 1920, ibid.

32. Greeley to District Forester, Apr. 22, 1920, ibid.

conclusions about Sierra Club policy and was responsible for the whole unfortunate incident. Greeley apologized for Carhart's behavior, describing him as "enthusiastic and aggressive" but as one who thought highly of the Park Service. It was Mather, Greeley suggested, who had misunderstood.[33]

Mather had publicly opposed Forest Service appropriations for recreation, believing that all such funds should go to his parks. When Carhart made a few extemporaneous remarks about Forest Service recreation programs during a National Parks conference in Des Moines, Iowa, in January 1921, Mather apparently felt it necessary to respond. A zealous newspaper editor gave full display to his reporter's account of the open dispute between agencies. Although Greeley was upset by Mather's opposition to Forest Service recreation, he did not permit "the controversy to go any further, as far as the Forest Service is concerned." Albert Fall's attempt to assume control of the Forest Service was occurring at this time, and the interior secretary proposed that the Park Service administer recreation on national forests. To Greeley, Fall's suggestion was "absurd and impossible."[34] The Carhart–Mather episode died down, but the scars remained.

Public demand for recreation continued to increase. Unfortunately, from the Forest Service point of view, the public was more familiar with national parks than national forests as recreational facilities and hence wanted more parks. Carhart had been trying to counter this lack of familiarity with national forest facilities when he offended Mather. Greeley now tried to divert efforts in the South for more parks by publicizing recreational aspects of national forests. He confided that so far he had headed off those "threatening movements" for parks. The diversion was short-lived, however; by 1926, Mather had won congressional authorization for three parks in the South.[35]

Still more aspects of recreation demanded Forest Service consideration. As the public's interest in recreation grew, so did its awareness of logging practices. When the issue of leaving scenic strips in sale areas to mask logging came up during a staff meeting in 1924, it revealed the

33. William F. Bade to Greeley, Feb. 17, 1921, drawer 192, ibid.; Greeley to Bade, Feb. 28, 1921, ibid.

34. Greeley to Henry C. Wallace, March 14, 1922, ibid.; Baldwin, *The Quiet Revolution*, pp. 61–71. Swain, in *Federal Conservation Policy*, insists that the basic disagreement was conservation versus preservation (see pp. 134–38).

35. Greeley to Philip W. Ayres, Nov. 6, 1923, drawer 193, RG 95–4, RFS; Swain, *Federal Conservation Policy*, p. 136.

perplexities of public image. J. F. Preston, forest management staff, spoke against scenic strips, arguing that "we are dedicating a small strip of land to National Park purposes and have passed up an opportunity to show the traveler what a National Forest cutting is like." He recommended, instead, using roadside logging as demonstration areas to better explain "what we are trying to do." Herbert Smith, responsible for public information, disagreed, pointing out that at the turn of the century when Fernow had showed the public what he was doing in New York State, he had reaped dismay, not support. Assistant Chief Roy Headley supported Preston and complained that when an artist went to Europe he painted man's work, "but when he goes to our woods he seems to be horror struck at anything which mars the original virgin condition of the stand." To Greeley it was not a question of "being ashamed of forestry" or trying to conceal any practices. Instead, it was a positive aspect of Forest Service policy to dedicate land to its highest use; therefore, areas best suited for logging should be logged. But E. E. Carter, assistant chief for forest management, pointed out that some "extremists" were demanding scenic strips along all roads. To that T. W. Norcross, head of engineering, added that it did not matter what they in the Forest Service believed; most of the public favored recreation. "We should be flying against present public sentiment," he insisted, "if we adopted a policy of cutting straight to the road."[36] The discussion failed to resolve the issue, although roadside screens became more and more common.

Whether timber, range, or recreation was involved, running the Forest Service like a business remained a goal. We have already seen computations on the financial legitimacy of wilderness use. In this vein, the secretary of agriculture wrote to Ovid Butler of the American Forestry Association: "Recreation in the National Forests is more than paying the costs of its supervision." He was referring to a report by Sherman, which made a strong case for recreation as an important use of the national forests. Sherman talked of coordinating recreation with other uses and advised designating all areas of recreational value. In most cases, he advocated, use should be free of charge. Even so, public benefits would exceed administrative costs. Sherman also supported continuation of the summer home lease policy. Others commented that summer homes remained in official favor because they

36. Minutes of the Service Committee, Nov. 26, 1924, box 115, RG 95–8, RFS.

provided a steady source of revenue — revenue being all-important to an agency striving for receipts in excess of appropriations.[37]

The secretary did not support all aspects of Forest Service recreation policy. During an unscheduled inspection of national forests in Montana, he had found conditions not to his liking. He thought that the local Forest Service officers were unrealistic. He gave a direct order for the Forest Service to stop building recreational roads, as they "constitute an unwise use of our funds and would bring us directly into competition with the National Parks." The secretary added that most Forest Service roads provided access to remote areas and therefore already were of recreational value.[38]

At the same time as Greeley and his staff were grappling with recreation policy, trying first to determine and then to respond to public sentiment, they were attacked on another flank. "Has Our Forest Service 'Gone Daffy'?" William C. Gregg asked the readers of *Outlook* in 1925. The Forest Service spent "millions" for recreation, he charged, "but how much for reforestation?" To Gregg, too much money was spent "entertaining visitors" and not enough on "the original but more somber work of forestry." Reflecting the magazine's influential status, Greeley soon received a letter of inquiry from Congressman Edward E. Browne, who described Gregg's article as "quite a serious arraignment of your Department." It should be refuted, the congressman insisted.[39]

Greeley responded to the congressman, charging that the editor of *Outlook* and Gregg wanted the Forest Service barred from recreation. Explaining that forestry is not timber alone or recreation alone, Greeley told Browne that "for a number of years there has been opposition, sometimes open, sometimes veiled, to Forest Service development of recreational use of the National Forests."[40]

In a later issue of *Outlook*, Greeley was able to respond to Gregg's attack. The Forest Service was "sticking to its job," Greeley maintained, and Gregg had quoted out of context from official reports

37. Jardine to Butler, May 26, 1925, box 72, SAF Records; E. A. Sherman, "Outdoor Recreation on the National Forests," n.d., ibid.; F. W. Reed to Philip W. Ayres, Feb. 19, 1926, box 38, RG 220. In 1916 Sherman had published "Use of the National Forests of the West for Public Recreation" (*SAF Proceedings* [July 1916], pp. 293–96), in which he emphasized increasing recreational demand and predicted that recreation would become an important use.

38. Leon Kneipp to Fred Morrell, Aug. 6, 1925, drawer 219, RG 95-4, RFS.

39. William L. Gregg, "Has Our Forest Service 'Gone Daffy'?" *Outlook* 139 (Jan. 11, 1925): 226–27; Edward E. Browne to Greeley, Feb. 17, 1925, copy in FHS clipping file.

40. Greeley to Browne, Feb. 27, 1925, copy in FHS clipping file.

when he implied neglect, incompetence, collusion, and graft within the Forest Service. Strong language from a man of even temperament; Greeley was obviously upset. Even more upsetting must have been the editor's handling of his response. In a note introducing Greeley's article, the editor denied that either he or Gregg had engaged in innuendo, but on the table of contents page the readers were told that "the storm of comment" had made *Outlook* even "more certain than ever" that the Forest Service needed full investigation. Readers were advised that there would be a whole series of editorials on the Forest Service.[41]

Although aggravated, Greeley confided to Graves that Gregg's efforts had not swayed congressional opinion. Gregg had, however, made Greeley uncomfortable about the propriety of the Forest Service's developing recreation as a major use. Some of Greeley's advisers thought that the whole affair had been brought about by the Park Service in another attempt to limit Forest Service recreational activity. Greeley predicted that the campaign in which *Outlook* was participating would delay advancement of recreation policy. He hoped that the National Conference on Outdoor Recreation could be induced to strengthen its support of the Forest Service. Not surprisingly, Graves, in his response to Greeley, thought that the ultimate solution would be to transfer the Park Service to the Department of Agriculture, a reiteration of his long-held opinion.[42]

As promised, *Outlook* continued to take the Forest Service to task. Arthur M. Baum, introduced to the readers as an ex-forest supervisor, strongly criticized Forest Service policy on fire control. The editor reminded the reader that Gregg had accused the Forest Service of being "daffy" over recreation instead of tending to business. Now Baum charged that the agency was failing at its primary task, had not grasped the importance of fire, had not worked hard enough to get adequate funding, had wasted money with its top-heavy bureaucracy, and had used its men inefficiently. Part of the problem, according to Baum, was too much emphasis on college education and not enough recognition of practical experience. Although Greeley was not invited to reply to Baum, as he had been to Gregg, he again sent in an article. This time *Outlook* did not print Greeley's defense because, the editor explained, it was too long and he could not expect his readers to make

41. William B. Greeley, "The Forest Service Is Sticking to Its Job," *Outlook* 139 (March 4, 1925): 336–39; Editor's note, ibid., p. 333; Table of Contents, ibid., p. 322.

42. Greeley to Graves, Feb. 24, 1925, drawer 202, RG 95–4, RFS; Graves to Greeley, March 2, 1925, ibid.

"detailed considerations of [opposing] statements" made several weeks apart.[43]

Controversies like those fostered by Gregg and Baum confront the head of any administrative agency. In theory, at least, administrators seek outside commentary and evaluation, always hopeful that the differences of opinion can be examined with fairness. In addition, administrators attempt to stave off criticism by educating and convincing potential dissidents that agency policy is appropriate. Pinchot had been particularly adept at garnering public support by a barrage of news releases. Although Congress perhaps was justified in slapping Pinchot down because he seemed to be using public funds to lobby, the public relations efforts continued but on a less blatant scale.

In 1915, Arthur Ringland had suggested to Henry Graves that the Forest Service needed a formal public relations department. World War I and other subsequent events delayed adoption of Ringland's proposal. On May 20, 1920, the Branch of Public Relations was established "for more careful planning of methods by which public interest may be increased in both the protection and use of the Forests. . . ."[44] Eventually, the name was changed to the Division of Information and Education, or I and E. Division head Herbert Smith, who began with Pinchot, had always served in this capacity, although his prior title had been editor.

Public relations has its more subtle aspects and cuts across all functions. When Leon Kneipp reported to staff on his efforts in recreational planning, he was emphatic about the "opportunity here to demonstrate in the next five years the superior ability of the Forest Service" to administer recreation. He pointed out that the general public looked to the Park Service as the "outstanding if not exclusive" authority on recreation. Kneipp insisted that the Forest Service should take advantage of the opportunity to publicize its recreational skills.[45]

Range

Recreation and public relations were obviously intertwined. Wartime emergency considerations had allowed grazing in national parks,

43. Arthur M. Baum, "Why Our Forests Are Burning Up: The Forest Service Has Lost Its Vision of Its Real Job," *Outlook* 146 (Aug. 10, 1927): 508–10; "The Necessity of Immediate Reform," ibid. (Aug. 24, 1927): 540, 542–43; Editorial, ibid. (Oct. 12, 1927): 167.

44. Ringland, *Conserving Human and Natural Resources*, pp. 110–11; Forest Service, *Annual Report* (1920), p. 9.

45. Minutes of the Service Committee, Feb. 9, 1928, box 115, RG 95–8, RFS.

a situation that must have tested Mather's own public relations ability. Grazing in national forests, too, at times overlapped into recreation policy, because high meadows were attractive to both hikers and stockmen. Each thought the other's use was lower priority. Since at that time game management was viewed as an aspect of recreation, keeping big game herds in balance with domestic stock was also related to recreation policy. A major responsibility for Smith and his Branch of Public Relations was to explain and, it was hoped, to convince the many user groups of the soundness of Forest Service policy. Their theme would be that uses were interrelated and all merited consideration. Gaining acceptance for grazing was as difficult a task as any.

Grazing policy had been a prime consideration in Pinchot's administration. The high points, from an administrative point of view, came when Supreme Court decisions on Light and Grimaud, which upheld federal authority over national forest ranges, were handed down over a year after Pinchot left office. Albert Potter, Pinchot's choice to develop Forest Service grazing policy, had continued in that capacity under Graves. The chief forester, who relied heavily on Potter's advice, asked whether their grazing regulations did not encourage the large owner at the expense of the small. Graves preferred favoring national forest residents. Potter assured his chief that the Forest Service did not favor the larger stockmen. Since local residents frequently sold their grazing privileges to stockmen, it only appeared that the large operator controlled the range.[46]

For the 1903 Public Lands Commission, Potter had sent questionnaires to stockmen using the public range. Most supported "reasonable" regulation of grazing, justified by the recognized need to protect the range. Potter went to the stockmen again, six years later, and found 75 percent to be in favor of Forest Service policy. An example of the cooperative programs they were supporting was a system of agreements where stockmen supplied fence posts and the Forest Service supplied wire and staples. Even though stockmen had to dig the post holes, the program offered an obvious encouragement to fencing. In 1914 the National Woolgrowers Association "for the first time in history" passed a resolution endorsing the work of the Forest Service,

46. Graves to Potter, Sept. 29, 1913, drawer 201, RG 95-4, RFS; Potter to Graves, Oct. 3, 1913, ibid. Allowing another stockman to use permit privileges was against Forest Service regulations. For Potter to complain about it, however, suggests that enforcement was very difficult.

rather than criticizing it.[47] General policies were acceptable, although more to cattlemen than to sheepmen.

How many stock could be grazed without damaging the range posed a difficult question for the Forest Service to answer, but how much to charge proved tougher yet. The court decision upholding the right to levy fees made it a little easier, but stockmen staunchly resisted paying more than token amounts.

Traditionally, the Forest Service had sold its forage resources on a different basis from its timber resources. Timber appraisal was based on current market conditions and sold by competitive bidding. Forage, on the other hand, was sold based upon a "reasonable fee," the amount to be determined locally within national guidelines. In addition, Forest Service fees were lower than market value. Pinchot had justified these policies by pointing out that the small stockowner, who often lived in or adjacent to a national forest, could not survive competitive conditions.[48]

Leon Kneipp, at a weekly staff meeting in 1913, had announced that the National Woolgrowers Association was beginning an active campaign to have grazing fees reduced. Sheepmen felt that they were paying too much in comparison with cattlemen. Months later the matter of grazing fees was brought up again at a staff meeting, this time by Potter. He thought that fees were of "considerable importance." Stressing that the relations between the Forest Service and stockmen were finally harmonious, Potter questioned the wisdom of jeopardizing the situation by asking for a change in fees. He reiterated existing policy: charge enough to cover administrative costs, plus "reasonable compensation" for benefits derived.[49]

Annual fees in 1910 had ranged from thirty-five to sixty cents for cattle and ten to eighteen cents for sheep. Five years later, rate increases boosted the fees to forty cents minimum and $1.50 maximum for cattle. Sheep were levied at 25 percent of the cattle rate. Then in 1917 the Forest Service moved to increase grazing fees again. Potter told the stockmen that fees would be doubled over a three-year period.

47. U.S. Congress, Senate, *Report of the Public Lands Commission*, Sen. Doc. 189, 58 Cong. 3, 1905, pp. 5–8; memo for week of Sept. 18, 1909, Sept. 25, 1909, copies in drawer 152, RG 95–4, RFS; Service Order no. 2, July 22, 1907, drawer 201, RG 95–4, RFS; Potter to Graves, Jan. 19, 1914, drawer 201, RG 95–4, RFS.

48. USDA, Forest Service, *Use Book: Grazing*, 1910, pp. 41–42; Overton Price to district forester, D–2, Sept. 28, 1909, drawer 165, RG 95–4, RFS.

49. Minutes of the Service Committee, Aug. 13, 1913, Nov. 20, 1913, box 115, RG 95–8, RFS.

As far as possible, fees would be based on real value, an amount difficult to determine since forage was not sold on a competitive basis. The Forest Service, however, would continue to charge less than full value, because permittees provided a first line of defense against fire and deserved some reward for these services. Potter tried to rationalize the increase by pointing to the large number of applications for grazing permits. Obviously, he claimed, the stockmen themselves thought present rates were a bargain.[50]

The unimpressed cattlemen responded by insisting that current fees already more than covered administrative costs. One stockman accused the Forest Service of substituting commercialism for conservation. Objections and wartime priorities for increased stock production prompted Secretary of Agriculture Houston to defer higher fees for the duration of the war.[51] In fact, war needs had dictated that overgrazing of national forest ranges would be tolerated.

Stockmen wanted stability, a condition that the annually renewed permits denied them, and following the war they negotiated successfully with Potter for five-year agreements. In conjunction with the new permit schedule, however, the agency doubled fees. Sherman supported stable fee structure during the life of a permit and warned that the House Committee on Agriculture was considering tripling fees, hinting that stockmen should be satisfied with a mere doubling. Postwar market declines caused the Forest Service to repeatedly defer announced fee increases, at least by liberal granting of extended payment dates. Congress was sympathetic, too. In addition to deferred payments, stockmen then pressed for ten-year permits to assure even greater stability. They again asked that fees be based on cost of administration rather than the commercial value of the forage.[52] Forest Service officials saw that a total policy overhaul was needed to avoid perpetual piecemeal changes.

C. E. Rachford, in charge of grazing for the California district, agreed with Greeley in December 1920 on the need for a comprehen-

50. W. L. Dutton, "History of Forest Service Grazing Fees," *Journal of Range Management* 6 (Nov. 1953): 393–98; American National Live Stock Association, *Proceedings* (1917), pp. 115–23.

51. American National Live Stock Association, *Proceedings* (1917), Resolution no. 6, "Protesting against Proposed Advance in Grazing Fees on National Forests," pp. 124–25; Statement by W. S. Whinnery, p. 126, ibid.; Houston to American National Live Stock Association, printed in Forest Service news release, Nov. 20, 1917, FHS clipping file.

52. Minutes of the Service Committee, Jan. 30, 1919, March 18, 1920, box 115, RG 95–8, RFS; Henry Wallace to George Norris, Apr. 1, 1922, RG 95–4, RFS; Minutes of the Service Committee, March 22, 1923, box 115, RG 95–8, RFS.

sive range study. Rachford was alarmed by the lack of solid data upon which to base decisions. What was needed, he explained, was a measure of the economic relationship of range to public welfare. Attached to Rachford's memorandum was a summary of Forest Service grazing policy, existing and proposed. Most significant were proposals to classify rangeland according to commercial values and to base fees accordingly. The summary emphasized that since the Forest Service should be operated as a business, "we are going to charge for the National Forest ranges what they are actually worth."[53]

Will Barnes, chief of grazing since 1916, when Potter's increasing administrative responsibilities as associate chief forester had forced him to give up the post, offered Rachford suggestions on his study. Acknowledging that the fee system was based on expediency, he warned that the public was demanding better care of the range. Barnes described Forest Service grazing policy as inconsistent, in fact, discriminatory. It increased speculation among permittees, and it deferred proper range management. Barnes ended his indictment by charging Rachford with the responsibility of determining how much a stockman could pay in fees and still make a profit. Barnes wrote to Greeley praising Rachford's proposal, particularly Rachford's idea of reducing the numbers of stock allowed under each permit in order to distribute grazing privileges more widely.[54]

It took Rachford four years to complete his study, based largely on frequent exchanges of thought and data with range men throughout the Forest Service. Under investigation were 110 million acres of national forest range used by approximately 38,000 permittees. Ultimately Rachford defined forage as a commodity to be sold or traded like any other. Most difficult for him to decide was whether the government was like an ordinary business and therefore should charge whatever the market could bear. Rachford reasoned that the national forests had been created for benefit of all people, equally. Allowing stockmen a large profit would be conferring unequal benefits. To submit to growing congressional pressures to base grazing fees on fair market value rather than on administrative costs would mean raising fees 60 to 70 percent. The proposed increases would have raised cattle

53. C. E. Rachford to Greeley, memo dated Dec. 10, 1920, copy in drawer 261, Records of the Division of Range Management, General Correspondence, 1905–52, Record Group 95, RFS.

54. Barnes to Rachford, Nov. 26, 1920, ibid.; Barnes to Greeley, Dec. 13, 1920, ibid.

fees from an average of $1.25 to $2.17. Rachford's recommendation for ten-year permits was implemented in 1926. Fee increases, although acknowledged as just, were deferred because of the fluctuating economic conditions of the mid-twenties.[55] Rachford's efforts made a favorable impression, for in 1928 he became chief of grazing, replacing Will Barnes.

Meanwhile, stockmen flatly opposed any increase in fees. Their opposition prompted Secretary of Agriculture William M. Jardine to ask Dan D. Casement, a Kansas stockman, to review Rachford's proposals. Casement recommended substantial reductions. Senator R. M. Stanfield of Oregon, who had recently lost his grazing permit on account of misrepresentation, introduced legislation which would have maintained the status quo on fees and given contract status to the ten-year permits—the vested-rights issue again. Stanfield also wanted local committees to administer grazing and to act as boards of appeal.

Ovid Butler of the American Forestry Association was appalled. "Our big fight is on," he wrote to E. T. Allen. The Stanfield bill would be "a destroying wedge in our National Forests . . . it must not pass." Earlier, the AFA had received a ten-thousand-dollar contribution from a private donor "to head off any drive" by either lumbermen or stockmen that would disrupt the Forest Service. Henry Graves at Yale had urged the conservation group to get into "the thick of the fight." The American Forestry Association did so, and was joined by the Society of American Foresters. Stockmen supporters in Congress backed away from an all-out battle and took no action. The secretary of agriculture met with the westerners and agreed on only modest fee increases spread over a long period.[56] Grazing issues were resolved, for a time.

Routine

All was not legislative battles and confrontation with user groups during Greeley's administration. Routine affairs of day-to-day opera-

55. C. E. Rachford, "Range Appraisal Report," pp. 8–10, Nov. 5, 1924, 64 pp., ms. copy in box 89, ibid.; Dana, *Forest and Range Policy*, p. 229; W. B. Greeley, "Grazing Administration and Range Fees on National Forests," in American National Live Stock Association, *Proceedings* (1924), pp. 133–41.

56. Ovid Butler to E. T. Allen, Feb. 5, 1926, box 12, WFCA Records; George D. Pratt to Butler, Sept. 29, 1925, box 5, AFA Records; Board Minutes, Dec. 28, 1925, box 6, AFA Records; Dana, *Forest and Range Policy*, pp. 230–31.

tion consumed the bulk of official time. Shortly after he became chief forester, Greeley took steps to build closer ties with his staff. He invited many more men to the weekly Service Committee sessions, as a way to familiarize them with Forest Service programs. The notion of the agency as a family, so much in evidence today, began with Greeley. A staff meeting meant calling the "whole family" together.[57]

Despite his wishes to be closer to staff, to have them speak their minds in his presence, the men did not fully respond. Throughout his career Greeley found it difficult to get along with both subordinates and superiors. His inclination to argue with authority had been evident during the Pinchot years, when Washington office staffers saw Greeley acting entirely on his own initiative, with apparent lack of consideration for other views. His men accordingly were reluctant to offer advice. Despite a perhaps overly aggressive nature, Greeley was able to command the respect of his men because no one challenged his abilities. It did seem, though, that he would have to "suffer some bruises" as he advanced through the ranks. Fifteen years later and now chief, Greeley's personality continued to unsettle many around him.[58]

The New England Section of the Society of American Foresters requested an investigation of Forest Service administration. As SAF president, Samuel T. Dana informed Greeley of the action by the section and told him frankly of critical reactions toward him personally and toward his administration of the Forest Service. Dana later put the substance of this conversation in writing. Among other things, he repeated the common belief that morale had suffered from overly critical inspections, emphasis on efficiency without regard to the "personal equation," and failure to delegate responsibility to the field.[59]

Greeley must have suspected that all was not well, but still he was hurt. Realizing that Dana was one of the few who would be candid, Greeley asked him to "feel perfectly free at all times to bring things to my attention." He told Dana that he was "grieved" that his staff had been talking about him among themselves, rather than speaking directly. Never had he supposed that any of the men would hesitate to discuss problems openly, as "one of the family." He wanted Dana to spread the word that he was fully accessible.[60]

57. Minutes of the Service Committee, Sept. 23, Oct. 7, 1920, box 113, RG 95–8, RFS.

58. John B. Adams to C. S. Chapman, Aug. 24, 1909, box 576, Gifford Pinchot Papers, Library of Congress.

59. S. T. Dana, personal communication, May 27, 1975; Dana to Ward Shepard, March 8, 1925, Samuel T. Dana Papers, FHS.

60. Greeley to Dana, Apr. 10, 1925, ibid.

Meeting the problem frontally, Greeley set up a two-week session of district foresters and assistant chiefs to be held in Denver in November 1925. Allotment of that much high-level time demonstrates Greeley's obvious concern. He told those attending that he saw need for "a fresh study and grasp of our common problems." Greeley's action elicited approval from Dana and stilled any further consideration of the New England SAF section demand for an investigation.[61]

Greeley predicted that "the permanent value of the Denver conference will be largely its contribution to the unwritten code of the Service." The two-week conference included discussions of many facets of Forest Service operation, and Greeley commented on each recommendation. Decentralization, the keystone of Forest Service administration since Pinchot, came under close scrutiny. "Genuine" decentralization, the field men decided, was achieved not by administrative theory but by increasing qualifications of men not headquartered in Washington, D.C. The tendency of the Washington office to acquire the best men could easily "result in more leadership and pressure than the doers can absorb."[62]

The group decided that morale measured as high as ever, but a committee on personnel focused on an issue that had initially prompted many to be critical of Greeley, that of paying men less who were not college trained. By a narrow twelve to ten vote the men suggested that no salary distinction be made. Greeley disagreed, as he believed that college-trained men were worth more in the long run.[63]

Greeley was satisfied that the conference had been beneficial. Sending a copy of the conference report to each forestry school, he explained to the deans that the meeting had shown both centrifugal and centripetal forces—"the Forest Service must have both the spirit of order and the spirit of initiative." After mulling the episode over, Greeley saw that his campaign to tighten up administrative procedures had been at its root. When he became chief, he thought all too frequently inefficiency was blamed on the aftermath of the war, a tough fire season, or some other major distraction. Rationalization of inefficiency had almost become habitual. His much-needed corrective actions had obviously caused much unhappiness. But the inefficien-

61. Greeley to District Foresters and Assistant Foresters, Apr. 7, 1925, ibid.; Dana to Greeley, June 18, 1925, ibid.

62. Greeley to District Foresters, Jan. 26, 1926, ibid.; "Defects and Needed Betterments in the Handling of Physical Resources," Nov. 2–14, 1925, copy of committee report, ibid.

63. Denver Conference Report, 1925, ibid.

cies had been eliminated, and the Denver conference had soothed the bad feelings.[64]

Decentralization, in some sense the main issue of the Denver conference, always proved difficult to maintain. There was a constant tendency toward Washington. Henry Graves had noted, for example, that the routine personnel action of transferring a forest supervisor had seven sets of Washington office initials on it, hardly an example of delegating authority downward. During staff meetings, the subject of decentralization came up repeatedly. Most agreed that in the Forest Service authority was more dispersed than in other federal agencies, but some felt not enough. The complaint was heard that forest supervisors failed to train their rangers adequately to handle more responsibility; whenever a ranger showed ability, he was promoted and transferred out of the field into a central office. Good men were needed at all levels. The promotion system tended to defeat this goal unless supplemented with adequate recruitment.[65]

Recruitment was not always easy, and keeping the men could be even harder. Low salaries, fought so valiantly by Graves and universally remembered in foresters' memoirs years later, were hard to tolerate for long periods. Constant abrasion between technical foresters and practical men at promotion time had no graceful solution. Living conditions, generally primitive, presented another chronic source of dissatisfaction. This latter problem stemmed from a congressional edict that no headquarters could cost more than 800 dollars to construct. An inspection of 210 ranger stations in 1920 showed only 46 with running water and 3 with bathtubs.[66]

After the men were on the job, they had to be trained. In the Southwest, newcomers were required to study the manual and answer questions. Inman Eldredge, inspector for the Division of Timber Management, emphasized the need for still more knowledge; increased timber values, increasing interest in recreation, and increased public awareness—all reduced the ranger's margin for error. Decentralization, too, demanded training. If district rangers, instead of a specialized staff of experts, were to administer timber sales as policy directed, then rangers had to have the capability.[67]

64. Greeley to Forestry Schools, Feb. 19, 1926, drawer 199, RG 95–4, RFS; Greeley to Dana, Apr. 6, 1926, Dana Papers.

65. Minutes of the Service Committee, May 10, 1917, Feb. 17, 1922, box 115, RG 95–8, RFS.

66. Minutes of the Service Committee, Nov. 24, 1920, ibid.; "My Landlord, Uncle Sam," *The Federal Employee*, Jan. 1920, p. 17.

67. Minutes of the Service Committee, Jan. 20, 1921, March 26, 1925, Nov. 16, 1922, box 115, RG 95–8, RFS.

The ranger was responsible for his district. The forest supervisor had charge of several rangers and had to make inspections to keep track of his responsibilities. And so on up the ladder. Local discretion was acceptable, but basic policies, to Greeley, were inviolate. The Forest Service, he ordered, would support policies of the Budget Bureau and the president. In private industry, he reminded his staff, the boss called the shots. He wanted it clear that the Forest Service supported the administration's policy "as laid down." To Greeley, "that is the way to play the game." Also typical of his forthright methods, Greeley presented well-trimmed budgets, believing that it was bad policy for an agency to ask the Bureau of the Budget for more than actually needed.[68]

To test the success of policy, recruitment, training, and to measure the success of decentralization, Pinchot had created an elaborate system of inspections. Graves, Greeley, and all chiefs to date continued the inspection program. Initially, being an inspector was a full-time job for certain staff officers. In later years the Forest Service developed a range of inspections, conducted by teams of line and staff. Inspection reports showed the field men how well they carried out Forest Service policy. Perhaps even more valuable, inspections offered an effective training program for high level staff, who only infrequently got into the field. Effectiveness of inspection was relative. Obviously much of value was gained, but as shown in events leading up to the Denver conference in 1924, tact was not always an inspector's hallmark.

Greeley was a tough-minded administrator. Those who were close saw a highly likable man with a well-developed sense of humor. Those outside his immediate circle often saw what they believed to be a somewhat conceited and intolerant man. He was a teetotaler and avoided profanity. Once he admonished district foresters against excessive poker playing, warning them not to play so late at night that they could not arrive punctually at work the next day. During Prohibition, Greeley ordered his men to set a good example for the public.[69] Yet, his puritanical conscience had in no way prevented close association with men either in or out of the Forest Service.

Greeley's main contribution, one that set the tone of Forest Service

68. Minutes of the Service Committee, Feb. 4, 1922, July 13, 1921, ibid.

69. Morgan, *William B. Greeley*, pp. 14–15; Inman Eldredge, Oral History Interview by Elwood R. Maunder, 1959, FHS, p. 25; Greeley to District Foresters, Nov. 27, 1922, drawer 192, RG 95–4, RFS; Greeley to Flory, Feb. 2, 1927, ibid.

policy to date, centered on federal fire protection legislation. When one considers Greeley during his Forest Service years, the Clarke–McNary Act and all that it represents comes to mind. As landmark legislation for the agency and for the conservation movement, it merits thorough examination.

CHAPTER VII

Fire and Taxes: A Cooperative Solution

"IT WAS A GREAT THRILL to be in at the kill—even if the victory was bloodless."[1] William Greeley was remembering his feeling on that June day in 1924 when Congress approved the Clarke–McNary Act. Greeley ranks second only to Pinchot in stamping his personal philosophy on American forestry, and the law was his great personal victory after four years of effort.

The Clarke–McNary Act, in sum, substantially expanded Weeks Law programs and added others. Cooperation and incentives—and, significantly, not force—would be used to improve conditions on private forest land. Fire and taxes, believed the two worst deterrents to good forestry, would be faced together by the federal, state, and private sectors. It was hoped that reducing risks would prompt landowners to adopt less destructive cutting practices, because then they could better afford to hold timber for future use.

The essence of the Clarke–McNary Act—cooperation to inspire voluntary action—has been the essence of the Forest Service. The timber management and protection policy of the National Forest Administration, the thrust of State and Private Forestry, and the orientation of Research have reflected this cooperative philosophy. To be sure, at times some would run out of patience with cooperation and advocate controls. Overall, however, cooperation dominates the relationship between public and private forestry.

FIRE

From land clearing, from railroad engines and red-hot brake shoes, from logging machinery or lightning—natural and man-caused fire

1. William B. Greeley, *Forests and Men* (Garden City, N.Y.: Doubleday, 1951), p. 110.

had burned vast areas of American forests by the turn of the twentieth century. While causes of fire were omnipresent, means of combatting it were few. Lumbermen, believing that fires were inevitable, thought that logging rapidly and wastefully and then abandoning the cutover land was justified. Only the foolhardy would husband timber for some future need, because in all likelihood fire would destroy whatever they had saved.

Then the lumbermen's westward migration ended. The blue Pacific Ocean, not another ridge of green forests, now met their gaze. Thoughts shifted toward permanence. Enormous supplies of West Coast timber were acquired, enough to feed sawmills for decades, perhaps generations. But supplies failed to free the lumbermen from the future. Indeed, because of these supplies, the future with its potential threats at times seemed ominous.

For example, in 1902 the Yacolt Burn in southwestern Washington offered tragic proof that major capital investments in timber could be wiped out in a flash. Over 400,000 acres containing an estimated 5 million dollars' worth of timber were burned. A shaken industry, realizing that individual companies were extremely vulnerable to fire, banded together and lobbied at the state capitol in Olympia, rushing through legislation which created the office of state fire warden. When appropriations for this new agency proved niggardly the lumbermen assessed themselves to raise necessary funds. The timber industry's support for the state fire warden at first exceeded public appropriations, a sure indication that forest landowners now believed in the value of fire protection.

Threat of fire remained grave in the Pacific Northwest, in a substantial degree because of expanding logging activities, and in 1909 the lumber industry formed the Western Forestry and Conservation Association. Headed by E. T. Allen, who had been Pinchot's district forester in the Northwest, the industrial group coordinated fire protection on private land to supplement growing state efforts. As an added measure, the industry sponsored successful legislation requiring all forest landowners in Washington to participate in fire protection, thus relieving a progressive few from bearing the burden for all.[2]

2. For a fuller view see George T. Morgan, Jr., "The Fight against Fire: Development of Cooperative Forestry in the Pacific Northwest, 1900–1950," Ph.D. diss., University of Oregon, 1964; and Harold K. Steen, "Forestry in Washington to 1925," Ph.D. diss., University of Washington, 1969. Stewart Holbrook, *Burning an Empire: The Story of American Forest Fires* (New York: Macmillan, 1952), provides a useful view of the nation's most important forest fires.

During the first two decades of the twentieth century, the Forest Service supported state and private protection programs. Forestry agencies, at all levels of government, joined with industrial protective associations and individual companies in an effort to reduce fire loss to acceptable levels. Since all of society benefited from forests, everyone must help with the problem.

The Weeks Act yielded a partial solution. Federal matching funds up to ten thousand dollars per year to state fire protection agencies provided an incentive for local programs, even if limited to patrolling watersheds of navigable streams. But more was needed. Increasing property values enlarged the risk of retaining standing timber, and the need for larger efforts grew evident.

Public timber, of course, was just as combustible as private. On its own lands, the Forest Service suffered disastrous fire seasons, the most spectacular being the Idaho and Montana infernos during the summer of 1910. Extensive forests in either state or federal ownership received primarily custodial or maintenance management against the day that this timber would be needed to replenish dwindling private supplies. Until then, fire protection ranked as the dominant activity on national forests. Forest Service permittees—grazing, timber, or power—were obligated to fight fire without compensation whenever their permit area was threatened. In fact, officials often listed fire protection as a major justification for issuing permits.

One could learn from fires. Graves thought that the 1910 season taught the need for more trails, telephones, and patrols. He cited Supervisor Elers Koch of the Bitterroot National Forest in Montana as an outstanding example of an administrator who had been able to hold fire loss to an acceptable level by having constructed a large number of trails.[3]

Forest Supervisor Burt Kirkland of the Snoqualmie National Forest in Washington doubted that the cost of protecting much national forest land could be justified on the basis of current values. The enormous private timber inventory, he claimed, had eliminated the market, and thus the value, for the federal timber, except as watershed protection. Future values were obviously substantial, but Kirkland recommended caution in assigning unrealistically high figures. He advocated a "scientific basis" for allocating fire protection funds.[4]

3. Minutes of the Service Committee, Sept. 2, 1908, Sept. 14, 1910, RG 95–8, RFS.

4. Burt Kirkland, "The Place of Fire Protection in Forest Management," March 1, 1911, copy in drawer 451, Research Compilation File, 1897–1935, Record Group 95 (hereafter cited as RG 95–115), RFS.

Kirkland's proposal was too sophisticated for the time; it would be many years before the Forest Service could justify more than a theoretical concern about too much emphasis on fire suppression. Instead, the problem was too little recognition of the awesome responsibilities, and the agency struggled with inadequate appropriations, small staff, and vast areas needing protection. It inaugurated education campaigns with fire prevention slogans displayed in strategic sites. As aircraft became capable of operating over forested areas, the Forest Service began to augment lookouts and ground patrols with aerial surveys.[5] Other technological advances, such as the Osborne firefinder, portable pumps, and then radio, combined with increased public awareness and cooperation to improve prevention and suppression. To foresters, however, and especially to the Forest Service, the major fire problem lay with protecting private timber supplies.

Cooperation or Regulation. Fernow's research program during the 1890s had been popular with the wood industry. Pinchot's cooperative management efforts during the first decade of the twentieth century, with his staff writing management plans for millions of acres of private land, met with broad support from forest landowners. Pinchot next had testified in favor of lumber tariffs, because he believed that a profit-making company would be more likely to adopt less wasteful logging methods. Henry Graves continued his predecessors' positive approach. During his administration the Forest Service sponsored studies to show that the Bureau of Corporations' *Report on the Lumber Industry* had treated lumbermen unfairly. In the latter part of the Graves administration, however, the chief forester decided that some degree of public regulation of private cutting practices was necessary to safeguard future timber supplies.

In 1919 Graves explained his proposal to Royal S. Kellogg, formerly with the Forest Service but now an industrial spokesman. Inclusion of private timberlands in any comprehensive forestry plan was essential to Graves. He added that the Forest Service had tried for twenty years to get effective industrial participation in cooperative programs; but although cooperation produced much industrial praise, Graves thought that it had showed too few tangible results. He now proposed federal legislation to allow fuller Forest Service cooperation with state agencies for fire protection. Included in this protective

5. Minutes of the Service Committee, Oct. 1, 1913, Apr. 21, 1916, March 27, Apr. 3, Nov. 13, 1919, RG 95–8, RFS.

Hauling logs by rail, Georgia, 1903. Courtesy of the U.S. Forest Service

Hauling logs by truck, Gifford Pinchot National Forest, Washington, 1951. Courtesy of the U.S. Forest Service

Loading logs with steam power, Snoqualmie National Forest, Washington, 1911. Courtesy of the U.S. Forest Service

Cable yarding system powered by a steam donkey engine, Snoqualmie National Forest, Washington, 1943. Courtesy of the U.S. Forest Service

Yarding logs by helicopter, California. Courtesy of the U.S. Forest Service

Yarding logs with tractor, Willamette National Forest, Oregon, 1957. Courtesy of the U.S. Forest Service

Yarding logs with horse-drawn bigwheel, Deschutes National Forest, Oregon, 1928. Courtesy of the U.S. Forest Service

legislation would be regulation of logging by states, also necessary to reduce waste and to encourage reforestation. All commercial timber and logged off land would be subject to the act.[6]

Kellogg discounted most of Graves's plan, stating that "if forestry is a business proposition, it must pay dividends under business conditions." Kellogg thought that professional foresters should acknowledge what practical men had long maintained; long-term forestry programs were too hazardous and offered too low a rate of return to attract private capital. Federal or state legislation designed to force private investment in unprofitable ventures was doomed to fail. What was really needed, Kellogg insisted, was public purchase of cutover land and more federal funds for fire protection.[7]

Graves rejected Kellogg's arguments. He had talked to many lumbermen, and they accepted the idea of controls, provided they received fire protection in return. Agreeing with Kellogg that forestry must pay, Graves proposed a new accounting system whereby savings were considered payment. Kellogg complained that he had been misinterpreted. The main problem resulted from exaggerating the importance of timberland to the nation and using this alleged importance to justify public intervention. After all, wood was no more important than wheat, and no one was advocating "mandatory" wheat growing.[8]

The Society of American Foresters entered the fray. Playing its first major role in national affairs, the SAF established the Committee for the Application of Forestry. Gifford Pinchot was chairman, presiding over a distinguished group of professional foresters. The committee studied American forest conditions and submitted its report for publication in the *Journal of Forestry*, official organ of the Society of American Foresters.

The report summarized the forest situation. Forests were essential to public well-being, and the "beginning of timber shortage" had arrived. The shortage would soon be acute, predicted the committee. They estimated that supplies of mature timber could be exhausted by the 1970s and expected that second-growth volumes would be too small to meet the demand, particularly since export trade seemed certain to increase. Skyrocketing lumber prices would herald the impending timber famine.

6. Graves to R. S. Kellogg, May 29, 1919, box 143, SAF Records, FHS, Santa Cruz, Calif.
7. Kellogg to Graves, June 21, 1914, ibid.
8. Graves memo [July 5, 1919], box 143, SAF Records; Kellogg to J. Girvin Peters, July 8, 1919, ibid.

Arguing that "ownership of forest land carries with it a special obligation not to injure the public," the committee proposed federal legislation to regulate logging on private land. Regulation would reduce waste, promote reforestation, increase fire protection, and maintain production in line with economic cycles. Fire insurance and special borrowing privileges to provide capital were included in the SAF proposal. One committee member, J. W. Toumey, disagreed with other members and insisted upon issuing a minority report. Toumey's report objected to direct federal regulation, preferring instead federal standards for state legislation. Graves, not a member of the committee, also proposed a stronger state role, his only fundamental difference with Pinchot.[9]

The SAF executive committee met and approved publication of both reports. When Pinchot agreed to pay with personal funds the cost of reprinting fifteen thousand copies of both the minority and majority reports, the way was clear for publication.[10] Raphael Zon, editor of the *Journal*, predicted that "the feathers will begin to fly" when the December 1919 issue carried the report of Pinchot's committee. Graves meanwhile agonized to his diary that instead of helping the Forest Service advance its program, Pinchot had "undertaken a scheme of his own" and would try to push Graves out of the limelight and "become the hero in saving the forests of the country." Graves suspected that Zon had worked with Pinchot in drafting the committee report that contained "many socialistic features."[11]

"The continued misuse of forest lands privately owned has now brought about a critical situation in America." So began Pinchot's frontal assault on the lumber industry in the short essay with which he introduced the committee report to the readers of the *Journal of Forestry*. After an unduly long wait for voluntary compliance, now, in his words, "The lines are drawn." The title telegraphed its contents: "Forest Devastation: A National Danger and a Plan to Meet It." Acknowledging his differences with Graves, Pinchot in an open letter of transmittal to SAF President Frederick E. Olmsted maintained that their "purposes were the same."[12]

9. "Forest Devastation: A National Danger and a Plan To Meet It," 3 pts., *Journal of Forestry* 17 (Dec. 1919): 911–45; Henry Graves, "A Policy of Forestry for the Nation," ibid., pp. 901–10.

10. Zon to Fernow, Nov. 28, 1919, box 143, SAF Records; Fernow to Zon, n.d., ibid.; Zon to Fernow, Dec. 3, 1919, ibid.

11. Zon to Earle Clapp, Nov. 4, 1919, Raphael Zon Papers, Minnesota Historical Society, Saint Paul; Graves, Diary, Nov. 25, 1919, Henry S. Graves Papers, Yale University.

12. Gifford Pinchot, "The Lines Are Drawn," *Journal of Forestry* 17 (Dec. 1919): 899–900; Pinchot to Olmsted, Nov. 1, 1919, printed ibid., pp. 911–12.

The regulation issue was debated in the pages of the *Journal of Forestry* during the following year. A referendum seemed in order, and the returned ballots showed that the forestry profession supported public regulation of logging by a three to two margin. It was a substantial, if not overwhelming, endorsement of the committee recommendation. The committee that tabulated and analyzed the ballots noted in their report: "This is the first time the Society of American Foresters has, as a body, expressed itself in favor of legislation for the perpetuation of forests."[13]

Industrial Efforts. The lumber industry was unsettled by the trend of events. One lumberman raged that the Society of American Foresters should be "annihilated," then more calmly demanded that the SAF at least impose adequate censorship on its journal. Cooler heads prevailed, however, and the industry countered by forming the National Forestry Program Committee. This new group consisted of representatives from the American Forestry Association, American Newspaper Publishers Association, American Paper and Pulp Association, Association of Wood-Using Industries, National Lumber Manufacturers Association, National Wholesale Lumber Dealers Association, Newsprint Service Bureau, Society for the Protection of New Hampshire Forests, and the Western Forestry and Conservation Association. Royal Kellogg of the Newsprint Service Bureau served as chairman during the eight-year life of the committee.[14]

The committee was Greeley's idea. He proposed to Kellogg that several industrial associations form a united front if they wanted to achieve legislative success. He recommended that Kellogg get in touch with E. T. Allen, whom he considered a "master strategist." As a newly appointed member of the National Lumber Manufacturers Association committee on forestry, Allen would be valuable in preparing an industrial position. For his part, Greeley promised to cooperate in any way he could and enclosed a draft of a bill to deal with forestry problems. Kellogg reacted favorably and agreed with Greeley that forest fire protection was the most important facet of the program. The industry would help in Congress to get increased Forest Service appropriations for fire.[15]

13. "Report of the Committee on the Results of the Referendum Ballot on the Society's Plan for a National Forest Policy," ibid. 18 (Oct. 1920): 581–89.

14. Zon to Clapp, Nov. 4, 1919, Zon Papers; Samuel T. Dana, *Forest and Range Policy* (New York: McGraw-Hill, 1956), pp. 214–15; "Conference of Allied Organizations upon the Question of a National Forest Policy" [Oct. 12, 1920], NFPC Records, Cornell University.

15. Greeley to Kellogg, June 5, 1920, NFPC Records; Greeley, *Forests and Men*, p. 103; Kellogg to Greeley, June 23, 1920, NFPC Records.

Following Greeley's advice, Kellogg contacted Allen, who voiced reservations about the protection appropriation. There were regulatory strings attached, and Allen wanted to "convert the Forest Service to a quite new attitude" before he could approve. He wanted to work through Greeley as much as possible but to avoid publicity in order not to jeopardize the chief's integrity. He planned to see Greeley soon and would try to negotiate a Forest Service policy more favorable to the industry. Greeley had assured Allen that Forest Service policy would emphasize cooperation and "encouragement of local initiative," not direct federal control.[16]

J. W. Toumey thought of the Society of American Foresters' program as radical, that of the industry as reactionary, and that of the Forest Service as intermediate—perhaps as good a description as any.[17] Greeley labored to bring the extremes closer to a consensus, but he had to deal with strong personalities.

In September 1920 the Forest Service asked all state foresters to support its legislative proposal, which focused on protection. Pinchot, now state commissioner of forestry for Pennsylvania, bluntly rejected the overture. In his typical fashion, he said that the program was "fundamentally wrong in principle, can never be put through Congress, and if it could would be unworkable." He would not support legislation that depended upon state action for implementation; it was "a question of National control or no control at all." Pinchot saw the emphasis on fire protection "almost to the exclusion of forest devastation" as playing into industrial hands. To him, lumbermen had been hiding their unacceptable practices behind the specter of fire too long.[18]

Greeley stood firm. He reasoned that although federal regulation would indeed be more effective than state control, Congress would not accept federal intervention. Always the pragmatist, Greeley believed it was better to go with what would work, rather than lose everything by insisting upon an ideal but unworkable solution. He also disputed Pinchot's contention that the fire problem was overrated. In fact, according to Forest Service data, logging devastation was "insignificant" compared to destruction by fire. The chief forester

16. Allen to Kellogg, Sept. 15, 1920, NFPC Records; Greeley to WFCA, Nov. 20, 1920, drawer 380, RG 95–115, RFS.

17. Minutes of the Service Committee, Aug. 27, 1920, box 112, RG 95–8, RFS.

18. Pinchot to J. Girvin Peters, Sept. 21, 1920, NFPC Records. Much of this correspondence can also be found in the E. T. Allen Papers, Oregon Historical Society, Portland.

reiterated that his main concern was with the "most direct route to results." He chastised the ex-chief for opposing the Forest Service program; even with their differences of opinion, more was to be gained by state regulation than lost. Greeley asked Pinchot to consider whether blocking each other's efforts would benefit American society.[19]

Pinchot concurred on the need to avoid controversy but characteristically agreed only to "concede anything that is not vital." Fire was indeed the number one problem, but treatment of devastation could not be deferred until fires were no longer a major concern. The matter of judgment was all-important; each had failed to convince the other of the merits of his argument. Pinchot was willing to let Congress decide, but he would not withdraw his objections to Greeley's program.[20]

The two soon debated again, this time in public. The audience was the National Association of State Foresters at its annual meeting in November 1920. Greeley told the assemblage that he thought federal control of logging would be impracticable. He favored letting the states do as much as they could, with the federal government supplementing their activities where necessary. He assured them that the Forest Service would approach the lumbermen in the "spirit of assistance rather than of regulation."

With his flair for the dramatic, Pinchot then asked those state foresters who felt free of political control to raise their hands. He counted only twelve. Pinchot then described Greeley as "vigorous and energetic," but he reserved the right to comment whenever the Forest Service was wrong. It was wrong now. Who, Pinchot wanted to know, was opposed to the Interstate Commerce Commission? He compared regulation by the ICC to his proposal for the Forest Service. He argued that although fire and taxes were state questions, forest devastation was of national importance. Federal regulation was necessary.[21]

In line with the debate between Pinchot and Greeley, the Senate in February 1920 had asked the secretary of agriculture for a report on the timber situation. As frequently happens, the Senate did not provide funds for the project, so the Forest Service had to pull together

19. Greeley to Pinchot, Oct. 6, 1920, NFPC Records.
20. Pinchot to Greeley, Oct. 22, 1920, ibid.
21. NASF, "Minutes," 1920, pp. 7–14, 18, 26–29, NASF Records, National Conservation Library, Denver.

existing information by using staff committed to other tasks. Known as the Capper Report after Arthur Capper of Kansas, the Senate sponsor of the original request, the study showed that there had been serious timber depletion, causing record high prices. With the Capper Report calling for increased cooperation in fire protection, Greeley hoped for one million dollars to begin the new program.[22]

The industry was satisfied. George S. Long of the Weyerhaeuser Timber Company thought it "gave the lumber industry all the fairness it was entitled to." E. T. Allen judged it to be "moderate, constructive, and workable," believing the report showed Greeley to be firmly in favor of cooperative forestry and opposed to "compulsive" forestry. The report was "notably fearless," and Allen thought that most partisans would be disappointed. Judging the report's effectiveness, a disappointed Allen revealed his own partisanship by claiming that Senator Capper himself had then "virtually repudiated" the moderate, accurate findings of the report bearing his name by introducing the "Pinchot measure."[23]

The "Pinchot measure" was a bill introduced by Senator Capper on May 20, 1920, to regulate timber cutting on private forest lands. Regulation would be by a federal forest commission, as advocated by the Pinchot-chaired SAF committee. Although the Capper bill stood on weak constitutional ground, following a Supreme Court decision concerning interstate commerce, it posed a serious threat to the lumber industry and galvanized action. At Greeley's suggestion and encouragement, the industry had formed the coalition named the National Forestry Program Committee. The committee now worked to prepare a bill favorable to their interests, so that it could be introduced before December 1.[24]

Greeley and Kellogg thought that the bill should be given wide publicity. A committee news release on November 15 announced that "for the first time in history a united campaign will be behind a national forest policy. . . ." The American Forestry Association made the preliminary statement, having offered its publicity department as the best means to place the program before the public. AFA

22. U.S. Congress, Senate, Resolution no. 311, "Directing the secretary of agriculture to furnish information in regard to the alleged depletion of the forest resources of the United States," 66 Cong. 2, Feb. 21, 1920; Minutes of the Service Committee, Feb. 26, June 10, 1920, RG 95–8, RFS.

23. George S. Long, "State Lands and Purchase of Land," speech to WFCA, Dec. 1920, copy in Allen Papers; E. T. Allen, "The Capper Report," June 12, 1920, copy ibid.

24. Allen to George S. Long, Oct. 29, 1920, NFPC Records.

action gave the measure the semblance of support of conservationists, and Greeley also expressed his approval. The release explained that the new bill would use cooperation between public and private forestry groups as prime means to protect the forests of America.[25]

Members of the National Forestry Program Committee found it necessary to set up divisions to deal with conservation affairs. When the U.S. Chamber of Commerce, as a member of the NFPC, moved to establish a natural resources department, Committee Secretary Kellogg congratulated the chamber for taking the significant step and expressed interest in who would be named head of the new department. Two months later Greeley was asked to leave the Forest Service and take over the chamber's natural resources department, at a salary of twelve thousand dollars per year. Greeley, chief only since April 16, 1920, and earning five thousand per year, declined. He wanted more time to work on Forest Service programs or "I would join you without any hesitation."[26] How different the course of history might have been had Greeley been swayed by the lucrative offer, we cannot know. But at the very least, the incident substantiates Graves's deep concern for the inequitable rewards of public service.

Representative Bertrand Snell of New York introduced the industrially supported measure in the House on December 22, 1920, in keeping with Kellogg's schedule of the previous June. Attempts to have Wisconsin's I. L. Lenroot introduce a companion bill in the Senate failed, as he was opposed to the total of appropriations involved. Federal funding to support state regulatory legislation was the essence of the Snell bill, a condition Graves and Toumey had advocated earlier. Snell received full backing from the National Forestry Program Committee, which sent promotional mail to a lengthy list of men affiliated with the forest industries. This list included ex-newspaper publisher and President-elect Warren G. Harding, who had been worried about supplies of newsprint. Pinchot, too, tried for presidential support, but for the alternative Capper bill. E. T. Allen gleefully reported that Harding viewed Pinchot's overtures as "pestering him with crank theories." Harding would take his "dope" from

25. Greeley to Kellogg, Nov. 10, 1920, ibid.; Kellogg to Greeley, Nov. 15, 1920, ibid.; news release, Nov. 15, 1920, ibid.; "Organization of the National Forestry Program Committee," n.d., ibid.

26. Kellogg to E. W. McCullough, Nov. 24, 1920, ibid.; Elliot Goodwin to Greeley, Jan. 20, 1921, William B. Greeley Papers, University of Oregon; Greeley to Goodwin, Jan. 29, 1921, Greeley Papers.

people who really dealt with forests. Allen was sure that Pinchot had little influence with Harding.[27]

Twenty state foresters, eighteen state forestry associations, and twenty-three industrial organizations endorsed the Snell bill. Pinchot maintained his opposition, asserting that regulation by state agencies would be inadequate and that direct federal regulation was crucial. Although Greeley fully supported the Snell bill, Secretary of Agriculture Wallace seemed to be wavering. Kellogg was dispatched to measure the strength of Harding's support, because the new president could obviously help with the secretary.[28]

More concerned about winning congressional acceptance of principles included in the Snell bill than securing appropriations to implement it, Kellogg kept up the pressure. His main job was to coordinate efforts by the National Forestry Program Committee members and to obtain support from as many other organizations as possible. Using the slogan "continuous forest production," he guided the efforts along "constructive" lines, that is, "no attacks on nor mention of antagonistic views."[29]

Greeley, still having problems with Secretary Wallace on the matter, asked Kellogg to call fewer men than planned to testify for their chosen bill. Perhaps Wallace's attitude explains why Greeley began to change his public posture on regulation and cooperation. Caught "totally unprepared" by Greeley's apparent shift, Allen wrote him privately asking whether his viewpoint had changed. Unless Greeley said otherwise, Allen would assume that the Forest Service still supported the principle of cooperation and was not beginning to view the forests as public utilities. On a carbon copy of an official letter to the chief forester written the same day, Allen scribbled to Kellogg, "You better find out what Bill is up to." Kellogg tried to reassure Allen about Greeley's loyalty to the Snell bill: "I think the real trouble if any lies with the present Secretary of Agriculture, who is anxious to do noth-

27. Greeley and Kellogg first hoped that Guy U. Hardy would introduce the House bill, but Snell did instead. Lenroot to Hugh P. Baker, Dec. 14, 1920, NFPC Records; Kellogg to C. A. Babcock, Jan. 8, 1921, ibid.; Kellogg to Warren G. Harding, Jan. 10, 1921, ibid.; Allen to Kellogg, Jan. 1, 1921, ibid.; Snell to Kellogg, May 4, 1921, ibid. For a counterview of Pinchot's influence on Harding, see Burl Noggle, *Teapot Dome: Oil and Politics in the 1920s* (New York: W. W. Norton, 1965), pp. 7–8.

28. "Endorsements of the Snell Forestry Bill to March 10, 1921," n.d., copy in NFPC Records (the list cites the AFA as an industrial organization); Dana, *Forest and Range Policy*, pp. 215–16; E. A. Sherman to Kellogg, Apr. 29, 1921, NFPC Records.

29. Kellogg to D. C. Everest, Apr. 19, 1921, NFPC Records; NFPC memo, n.d., ibid.

ing to displease G.P."[30] Obviously the ex-chief was still a potent conservation voice.

The legislative program seemed to falter, and industrial support began to dwindle. Kellogg believed that Pinchot's endorsement of the Capper bill and federal regulation was not impressing congressmen at hearings, describing his testimony as "pretty well punctured." When Greeley testified in favor of the Snell measure, some members of the House Committee on Agriculture claimed the bill was an invasion of states' rights. Pinchot heard similar protests when he testified to the same committee. Lumbermen were strongly divided on the matter of public regulation, although the National Lumber Manufacturers Association officially supported Snell. A committee minority report from the U.S. Chamber of Commerce branded the whole idea as "one of the most dangerous tendencies in American government at this time," arguing that federal aid led inevitably to subordination. With Pinchot making little headway for federal regulation and the state regulation aspects of Snell's proposal losing support, Greeley proposed a compromise: drop regulation and emphasize cooperative fire control. He reported back to his staff that Pinchot agreed. Secretary Wallace tipped off the president about the shifting strategy, pointing out that neither the Capper nor Snell bill was likely to pass and fire control, after all, was the number one concern.[31]

The Clarke–McNary Act. On January 3, 1923, the Senate adopted a resolution calling for a committee "to investigate problems relating to reforestation, with a view to establishing a comprehensive national policy." Senator Charles L. McNary of Oregon was named chairman of the Select Committee on Reforestation. He scheduled twenty-four hearings in sixteen states plus one in Washington, D.C. Dean Franklin Moon of the New York State forestry school wired Greeley

30. Greeley to Kellogg, June 1, 1921, ibid.; Allen to J. C. McLaughlin, Feb. 21, 1922, ibid.; Allen to Greeley, Nov. 14, 1921, ibid.; Allen to Greeley, Nov. 14, 1921, ibid.; Kellogg to Allen, Nov. 21, 1921, ibid. The Wallace and Pinchot friendship is well described in Donald L. Winters, *Henry Cantwell Wallace as Secretary of Agriculture, 1921–1924* (Urbana: University of Illinois Press, 1970), chap. 8.

31. Kellogg to A. L. Osborne, Jan. 16, 1922, box 59, NFPA (formerly NLMA) Records, FHS; Kellogg to Snell, July 6, 1921, ibid.; U.S. Chamber of Commerce, "On the Report of the Committee on National Forestry Policy," Referendum no. 42, Oct. 3, 1923, ibid.; Minutes of the Service Committee, Jan. 19, 1922, RG 95–8, RFS; Wallace to Harding, June 8, 1922, drawer 192, Records of the Office of the Chief, 1908–47, Record Group 95 (hereafter cited as RG 95–4), RFS. Earle Clapp's suggestion that minimum silvicultural requirements and protection should be included in the pending legislation bothered the industry. Samuel T. Dana suggested keeping the plans confidential. Minutes of the Service Committee, May 20, 1921, RG 95–8, RFS; Dana to P. T. Coolidge, July 7, 1921, Samuel T. Dana Papers, FHS.

asking if he supported the resolution or thought it "loaded with dynamite." Suggesting that the Senate action was not coordinated with others, Associate Chief Sherman answered for his chief that the resolution seemed to have been introduced in good faith and would provide important information.[32]

McNary's committee would soon demand a cogent statement of Forest Service policy. On the day following the Senate resolution, Greeley and his staff discussed the lumber industry at length. Greeley led off by questioning their traditional perspective of forestry. They had always agreed that "the growing of timber is an economic process [following] the law of supply and demand." The Forest Service had never expected a property owner to do things not in his best interest. Now Greeley saw criticisms "creeping in" to Forest Service statements about landowners seeking a profit. He insisted that Forest Service policy be consistent and support the economic basis of forestry. Greeley chided his men for being no more altruistic than the lumbermen when it concerned their own pocketbook. Why expect others to be more generous? Greeley suggested that the Forest Service had "called wolf" too often and he was opposed to "preaching disaster."

Raphael Zon challenged his chief, agreeing that the Forest Service should not view with alarm "situations that were not alarming," but warning that "we should not view with pride when there was nothing in particular to be proud of." Tactfully, Zon suggested that Greeley's position might be "misunderstood." To Zon, de-emphasizing bad practices and stressing the good was not acceptable. Austin Cary, who had been working with a missionary's zeal to spread industrial forestry in the South, agreed with Zon that it would be dangerous to exaggerate industrial progress. Herbert Smith also supported Zon. Smith asked Greeley why, if there had been so much progress, the chief was pressing for more legislation. Leon Kneipp then observed that forestry had begun as an emotional movement, but now it was economic and warranted full disclosure of all the facts.

Greeley, taken aback by solid staff opposition, feared that indeed he had been misunderstood. He responded that he did favor vigorous pursuit and exposure of unacceptable practices but then added somewhat lamely that pointing out good examples of reforestation would influence other landowners into trying it for themselves.[33] The

32. Senate Resolution no. 398, 67 Cong. 4; Moon to Greeley, Jan. 31, 1923, drawer 182, RG 95–4, RFS; Sherman to Moon, Jan. 31, 1923, ibid.

33. Minutes of the Service Committee, Jan. 4, 1923, RG 95–8, RFS.

matter of Forest Service posture on the lumber industry was not resolved, but Greeley clearly understood that he would have to be moderate in his praise of industrial progress.

At the McNary committee hearings in New Orleans, the National Lumber Manufacturers Association proposed that reforestation was a public concern, as too little was known about actual reforestation methods to expect the private sector to absorb the risk. J. E. Rhodes of the Southern Pine Association, speaking in behalf of the NLMA, demanded minimal federal regulation or requirements. McNary countered that it would amount not to "Government leadership" but to "State dictatorship." The senator saw no difference between federal road standards and the accompanying federal funds which states gladly accepted, and maintaining federal fire standards with federal funds. He rejected the notion that states could logically accept federal money without federal rules. Greeley attempted to rescue the lumberman by assuring him that the program would not be forced upon the states; voluntary acceptance of funds would mean voluntary acceptance of rules.[34]

Greeley, in *Forests and Men*, remembered "packing" the reforestation hearings with witnesses who testified that fire was the number one problem facing forest landowners. Nearly fifteen hundred pages of printed testimony show Greeley's effectiveness. The Select Committee on Reforestation heard witness after witness explain how risk from fire made long-term forest management impractical. Greeley deservedly was pleased.[35]

After completing its hearings, the committee reported back to the Senate. Following its regional summaries of forest conditions, the committee described the national situation. The more than 800 million acres of forests originally standing in the United States had been reduced by land clearing, fire, and logging to less than 140 million acres of virgin timber, plus 81 million acres of barren logged-off land that merited reforestation. The report predicted that the present rates of consumption and replacement assured future shortages.[36]

Pointing out that the 81 million acres of logged-off land was a

34. Minutes of the Service Committee, March 29, 1923, ibid.; U.S. Senate Select Committee on Reforestation, *Hearings*, 67 Cong. 4, March 26, 1923, pp. 236–37; Greeley to Wilson Compton, March 28, 1923, WFCA Records, Oregon Historical Society, Portland.

35. Greeley, *Forests and Men*, p. 107; U.S. Senate, Select Committee on Reforestation, 1923, *passim*; Greeley to Allen, Apr. 7, 1924, WFCA Records.

36. U.S. Congress, Senate, *Reforestation*, Senate Report no. 28, 68 Cong. 1, Jan. 10, 1924, p. 13–14.

substantial tax burden for local governments, the report claimed that fire "more than any other cause" had prevented reforestation. Not only did fires destroy, but the risk of fire prevented landowners from investing in reforestation. Adequate protection would create "a tremendous impetus" for increased timber production. The committee proposed two solutions: to place under public ownership especially hazardous areas or areas of special value to the public, such as watersheds of navigable streams; and to remove the risks that were deterring reforestation.[37]

In the meantime, congressional maneuvers set the stage for the final act. Congressman John D. Clarke of New York introduced in February 1923 a version of the Snell bill from which he had removed items referring to regulation. This deletion made Clarke's bill basically one dealing only with cooperative fire control. Meanwhile, Capper had reintroduced a revised bill that would levy a federal excise tax of five *cents* per thousand board feet of timber logged "properly" and five *dollars* per thousand for logging in disregard of federal guidelines. A Supreme Court decision in 1922, however, had ruled out the use of federal taxes for regulating industry. The proregulation forces then tried another version of the bill, which levied an excise tax of $5.00 per thousand on all logging but offered a payment of $4.95 per thousand for logging conducted by federal standards. But regulation was unacceptable to Congress; cooperation was the watchword.[38]

Greeley credits McNary's parliamentary finesse and Clarke's support for the ease with which the bill moved through Congress and received presidential approval. On the day of the vote in the House, Greeley had waited in the cloak room where he could hear the question raised in debate, passing in penciled replies to Clarke and other supporters. In the Senate, McNary had felt that Snell asked for too much and had gone to Greeley to find out what could be eliminated from the bill. Everything but fire, the senator was told, and he drafted it accordingly. With amazing ease, the bill cleared each hurdle. After introducing the bill in January 1924, McNary wired Allen in Portland on June 6, "SENATE JUST PASSED MY REFORESTATION BILL." President Coolidge signed it into law on the following day. Pinchot's preoccupation with his successful race for the Pennsylvania governorship had eliminated effective opposition. The only important item deleted from

37. Ibid., pp. 15, 23–24.

38. Dana, *Forest and Range Policy*, pp. 216–17; Minutes of the Service Committee, Nov. 30, 1923, RG 95–8, RFS; Warren G. Harding to Clarke, Jan. 24, 1923, copy in FHS clipping file.

the final version of McNary's bill was the restoration of presidential authority to create national forests from public lands.[39]

A triumvirate of cooperative forestry now existed—federal, state, and private. Expanding the 1911 Weeks Law, Clarke–McNary authorized federal participation in programs without being restricted to navigable streams. Section 2, perhaps the most important, made federal matching funds available to qualified state protection agencies, including private efforts as part of the state share. Federal contributions in any year could not exceed the amount provided by state and private. Although Congress authorized the secretary of agriculture to spend up to $2.5 million to achieve the cooperative goals, actual appropriations have traditionally been substantially below that amount. The Forest Service openly expressed its disappointment over reduced appropriations, charging that the Bureau of the Budget was responsible for the pared-down budget requests sent to Congress.[40]

The act authorized up to one hundred thousand dollars per year to be used cooperatively with states to establish nurseries. Nursery stock could be used for shelterbelts and reforesting farmland. To implement the reforestation, Congress earmarked another one hundred thousand dollars per year for providing technical advice and services to farm woodlot owners. This program, too, was to be handled on a fifty-fifty cooperative basis with the states.

The Clarke–McNary Act also dealt with land acquisitions for national forest purposes. The Weeks Law had limited purchase to headwaters of navigable streams. Now the National Forest Reservation Commission could recommend the purchase of land for timber production as well as protection of stream flow. Additionally, Clarke–McNary provided for accepting gifts of land and exchanges with other federal agencies.

Taxes

Another section of the Clarke–McNary Act dealt specifically with a serious obstacle to good forestry practices, second only to fire. Section

39. Greeley, *Forests and Men*, pp. 106–10; McNary to Allen, June 6, 1924, WFCA Records. The Clarke–McNary Act authorized presidential creation of national forests from federal reservations that were mutually acceptable to the secretary of agriculture and the secretary of the administrating department.

40. 43 Stat. 653–55; E. A. Sherman to Kellogg, Sept. 24, 1924, box 2 *passim*, WFCA Records; Forest Service, *Annual Report* (1928), pp. 4–5. Verne L. Harper, retired deputy chief for

3 authorized expenditure of a portion of the 2.5 million dollars of fire control funds to study tax laws and their effects on forest land management. Risk from fire was easy to explain, easy to understand. Risk from taxes was much more subtle, if only of slightly less concern for forest landowners. Although the taxes in question were not levied by the federal government, the Forest Service assumed a responsibility in the matter as part of its program to assure future supplies of wood.

The tax situation was, and still is, exceedingly complex. Forestry is a long-term venture. Forest landowners are particularly vulnerable to substantial tax increases during the course of an operation that may span generations. Owners of logged-off land faced uncertainty—the greatest of all threats to capital investment—because the local governments might raise taxes after reforestation. Taxation at a level acceptable at the time seeds are sown or seedlings planted might be increased after the owner committed himself to another rotation. Many lumbermen felt that it was more prudent to dispose of logged-off land and to purchase standing timber than to risk future tax levies, and of course fire, on their forest plantations.

Foresters generally accepted lumbermen's claims that uncertainty over future taxes deterred reforestation and other long-term investments in forest land, while current taxes provided an incentive to speed up logging rates. If timber was taxed at its market value, and bare land usually was assessed at low rates, then owners of timberland could reduce their tax burden by logging. To tax the same tree every year, year after year, at its current market value even though it could only be sold once—at the time of logging—was a concept unacceptable to lumbermen and foresters alike.

Professor Fred R. Fairchild of Yale University, noted authority on forest land taxation, began a decade-long study to provide the necessary information. Earlier studies by Fairchild for the 1909 National Conservation Commission brought into question the assumption that many lumbermen had actually adjusted logging schedules because of taxes, but many had agreed that it could happen. In reality, however, forest land taxes were extremely low, and not infrequently large tracts of timber were not taxed at all because of inadequate administrative

Research, strongly suspects that the blame for reduced Clarke–McNary appropriations must be shared by the Division of State and Private Forestry for not selling its cooperative program. He also feels that the industry and state foresters failed to work vigorously for congressional support. Personal communication, June 2, 1975.

machinery to deal with areas remote from the assessor's headquarters. Even though property taxes increased, the increase was often moderate compared to increases in stumpage, lumber prices, and the economy in general. But in the minds of the decision-makers, the forest landowners and their forestry advisers, they were vulnerable to taxes, and as a matter of principle they needed protection before making long-term investments. Taxes were both financial and psychological burdens.[41]

Witnesses had testified to McNary's committee that taxation was a major deterrent to forest land management. Even though Greeley carefully selected witnesses to hold up fire as the number one problem, testimony frequently focused on taxes instead. Emphasis on taxation was particularly intense in the Pacific Northwest, perhaps reflecting the large holdings found in that region. At one point in the hearings, the senators considered the advisability of withholding cooperative fire protection funds from states following undesirable taxing practices.[42]

Even if there was a consensus on the solution to the tax problem—probably some sort of yield tax rather than ad valorem—state constitutions generally prevented preferential treatment for any one class of land. Constitutional amendments, then, would have to be part of the tax-reform program. During the 1920s, many states did amend their constitutions to allow deferred tax payment for forest landowners. Taxes became due at the time of logging, when the owner realized income from his property.[43]

McNary's committee report to the Senate called attention to the tax problem, and Section 3 of the Clarke–McNary Act authorized a study. The study was needed, stated the Forest Service officially, because "heavy taxes have forced the owner of old-growth timber to cut his timber as rapidly as possible. . . . This vicious race between forest destruction and mounting taxes has raised a fear that managed forests may be subjected to confiscatory taxation."[44]

41. Fred R. Fairchild, "Taxation of Timberlands," in *Report of the National Conservation Commission*, vol. 2, Sen. Doc. 676, 60 Cong. 2, 1909; "Forest Taxation Instructions to Special Investigators," Apr. 1, 1910; Fairchild to D. Blakely Hoar, March 19, 1910, RG 95–115, RFS; *Oregon Forester* 1 (March 1903): 1–7; Forest Service, *Annual Report* (1924), p. 2.

42. U.S. Congress, Senate, Select Committee on Reforestation, *Hearings*, 67 Cong. 4, pp. 803–83 *passim*; Minutes of the Service Committee, Apr. 6, 1923, RG 95–8, RFS.

43. Senate, *Reforestation*, pp. 17–18, 26–27.

44. Forest Service, *Annual Report* (1925), p. 9.

Fairchild formally brought together his staff at Yale in April 1926. Seventeen technical personnel, mainly foresters and economists, began to investigate state and local forest land taxing processes. After eight years of studying all aspects of forest taxation, including comparisons with European methods, Fairchild concluded the project. The resulting 681-page report, published in 1935, represents the single most valuable forest tax study to date.[45]

Fairchild found that taxes contributed only slightly to the rate of logging, refuting for the second time the idea that lumbermen were forced to liquidate their holdings by excessive tax burdens. Reforestation, however, seemed to be substantially deferred by taxes. Fairchild believed that because many landowners had no interest in reforestation anyway, regardless of taxes or any other factor, taxes were probably an even more significant deterrent than the data showed. Some lumbermen believed that forestry was unprofitable with or without taxes. Others believed that taxes were important but only a part of the larger problem of risk, which included fire.[46]

Years later Fairchild's chief assistant recalled that they had found little evidence that taxes caused destructive lumbering and had accused the industry of exaggeration. R. Clifford Hall believed that overcapitalization in land and mills, not taxes, was the major cause of liquidation. In proportion to the total cost of doing business, taxes had not been a significant factor but provided a most convenient whipping boy.[47]

Fairchild's recommendation for equitable forest land taxation was not particularly innovative—a yield tax on the standing timber with the land itself taxed conventionally as bare land. This combination of taxes would protect local tax bases and treat timber as a crop rather than as property. The owner would pay taxes on his timber only at time of logging, therefore having no incentive to reduce taxes by logging and having cash in hand to meet the assessor's bill. Fairchild also recommended improving tax assessment and collection procedures to assure equitability. He was opposed to esoteric tax methods that would be vulnerable to passage of time. All in all, it was a straightforward report.[48]

45. Fred R. Fairchild, *Forest Taxation in the United States*, USDA Misc. Pub. no. 219, Oct. 1935.

46. Ibid., p. 267.

47. R. Clifford Hall, *Forest Taxation Study, 1926–35*, Oral History Interview by Fern Ingersoll, 1967, Regional Oral History Office, Bancroft Library, pp. 19–20 and *passim*.

48. Fairchild, *Forest Taxation*, pp. 635–40.

Perhaps because Fairchild did not find property taxes to be the overwhelming villain as portrayed in forestry literature, his study quickly dropped into obscurity. Rarely cited in subsequent tax literature, except as a curiosity, this monumental investigation yielded few tangible results, thus negating the effectiveness of Section 3, Clarke–McNary. In any regard, the Forest Service made a major effort to deal cooperatively with property taxes, an issue of prime concern to forest land managers.

The Clarke–McNary Act is undoubtedly one of the most important pieces of forestry legislation in American history. It assured that the cooperative relationships with nonfederal forestry programs formalized by the Weeks Law would be retained. In 1955, three decades after enactment, the chief of the Forest Service would report nearly 39 million acres protected by 9 million dollars of federal money and 30 million dollars from state, county, and private sources. Fifty million tree seedlings were shipped in 1955 to reforest cutover acres; nonfederal funds were tenfold larger than the Forest Service contribution.[49] Greeley died that same year; the Clarke–McNary Act was certainly the capstone to his Forest Service career.

Greeley resigned from the Forest Service on April 30, 1928, to become executive secretary of the Seattle-based West Coast Lumbermen's Association. To those who knew Greeley intimately, the change must not have come as a surprise, for he had always intended to leave the Forest Service at the right time. As noted, shortly after he became chief forester in 1920, the U.S. Chamber of Commerce had offered Greeley nearly three times his government salary to work for them. He declined, but four years later he was apparently tempted by a Michigan firm. The company president wrote Greeley that they could wait a year for him to decide, then they would have to hire someone else. Greeley declined with regrets as he was "not yet able to determine when I ought to leave the Forest Service." It would be "some time" before he was satisfied with his accomplishments, and then he wanted to work for a private company that was engaged in "continuous timber production." The University of Michigan offered Greeley the deanship of its forestry school. First he declined by saying he wanted to stay with the Forest Service longer, then he admitted

49. Forest Service, *Annual Report* (1955), pp. 24–25.

that it was doubtful he would ever get into education, because industry held the greatest attraction.[50]

After eight years as chief and twenty-four years with the agency, the time to leave was at hand. The General Exchange Act of 1922, providing for more logical ownership patterns, the cooperative Clarke–McNary Act of 1924, having forestry research formalized and strengthened by the McSweeney–McNary Act of 1928, battling Albert Fall to a standstill on transferral of the Forest Service, productive examination of range and recreation policy—these achievements satisfied Greeley and he yielded to the lure of private industry.

Greeley credited George S. Long of the Weyerhaeuser Timber Company for being the one who had convinced him. Long's foresight and optimism about the potential of industrial forestry caught Greeley's fancy. He wanted to help as best he could. In an official statement, Greeley acknowledged that he would have a lot to learn. Forestry was "inseparable from the stability and sound functioning of timber-using industries." The public and private sectors had a mutual interest in forestry. He claimed to feel "somewhat like a minister who steps down out of his pulpit and tries to practice what he has preached." The West Coast Lumbermen's Association gathered praises for Greeley from high officials to smooth the transition from public service to private enterprise in the public mind. The American Forestry Association passed a resolution regretting Greeley's resignation as an "irreparable loss." Referring to the "unreasonable heavy personal sacrifices" he had made in order to be chief, the resolution supported Greeley's decision to resign "at the height of his powers."[51]

Some, like Fred Morrell, district forester in Missoula, supported Greeley's decision with regrets at losing him as chief. To others, it was proof that Greeley had been more sympathetic to the needs of industry than the welfare of the Forest Service could justify. The Clarke–McNary Act, Greeley's major achievement, obviously was a disappointment to those who favored strict federal regulation. Their bitterness in defeat was evident. Pinchot would rate Greeley's per-

50. James W. Blodgett to Greeley, June 4, 1924, Greeley Papers; Greeley to Blodgett, June 6, 1924, ibid.; Blodgett to Greeley, June 13, 1924, ibid.; Greeley to Marion L. Burton, Jan. 28, 1925, ibid.

51. Greeley to Reginald H. Parsons, March 4, 1953, Greeley Papers; statement dated Feb. 25, 1928, ibid.; J. D. Tennant to Herbert Hoover, Feb. 20, 1928, Commerce Papers, Hoover Presidential Library, West Branch, Iowa; Resolution in Minutes, Board of Directors, March 21, 1928, AFA Records.

formance as "pitiful." Judson King implied that Greeley's new position was a payoff for services rendered while chief.[52]

After leaving the Forest Service, Greeley frequently served as negotiator between public and private foresters. His contributions to American forestry by no means ended in 1928. Until his death in 1955, Greeley was an important spokesman on forestry issues of national interest.

52. Fred Morrell to Greeley, Feb. 28, 1928, RG 95–4, RFS; Pinchot to H. H. Chapman, July 12, 1934, SAF Records; Judson King in National Popular Government League Bulletin no. 155, June 21, 1932, copy in SAF Records.

CHAPTER VIII

New Deal Planning and Programs

WILLIAM TERRY worked as a messenger for the Forest Service in Washington, D.C. At ten past eight the morning of October 23, 1933, Terry was parking his car in the lot behind headquarters, the red brick Atlantic Building on F Street, when, horrified, he saw a plummeting body crash onto the roof of a parked car. Robert Y. Stuart, aged fifty and chief of the Forest Service since replacing Greeley in 1928, was dead.[1]

Accident or suicide? Records of the coroner's autopsy inquest and police records of Stuart's death have been lost as part of routine file destruction, but that it was an accident was apparently not officially questioned. An investigation by the Society of American Foresters led to the theory that Stuart had suffered an attack of vertigo and had fallen through an open window to the parking lot seven stories below. W. Ridgely Chapline, in charge of the Division of Range Research, worked directly below the office from which Stuart fell. Chapline always felt certain that Stuart must have been leaning out the window to see if the Forest Service car was available in the parking lot and accidentally fell. Chapline and others recall that the windowsill was unusually low, making falling out of the window feasible, indeed. Questions remained unanswered, however. The doubts were sufficient to prompt at least two later investigators to decide that Stuart had taken his own life.[2]

1. Washington, D.C., *Evening Star*, Oct. 23, 1933; Washington, D.C., *Post*, Oct. 24, 1933.
2. Herbert A. Smith, "Robert Young Stuart," *Journal of Forestry* 31 (Dec. 1933): 885–90;

The chief forester arrived at his office that morning around eight. He hung up his hat, walked into an adjoining office, opened the window, and either fell or jumped. To some minds, the fact that Stuart's own office window was enclosed by a fire escape platform and railing while the window of the other office offered a clear path to the parking lot was suspicious. Why open a window in someone else's office, anyway? Perhaps he only sought cross-ventilation to air a stuffy room, or perhaps Chapline's hypothesis is correct. But leaning against Stuart's inkwell was a life insurance prospectus, adding further doubt about the accidental nature of his death.[3] It added up to pretty flimsy evidence, except when the state of his mental health was included.

A year and a half earlier, Stuart had suffered a nervous breakdown, from which he seemingly had recovered. In the spring of 1933, he again began to suffer from nervous fatigue. It was obvious to his coworkers that he needed rest. Stuart insisted that he could not stay away from his desk, that his work was too vital.[4] The good-natured chief who enjoyed laughing at a joke broke under the strain. Accident or not, the burden of office killed Robert Stuart.

Franklin D. Roosevelt

October 1929 is generally accepted as the beginning of the Great Depression. Farmers had complained of falling prices, however, throughout the Roaring Twenties. The lumber market began to spiral downward in 1926 and stayed down until World War II. Herbert Hoover, thought by many to be the best-equipped man ever elected president, grappled cautiously with the disintegrating domestic and international situation. Sadly, the man who had achieved world prominence for his humanitarian success in bringing relief to famine-plagued Russia in the aftermath of war and revolution now was

Chapline to Frank Harmon, Nov. 20, 1975, U.S. Forest Service History Office files; Bernard Sternsher, *Rexford Tugwell and the New Deal* (New Brunswick, N.J.: Rutgers University Press, 1964), p. 216; Arthur M. Schlesinger, Jr., *The Coming of the New Deal* (Boston: Houghton Mifflin, 1959), p. 340.

3. Washington, D.C., *Post*, Oct. 24, 1933; Washington, D.C., *Evening Star*, Oct. 23, 1933. Stuart's death certificate lists skull fracture and ruptured heart, lung, and liver as cause of death. The certificate notes that an autopsy inquest was held. Accounts differ on the time he arrived at his office.

4. Smith, "Robert Stuart," p. 885; Christopher M. Granger, *Forest Management in the United States*, Oral History Interview by Amelia R. Fry, 1965, Regional Oral History Office, Bancroft Library, p. 87.

blamed for the excesses of all. With millions out of work and optimism extinct, makeshift shantytowns sheltering desperate and despairing men sprang up and became known as Hoovervilles. Federal employees saw their vacations canceled and replaced by "Hoover Holidays"—two days off each month without pay. The administration that had been inaugurated with such optimism in the spring of 1929 was overwhelmingly rejected at the polls four years later. The nation mandated the wheelchair-bound, patrician governor of New York, Theodore Roosevelt's cousin Franklin, to find a way out of the calamity.

Franklin Delano Roosevelt had long been a devotee of forestry. Once when filling out a questionnaire for *Who's Who*, the president listed his occupation as "tree grower." He had planted trees to rejuvenate the exhausted soil of his Hyde Park estate, so that his grandchildren could harvest bumper food crops from the overworked farm. FDR thought forests to be lungs of the land, "purifying our air and giving fresh strength to our people."[5] In 1935 the Society of American Foresters would award Roosevelt the Schlich Memorial Award for his outstanding contributions to American forestry. The foresters had considered making the award to Pinchot, instead, but finally decided to offer recognition to presidential interest in their work.

Forestry and the broader concepts of conservation were inbred in FDR's personal as well as his political philosophy. The depression and his landslide election in 1932 gave him the opportunity to implement his views. Assembling a corps of planners—"braintrusters"—FDR swamped Congress with programs. As far as technical forestry was concerned, the New Deal offered little that was new; many of the programs either had been proposed or were in operation before Roosevelt became president. Massive funding was the one significant innovation pressed by the new administration. With the funds came the administrative burdens of doubling or tripling efforts practically overnight. Seven months after FDR's inauguration, Stuart was dead from overwork.

Ferdinand A. Silcox was picked to replace Stuart as chief. According to one observer, serious consideration had also been given to Aldo Leopold, a game management expert who had been influential in developing early policies on wilderness areas. Spokesmen for the

5. Schlesinger, *Coming of the New Deal*, pp. 335–36.

lumber industry and many in the Forest Service had tried to convince Secretary of Agriculture Henry A. Wallace to promote Assistant Chief Christopher Granger. The New Deal required new blood, however, and Silcox's relatively radical views gave him the nod over Granger, a member of the old guard.[6]

Silcox had begun his forestry career under Pinchot and Graves. Having been assigned to deal with labor problems during World War I, he stayed in labor relations—some believed because of differences with Greeley—until recalled in 1933. Thrust suddenly to the top after a fifteen-year absence, Silcox told his staff that he had accepted with reluctance, hoping someone else would be selected. He needed more time to think before discussing new policies but, he added, "we've got to consider forestry in its social relation to our industrial life."[7] The heavy burdens of office remained: six years later Silcox would be dead of a heart attack.

New Deal Planning

Copeland Report. Drafting of what eventually became the New Deal blueprint for forestry began a year before FDR's inauguration. In the spring of 1932, Stuart, who had long been urging a re-evaluation of the 1920 Capper Report, ordered Earle Clapp to study the forest situation nationally. Senator Royal Copeland of New York had asked by resolution for this information and wanted an answer within a few days. It was estimated that it would take a year until the report was complete, but Clapp hurried the men along, explaining that for political reasons all data had to be in Washington, D.C., by October 1, 1932. Some had hoped for more time in order to make a better economic study, but by utilizing existing sources whenever possible, the men complied as best they could.[8]

6. Tom Gill to Arthur Newton Pack, Oct. 30, 1933, box 4, Tom Gill Papers, FHS, Santa Cruz, Calif. Gill described Silcox as a liberal "but not of the G.P. type," and he thought that Silcox could reconcile idealism with practicality. Gill to Pack, Nov. 13, 1933, Raphael Zon Papers, Minnesota Historical Society, Saint Paul. Assistant Secretary of Agriculture Rexford G. Tugwell had left Columbia University for the opportunity to supervise the Forest Service. He claims to have "begged" Silcox to accept the appointment as chief. Silcox had flatly refused the first offer. See "A Forester's Heart," Washington, D.C., *Post*, March 7, 1940.

7. Tugwell, *The Stricken Land*, pp. 24–25; Granger, *Forest Management*, pp. 87–88; Minutes of the Service Committee, Nov. 16, 1933, RG 95-8, RFS.

8. Clapp to Regional Foresters, Sept. 24, 1932, box 59858, Records of Pacific Northwest Forest and Range Experiment Station, Seattle FRC; C. J. Buck to Clapp, Oct. 18, 1932, ibid.; U.S. Congress, Sen. Resolution no. 175, 92 Cong. 1; *Congressional Record*, Apr. 4, 1933, p. 1186.

The idea had come up during a senatorial discussion of unemployment. From the question whether reforestation might not be a source of jobs evolved the resolution asking for an analysis of the forest situation. Clapp asked each experiment station to participate. Believing that forestry was "altogether too much on the defensive," he thought that Copeland's resolution provided an excellent opportunity for a much-needed restatement of objectives. "The right kind of report" could result in the Forest Service's continuing on a "satisfactory footing."[9]

Clapp encouraged his fellow scientists to face up to the "full magnitude" of the national forestry problem, to re-examine it in terms of national welfare. He felt that, in the midst of the depression, forestry might be the only use for nonagricultural land. The Copeland Report could effectively show forestry to be a solution to certain social problems. He wanted the report to be realistic and impartial; all forestry agencies should be coordinated into a "single national enterprise."[10]

Clapp heard predictions that the report would show a breakdown in private ownership of forest land, as indicated by massive abandonments and sales. Cooperative forest protection programs were threatened, too. Moreover, the report would reveal that little forestry was practical on private land, and that overcapitalization in land and manufacturing plants had undermined industrial efforts. All in all, Clapp heard that the report would verify the necessity of major federal intervention in private affairs.[11]

As data and drafts poured into his office from experiment stations and regional offices, Clapp supervised analysis and revision. Many field men were detailed to the Washington office to assist with the report. Authors found their reports repeatedly returned for rewrite. Finally, either Clapp or a review board examined the manuscript.[12]

Sections dealing with ticklish problems received close scrutiny. For example, when A. B. Hastings, who was in charge of state cooperation, turned in his draft entitled "Federal Financial and Other Direct Aid to the States," Fred Morrell, who had been district forester in

9. "The Copeland Report," Jan. 29, 1946, memo in box 11, Earle Clapp Papers, Record Group 200, National Archives (hereafter cited as RG 200); Assistant Forester to Directors, July 9, 1932, Copeland Report Data, 1923–33, Record Group 95 (hereafter cited as RG 95–120), RFS.

10. Earle Clapp, "The Philosophy of the Copeland Report," unpublished ms. in RG 95–120, RFS.

11. SBS [S. B. Show?] to Clapp, memo dated Oct. 14, 1932, copy ibid.

12. Clapp, memo dated Oct. 25, 1932, copy ibid.

Montana, reviewed it thoroughly. Morrell objected to the critical tone of the article and to Hastings' lack of clarity. The reader should learn, Morrell insisted, how important cooperative fire protection was to forestry and how forestry was related to private ownership. He believed that resolution of the question of cooperation versus regulation was hanging in the balance, and Hastings' analysis ought to focus on reaching a sound conclusion. He added that conflicting philosophies had no place in the Copeland Report; sweeping condemnation of past lumbering practices was as serious a shortcoming as failure to mention them at all. Clapp agreed that revision was necessary. He wanted Hastings to rewrite the section on federal aid in such a fashion that readers would "understand the soundness" of the Forest Service position.[13]

After innumerable revisions, the Copeland Report was completed in the spring of 1933 and transmitted to the Senate. Clapp tried to achieve a "spontaneous" show of interest by advising field men to channel comments directly to Secretary Wallace or to the president himself. He also prepared draft legislation to implement the report's recommendations, hoping that the department would support its enactment.[14]

While Clapp was drumming up support for legislation, Stuart formally transmitted a draft bill to Wallace. With uncharacteristic overstatement, Stuart predicted that its enactment would cause FDR's administration to surpass that of Theodore Roosevelt "in the furtherance of forestry enterprise." He thought that immediate passage was crucial lest the public begin to believe that other emergency programs already in effect were adequate for forestry, too. The Forest Service was proposing to expand the Clarke–McNary Act in a variety of ways. The cooperation practiced in fire protection would be broadened to include forest insects, tree diseases, erosion, and flood control. Protective funding would be nearly tripled. Other modifications were offered, but the most dramatic change was a proposed land acquisition fund of 50 million dollars per year. The administration

13. Morrell to Hastings, memos dated Nov. 21 and Dec. 7, 1932, copies ibid.; Clapp to Morrell, Feb. 23, 1933, copy ibid. It is interesting to note that Morrell objected to the way his contribution to the report was handled, and he refused to sign it (see pp. 1329–41 of the Copeland Report). The author is grateful to E. L. Demmon for making this observation.

14. Clapp to C. J. Buck, May 9, 1933, box 59858, Records of the Pacific Northwest Forest and Range Experiment Station, Seattle FRC; "Through Official Channels," Jan. 31, 1946, memo in box 11, Clapp Papers, RG 200.

chose not to have the bill introduced, Clapp believed, because of legislative priorities elsewhere.[15]

The report itself was impressive. The two-volume, 1,677-page *A National Plan for American Forestry* described and evaluated virtually all aspects of forestry, public or private. Timber, water, range, recreation, wildlife, research, state aid, and fire protection were only some of the topics included.[16] The modern concept of multiple use appeared for the first time in a substantive way.

Since the Forest Service had long described its own work as superior, it was predictable that the agency would report that "practically all of the major problems of American forestry center in, or have grown out of, private ownership." Acknowledging some deficiencies in public management and proclaiming forest problems to be a "major national problem," the Copeland Report recommended a "large extension" of public ownership to be followed up by more intensive public forest management. Public acquisition of forest land would be only the first phase of a total program, the Forest Service reminded the reader; there was much more to the plan.[17]

In Clapp's view, the essence of the plan was to have public agencies assume responsibility for fully one-half of "the national enterprise in forestry." Private ownership, which held 80 percent of the commercial forest land and 90 percent of the growth capacity, had been expending only 10 percent of the "constructive effort." To Clapp, such a ratio of effort to responsibility was unacceptably out of balance, and the best solution was massive public acquisition. Advocating flexibility, Clapp offered regulation of private cutting as a quid pro quo for public assistance and protection.[18]

Frustrating to Clapp was the awareness that even within the Forest Service the Copeland Report fared ill. The report had been a "standing joke" around the Washington office, or so a disgruntled Clapp complained to Silcox in 1934. Even during preparation, Clapp had faced opposition. Some forest officers bitterly fought the fire protection recommendations, others were "indifferent or openly hostile to public regulation," and some men "seem to sabotage the effort for its

15. Stuart to Wallace, May 12, 1933, in Edgar B. Nixon, ed., *Franklin D. Roosevelt and Conservation, 1911–1945*, 2 vols. (Hyde Park, N.Y.: General Services Administration, 1957), 1: 163–65; Clapp, "Through Official Channels," Clapp Papers, RG 200.

16. U.S. Congress, Senate, *A National Plan for American Forestry*, Sen. Doc. 12, 73 Cong. 1, March 13, 1933.

17. Ibid., pp. v–x.

18. Ibid., pp. 76–78.

approval." Clapp thought that Silcox himself had never recognized the need for a comprehensive program, and that lack of dynamic leadership had kept the report from achieving its potential. Even so, Clapp was proud that the Forest Service had lived up to its responsibility. To the chief of research, the report was a good example of the quality of work that his division could do. History would be the ultimate judge of its worth.[19]

Despite Clapp's disappointment, the report was well received outside the agency. An American Forestry Association committee found it to be an "extraordinarily able analysis of the forest situation of the country." The committee thought that the Forest Service had presented "very convincing reasons" why national planning was essential. Proposals were judged to be in general agreement with existing policies of the conservation group.[20]

Strangely, the Society of American Foresters did not provide a formal evaluation, although many articles about various aspects of it did appear in the *Journal of Forestry*. The report perhaps was considered too sensitive for the organization. After all, a majority of members were Forest Service employees, bound to official policy. At least one professional forester, Alfred Gaskill, retired state forester of New Jersey, did voice strenuous objections. Gaskill submitted a critique to the *Journal of Forestry*, which Editor Franklin Reed sent out for review. Readers of the manuscript felt the article to be a "stimulating consideration" of the report and urged publication. Reed also sent galley proofs to Clapp, agreeing to delay publication to allow the Forest Service time to prepare a formal response, which would appear alongside Gaskill's review.[21]

Gaskill described the Copeland Report as a "thing of awesome proportions and disappointing substance." He accused the Forest Service of submitting its case to the people "by mass rather than by reason." The report stood as a "challenge to the intelligence and sincerity" of those who supported forestry. He found little evidence to substantiate the report's basic theme that public forestry was capable of solving the vast array of problems described. Gaskill's main fear was that since

19. "The Copeland Report," Jan. 29, 1946, box 11, Clapp Papers, RG 200; Clapp memo to files dated Feb. 13, 1946, ibid.; Clapp to Silcox, memo on research, March 19, 1934, box 1, ibid.

20. AFA Committee on Copeland Report, Oct. 17, 1933, box B-6, AFA Records, FHS.

21. Franklin Reed to Ralph C. Hawley, Nov. 20, 1933, box 51, SAF Records, FHS; Hawley to Reed, Nov. 23, 1933, ibid.; Reed to Clapp, Dec. 2, 1933, ibid.; Silcox to Reed, Dec. 11, 1933, ibid.; Reed to Silcox, Dec. 14, 1933, ibid.

public forestry could not be a panacea, it would be discredited after failing to live up to brave promises.[22]

Chief Silcox wrote an eloquent rebuttal, backed privately by H. H. Chapman of Yale, president of the Society of American Foresters. Both saw Gaskill's statement as suggesting that foresters should mind their own business—stick to timber management and stop trying to do too much, such as rehabilitating the range and tackling soil erosion. Angry and disgusted, Chapman considered writing a presidential editorial for the next issue of the *Journal of Forestry* on "defeatism."[23]

Dodging the occasional brickbat, the Forest Service began its campaign to implement the Copeland Report's numerous recommendations. In keeping with the concept of a single national plan for American forestry, strategy emphasized the need for one all-inclusive bill to coalesce the vast array of situations. Drafts circulated to industrial groups, conservation organizations, state foresters, and within the Forest Service itself. Staff were warned that as federal officers they must guard against political activity or lobbying, but they were "definitely responsible for the exercise of positive leadership" to encourage enactment of the bill. Setting 1936 as the target, staff were mobilized to achieve peak effectiveness.[24] The right time, unfortunately, failed to come. A testy Congress and continuing economic difficulties caused the bill to be pigeonholed. Congress would examine an omnibus forestry bill five years later, but then it would be World War II that shuffled priorities.

Yet, *A National Plan for American Forestry* was a useful and influential document. Throughout the New Deal years, it cropped up in report after report dealing with the full range of forestry problems. Not all references were complimentary, to be sure. Lumbermen were understandably disappointed by the scathing rebuke they suffered. Nonetheless, because of its broad scope, history must regard the report highly, as Clapp was certain that it would.

The Western Range. Following publication of the Copeland Report, the Forest Service elaborated on certain of its specific recommendations. The deteriorating condition of range land was one of the most

22. Alfred Gaskill, "Whither Forestry?" *Journal of Forestry* 32 (Feb. 1934): 196–201.

23. F. A. Silcox, "Forward Not Backward," ibid., pp. 202–7; Chapman to Reed, Jan. 6, 1934, box 51, SAF Records.

24. "Plan to Campaign for Enactment of the Omnibus Forestry Bill," Nov. 20, 1935, copy in box 4, Records of the Office of the Chief, 1908–47, Record Group 95 (hereafter cited as RG 95-4), RFS.

pressing problems. Poor range conditions and then enactment of the Taylor Grazing Act in 1934 prompted an investigation "somewhat comparable" to the Copeland Report, although on a smaller scale.[25] The 1924 Rachford Report had focused on grazing fee schedules; needed now was another study to investigate the state of American ranges.

In 1918 Henry Graves had dictated to his men that grazing was only secondary to the primary uses of the national forests, timber and water. Grazing must be made compatible with the dominant uses.[26] Then C. E. Rachford studied grazing fees; others wrestled with the problem of reducing the numbers of stock using national forest ranges. But now a broader issue came under consideration: did the condition of public range lands justify jurisdiction by a single agency instead of several?

During the Hoover administration, the states failed to respond to federal overtures for transfer of unappropriated public domain to local control, thus forcing consideration of federal regulation of public land use. Since vast portions of public land were used for grazing, the importance of range management loomed large. Congressman Don B. Colton of Utah had introduced a bill to create grazing districts in the public domain. Chief Stuart had offered his support, calling the bill "reasonable." The bill passed the House but failed to clear the Senate. In 1933 the next Congress saw Representative Edward C. Taylor of Colorado reintroduce the Colton bill as the Taylor bill. The bill proposed creating a grazing bureau in the Department of the Interior to administer range lands then under custody of the General Land Office.[27]

25. W. R. Chapline, memo to files dated Sept. 8, 1935, box 1120, Records of the Division of Range Management Research, General Correspondence and Reports, 1909–54, Record Group 95 (hereafter cited as RG 95–134), RFS. The Copeland Report had recommended correlation of grazing with other uses, improving quality of management, public acquisition of depleted range, and expanded range research programs. Senate, *National Plan*, pp. 1537–41.

26. Henry S. Graves, "Correlation of Forestry and Range Management," Policy Letters from the Forester, no. 2, Nov. 13, 1918, copy in box 209, RG 95–4, RFS.

27. During President Hoover's administration, a public lands commission recommended that federal range lands should be turned over to the states. It soon became obvious that states were unwilling or unable to assume this responsibility, inviting a federal solution. For study of the Taylor Grazing Act, most useful are E. Louise Peffer, *The Closing of the Public Domain: Disposal and Reservation Policies, 1900–1950* (Stanford, Calif.: Stanford University Press, 1951), pp. 214–24; and Wesley Calef, *Private Grazing and Public Lands: Studies of the Local Management of the Taylor Grazing Act* (Chicago: University of Chicago Press, 1960); also useful but flawed is Phillip O. Foss, *Politics and Grass: The Administration of Grazing on the Public Domain* (Seattle: University of Washington Press, 1960).

Taylor, long an opponent of federal controls, advocated local domination of the federal grazing bureau and thereby prompted opposition from both the American Forestry Association and the Forest Service. Silcox charged that the original bill, which the Forest Service had supported, had been so altered that it no longer was a conservation measure. He also believed, as did Ovid Butler of the American Forestry Association, that the bill required federal abdication of authority over the public range. To Silcox, what had been intended as a program to end range deterioration and to rehabilitate overgrazed areas had instead become a plan for siphoning federal funds at the local level. He implored Secretary Wallace to intervene with the president to obtain a veto should the bill reach his desk in its present form.[28]

SAF president H. H. Chapman, who approved of Forest Service grazing policy, shared Silcox's despair over the Taylor bill. The legislation would fix grazing rights, Chapman warned Silcox, and the value of these rights would be immediately reflected in the value of the permittee's holdings. He feared that proponents of recreation and other uses would view enactment of the Taylor bill as a "betrayal of public rights for the permanent benefit of the [range] industry."[29]

Despite opposition from the Forest Service and others, the Taylor grazing bill became law on June 28, 1934. The act authorized creation of 80 million acres of grazing districts, soon enlarged to 142 million acres. Two months later Secretary of the Interior Harold Ickes appointed Farrington R. Carpenter to be the first director of grazing. Officially, forest officers received instructions to support it, as it was the law of the land, but unofficially disapproval of the terms of the Taylor Act prompted open Forest Service retaliation.[30]

Feeling that grazing had too long been handled in piecemeal fashion, Secretary Wallace asked the Forest Service for a report covering all aspects of range management. W. Ridgely Chapline, one of the thirty-five authors assigned to the task, set out to find "concrete evidence" of poor management or lack of appreciation of grazing on the public domain by the Department of the Interior. Clapp predicted that grazing would be an important issue during the next session of Congress and a full study of range conditions would be timely.[31]

28. Ovid Butler to F.D.R., June 14, 1934, box F-11, AFA Records; F. A. Silcox, memo to Secretary of Agriculture, June 21, 1934, copy in box 37705, RFS, Denver FRC.

29. H. H. Chapman to Silcox, Jan. 4, 1935, box 45, SAF Records.

30. F. A. Sherman to Regional Foresters, June 30, 1934, box 37705, RFS, Denver FRC.

31. Clapp to Frank C. Pooler, Oct. 5, 1935, box 1119, RG 95–134, RFS; Chapline to Clapp,

The Western Range was sent to the Senate in April 1936, technically in response to a resolution by Senator George Norris of Nebraska. Unusually blunt for an official document, the report in no uncertain terms advocated that management of all federal ranges should be a Forest Service responsibility. It claimed that the Taylor Grazing Act was inadequate and stockmen were troubled by having to deal with more than one agency. Therefore, "the Department of Agriculture is . . . the logical and, in fact, the only well-equipped department for the administration of federally owned range." Consistent with past Forest Service views, the report was highly critical of private range conditions, estimating 85 percent to be deteriorating.[32]

As could be expected, those criticized by the Forest Service report responded in kind. Ickes complained to FDR that *The Western Range* was being used to "antagonize western livestock interests against your administration." He then complained to Wallace that Interior had been more concerned about proper range management than the Forest Service and ought to have been consulted in the preparation of or even in determining the need for the range report. Ickes also noted sarcastically that the more than 600-page report, "a thinly veiled attack on a sister department," was sent to the Senate only four days after Senator Norris' resolution. Obviously the Forest Service had made the study and prepared its recommendations, then had arranged for a senatorial sponsor to "request" such a study.[33]

Wallace responded to Ickes' charges with a lengthy defense. Range protection was an agricultural objective, the secretary argued, and the Forest Service was part of his department. Therefore, it was appropriate for that bureau to have made the study. After recounting the dreary history of Ickes' department in "protecting" public lands, Wallace lashed out that "the fundamental difference" between their views on grazing was the degree to which public rights were to be subordinated to the vested rights of a relative few. If Ickes would find the time to study Interior's record, then he would understand why the Department of Agriculture was concerned. Wallace scoffed at Ickes' complaint that *The Western Range* constituted an attack on Interior. Nothing in the report, he insisted, was as "biting" as Ickes' remarks

September 24, 1935, box 1120, ibid.; Clapp to Joseph C. Kircher, Sept. 28, 1935, box 1119, ibid.

32. U.S. Congress, Senate, *The Western Range*, Sen. Doc. no. 199, 74 Cong. 2, 1936.

33. Ickes to FDR, Aug. 19, 1936, in Nixon, ed., *Roosevelt and Conservation*, 1:550 (copy also in box 4, Clapp Papers, RG 200).

when proposing the transfer of several agricultural agencies to the Department of the Interior. Ickes' behavior had made self-defense necessary.[34]

Stockmen, singled out for criticism even more specifically than Interior, voiced outrage at *The Western Range*. "Utterly ridiculous," stockmen labeled the charge that most private landowners did not know how to conduct a proper business. Claiming that they had some rights and some voice in the matter, F. E. Mollin, secretary of the American National Live Stock Association, vowed to oppose increased regulation by the Forest Service, which was "a bigoted, conceited, bureaucratic setup." Mollin then wrote a defensive tract that portrayed drought as the real villain in range deterioration, for if it would only rain, "the grass grows again and all is well on the range."[35]

To the Forest Service, however, the report seemed both appropriate and accurate. Current and past deterioration of the range was a major problem, and evaluation of conditions could not in good conscience have been delayed any longer. The Forest Service had been working on the report since 1932; the Science Advisory Board to the president had stressed the need for information on the range, and the secretary of agriculture had complied by directing the Forest Service in August 1935 to complete its investigation. Clapp sent a list of suitable answers to forest officers explaining its background in order to make their defense of *The Western Range* more effective.[36]

The Western Range was very important to Clapp personally. His Division of Research had prepared it, as it had the Copeland and other major reports. Clapp's pride in the range study was exemplified by his transfer of nearly one hundred letters applauding *The Western Range* from official Forest Service files to his private papers. One letter was signed, "Yours for giving nature a chance."[37]

Issues raised by *The Western Range* remained unresolved and at times were obscured by larger events of the depression years, and Forest Service opinion of the Grazing Service remained at a low ebb. Forest officers, suspicious of actions taken by Interior's employees, fre-

34. Wallace to Ickes, Nov. 13, 1936, in Nixon, ed., *Roosevelt and Conservation*, 1:595–606.

35. F. E. Mollin, "On Western Grazing," *Successful Farming*, Oct. 1937; F. E. Mollin, *If and When It Rains: The Stockman's View of the Range* Question (Denver, Colo.: American National Live Stock Association, 1938), p. iii.

36. W. R. Chapline, "Answers to Questions Raised in the Western Range Report," n.d., copy in box 1119, RG 95–134, RFS; Clapp to Regional Foresters, Nov. 11, 1936, copy in A. F. Hough Papers, National Conservation Library, Denver Public Library.

37. Box 4, Clapp Papers, RG 200.

Secretary of the Interior Harold Ickes attempts to capture the Forest Service. Courtesy of the J. N. "Ding" Darling Foundation

President Franklin D. Roosevelt at first Civilian Conservation Corps camp, George Washington National Forest, Virginia, 1933. Courtesy of the U.S. Forest Service

Raphael Zon with map of Shelterbelt Project. Courtesy of the U.S. Forest Service

Acting Chief Earle H. Clapp signs New England hurricane timber salvage agreement, 1940, with Earl S. Peirce and Harry Joseph looking on. Courtesy of the U.S. Forest Service

quently charged that Grazing Service personnel encouraged stockmen to make difficulties for the Forest Service while being purposely lax in enforcing their own rules in order to curry favor. Gunnison National Forest Supervisor M. J. Webber raged from Colorado, "How much longer should the Forest Service continue to maintain an attitude of appeasement toward the Grazing Service? The time is ripe to adopt a militant policy." Within Congress, the Forest Service and the Grazing Service had their own supporters; most stockmen, however, overwhelmingly favored the Grazing Service. This split would continue into the following decades and cause a substantial drain of Forest Service administrative energy.[38]

Recreation. The Department of Agriculture had been involved with range administration since Pinchot became chief of the Division of Forestry in 1898. It was not until the New Deal years, however, that recreational use of national forests, always of some interest, began to receive truly serious attention. The Division of Recreation and Lands first appears in the 1935 Forest Service directory. Earlier issues give no recognition to recreation. Annual reports lumped it with game, presumably because hunting and hiking were both fun and ought to have uniform supervision.

From the Forest Service point of view, two recreational elements—wilderness and national parks—presented the most difficulties. Henry Graves had not been able to prevent creation of the National Park Service in 1916; consciously or not, the two agencies competed on recreational affairs. Chief Stuart in 1931 called for clearer distinctions between Park Service and Forest Service programs, acknowledging that his agency viewed recreation primarily from the standpoint of sanitation and fire prevention. Park Service Director Horace Albright concurred on the need for a joint study.[39]

Most irritating to Regional Forester Allen S. Peck in Colorado was the constant erosion of Forest Service land base whenever a new national park was created or an existing one expanded. Peck particularly bridled at claims that the Park Service alone recognized and appreciated the "higher social and spiritual values inherent in natural things." The Forest Service seemed to receive little recognition for its multiple-use management, which was "inseparably interwoven into

38. Allen S. Peck to Chief, Jan. 29, 1941, box 37703, RFS, Denver FRC; M. J. Webber to Regional Forester, Feb. 12, 1941, ibid.

39. AFA, "Board Minutes," Sept. 17, 1931, box B-1, AFA Records; Minutes of the Service Committee, Sept. 10, 1931.

the social and economic future of forest communities." Not only did the Forest Service furnish recreation to four times more people than did the Park Service, and had set aside much more land in the wilderness category than had the Park Service, but also it did not "bottle up" needed resources as did the "single-use" Park Service.[40]

Yet, despite what the Forest Service believed to be the indisputable logic of its arguments, when it became known that the agency had formal plans to authorize logging in the Olympic National Monument in Washington State, the Park Service acquired still another national park. Try as the agency would, it could not convince a substantial portion of the public that multiple-use management was an adequate safeguard for recreational values.[41] The conflict between recreation and multiple use would continue without satisfactory resolution.

Robert Marshall, author of the recreational portion of the Copeland Report, recommended a recreational survey so that decision-makers might know the preferred types of recreation and how much land ought to be set aside for these purposes. He thought all public recreation agencies, federal and state, should pool information. Marshall proposed seven categories of recreational areas and the acreage needed for each. His recreational spectrum ranged from wilderness ("a person may spend a week or two of travel in them without crossing his own tracks") to outing areas ("where one can get away from the sounds of highways"). Forty-five million acres would be needed for recreation; logging would be allowed in the 11 million acres of outing areas. Not a very radical proposal, since millions of acres of wilderness had already been reserved and campgrounds, as well as de facto outing areas, already existed. According to Marshall's arithmetic, however, only 11 million of the needed 45 million acres were reserved, leaving 34 million to go. What he proposed in the main was a unified approach to recreation, adding a sense of permanency.[42]

Imprecise terminology and muddy concepts contributed to the confusion. Forest Service leadership in wilderness preservation notwithstanding, the Park Service seemed to be getting all the credit. In the fall of 1935, Marshall told an assembly of regional foresters that the Forest Service needed a better definition of wilderness and to "see

40. Allen S. Peck to Chief, July 27, 1938, box 8, RG 95-4, RFS.

41. See Ben W. Twight, "The Tenacity of Value Commitment: The Forest Service and Olympic National Park," Ph.D. diss., University of Washington, 1971, pp. 85 ff.; Silcox to H. H. Chapman, Feb. 15, 1937, box 39, RG 95-4, FHS.

42. Senate, *National Plan*, pp. 1543-46.

that it is lived up to." Areas being evaluated for wilderness potential should be set aside until a final decision was reached. Marshall thought that grazing would be accepted in western wilderness areas, because livestock were important to the economy of that region. The big danger, he believed, was the "invasion" of wilderness by motorized vehicles. Marshall answered in the affirmative when Silcox asked whether varying degrees of wilderness would be acceptable. Evan W. Kelley, regional forester in Montana, pointed out—and Marshall agreed—that wilderness preservation was contrary to the concept of the greatest good of the greatest number, because only a few citizens could take advantage of wilderness. Ovid Butler of the American Forestry Association thought that the wilderness portion of Forest Service policy was its most vulnerable aspect. Not agreeing specifically, Silcox accepted the idea that unless the Forest Service formalized its wilderness program, the agency would lose jurisdiction over the reservations.[43]

In an attempt to clarify the issues, the Society of American Foresters distributed a questionnaire to its membership. The questionnaire was sharply slanted in favor of the Forest Service. Of seventeen questions on various aspects of recreation, six dealt with wilderness. All six queries had a similar tone, best summarized by the last: "Since the principle of inaccessibility is diametrically opposite to that governing National Parks, wilderness areas which are integral parts of National Forests should be permanently retained as portions of these National Forests."[44]

Aldo Leopold, authority on game management and wilderness recreation, returned his questionnaire, declaring that it was meaningless. In particular, he doubted the wisdom of "writing wilderness areas out of national parks." Others were even less complimentary, describing the effort as "diabolical," "the rankest thing I have yet seen," or "bigoted and unfair referendum." Opponents of Forest Service wilderness policies were in the minority within SAF membership, however. The vote came out over twenty to one against the Park Service. The same margin favored legislation to give wilderness full protection.[45]

43. "Record of Discussion and Action at Regional Foresters Meeting," Nov. 18–23, 1935, box 3, RG 95–4, RFS.

44. Copy of questionnaire in box 32, SAF Records; questionnaire published in *SAF Affairs* 4 (Dec. 1938): 207–8.

45. Aldo Leopold to Henry Clepper, Oct. 11, 1938, box 32, SAF Records; John P. Coffman to SAF Council, Nov. 1, 1938, ibid.; assorted letters to SAF Council, Nov. 1938, box 58, ibid.; *SAF Affairs* 4 (Dec. 1938): 209.

Meetings, reports, and questionnaires notwithstanding, Forest Service recreation policy and the agency's relation to the Park Service remained murky. Income from recreation was of key importance to the utilitarian foresters, but a bone of contention to many others. Silcox did believe, however, that the Forest Service should provide free recreation for low-income people. Despite all of the words and promises, John Sieker, chief of the Division of Recreation and Lands, felt called upon as late as 1939 to explain to state foresters that he thought it "perfectly logical that recreation should take its place beside the old and well established uses such as timber," a sure indication that recreation had not yet become an equal partner.[46]

The Forest Service tried to take its recreational case to the public in 1940, publishing *Forest Outings by Thirty Foresters*. Written to a large extent by Russell Lord, writer, editor, and one-time staffer in the Department of Agriculture, the book was only partially successful. Some members of Congress thought recreation frivolous and were critical of spending appropriations on such things.[47] For the Forest Service, it was a case of damned if it did and damned if it didn't.

An insider also voiced criticism of recreation practices, although from the opposite viewpoint. Shortly before his death from a heart attack in 1939, Robert Marshall, then chief of the Division of Recreation and Lands, made an inspection trip through the western states. He complimented Regional Forester Lyle Watts of the Pacific Northwest Region for an excellent recreation program. Except, Marshall wrote, "the one thing predominantly disappointing . . . was the lack of progress in establishing wilderness areas." In California, Marshall chastised Regional Forester S. B. Show for taking for granted that timber and range management required planning but apparently assuming that recreation had lesser needs. Marshall insisted that recreation affected more Californians than either timber or range, and it seemed to him that Show could "afford to substitute for the sporadic planning of the past a comprehensive plan for the entire area." On the train back to Washington, D.C., Marshall continued his report on recreation. Large-scale recreation was still so new on national forests

46. Forest Service, *Annual Report* (1938), p. 52; AFA, "Committee on Parks, Forests and Recreation," Sept. 21, 1938, box B-6, AFA Records; NASF, Minutes, 1939, NASF Records, National Conservation Library, Denver Public Library.

47. Forest Service, *Annual Report* (1939), p. 32; Verne L. Harper, *A Forest Service Scientist and Administrator Views Multiple Use*, Oral History Interview by Elwood R. Maunder, 1972, FHS, pp. 91–92; Russell Lord, ed., *Forest Outings by Thirty Foresters* (Washington, D.C.: GPO, 1940).

that he believed quality of recreational programs to be less consistent than almost any other "line of national forest activity."[48]

New Deal Programs

Planning, whether all-inclusive like the Copeland Report or specific like *The Western Range* or Marshall's proposals for recreation, consumed much staff time during the New Deal. In fact, projecting need has always been a major Forest Service activity. To manage many diverse resources more or less compatibly required careful planning, particularly when dealing with long-term ventures such as timber management, for it was unlikely that those who developed a timber plan would be able to observe the results. Moreover, flexible plans could accommodate the unexpected or abrupt shift in priorities. The New Deal fostered many shifts to revive and stabilize the sagging economy. Forest Service plans proved quite useful in implementing its New Deal assignments.

Civilian Conservation Corps. Pinchot had his problems when he tried to get messages to the public, and Congress eventually placed strict controls on the masses of news releases. Yet, agencies must keep their public informed. *Forest Outings* is a good example of the advance in public relations, and the Forest Service went on to utilize the latest techniques to gain support. In 1932, *Uncle Sam's Forest Rangers* began on the NBC radio network, as part of the "National Farm and Home Hour." Adopting the dialogue format used by the extremely popular *Amos 'n Andy*, the principal characters were an old-time ranger named Jim Robbins and Jerry Quick, a young "cub" recently out of forestry school. Housewives preparing their Thursday lunch could listen to the adventures of Jim and Jerry in the Pine Cone Ranger District, "at the edge of the forest." Jerry's first test comes quickly: he and Jim light up their smokes, and Jerry forgets to break his match in two. Fortunately, Jerry is indeed quick and learns rapidly. It was hoped that the housewives would, too.[49]

The weekly program lasted into the 1940s, each episode emphasiz-

48. Robert Marshall, "Inspection of Recreation and Lands Activity, Region 6," Nov. [10], 1939, box 13, RG 95–4, RFS; "Inspection of Recreation and Land Activities, Region 5," Nov. 2, 1939, ibid.; "Inspection of Recreation and Land Activities, Region 1," Nov. [10], 1939, box 12, ibid. Marshall died on Nov. 30, 1939.

49. Minutes of the Service Committee, Nov. 12, 1931, RG 95–8, RFS; *Uncle Sam's Forest Rangers*, episode no. 1, Jan. 7, 1932, Radio Scripts for "Uncle Sam's Forest Rangers," 1932–44, Records of the Division of Information and Education, Record Group 95, RFS.

ing a different mission or characteristic of Uncle Sam's forest rangers. Listeners learned that on the rare occasions rangers went to town, they removed their hats in the presence of ladies, a custom seldom observed by callous city dwellers. Etiquette aside, range, logging, and fire policies were illustrated, too. An episode on September 21, 1933, portrayed one of the most popular of all the New Deal programs.

The announcer's cultured voice introduced the theme. "This summer, many of our Rangers have been called upon to handle another big job—supervision of the work of thousands of young Civilian Conservation Corps boys." He explained that over fourteen hundred camps had been established with nearly half of them on national forests. As luck would have it, there was a camp on the Pine Cone Ranger District.[50] Jim and Jerry had their work cut out.

FDR had alluded to a million-man conservation work force in his acceptance speech to the Democratic convention. He had utilized young men in conservation tasks as early as 1925, while president of the Boy Scouts Foundation of Greater New York. Ten days after inauguration, the president of the United States directed the secretaries of agriculture and the interior to "coordinate the plans for the proposed Civilian Conservation Corps." Enabling legislation had been drafted in less than ten hours. To avoid opposition from labor, the CCC would work only on projects not already covered by public works relief. In addition, Roosevelt appointed Robert Fechner, an executive of the International Association of Machinists, as director.[51]

Speed was essential. In late March the CCC bill was still languishing in committee, and FDR wanted a quarter of a million men at work in the forests by early summer. The feat would require mobilization unmatched in American peacetime history. At first, Stuart doggedly insisted that the Forest Service was equal to the task, but soon he asked for Army support to deal with the masses of manpower and accompanying logistical support. By mid-May, with only fifty thousand men enrolled, the July 1 target of five times as many was abandoned. Somehow most of the technical and administrative problems were solved, and in August Roosevelt triumphantly toured five

50. *Uncle Sam's Forest Rangers*, episode no. 80, Sept. 21, 1933, ibid.

51. Ovid Butler to Arthur W. Proctor, Sept. 1, 1935; FDR to Secretaries of War, Interior, Agriculture, and Labor, March 14, 15, 1933, in Nixon, ed., *Roosevelt and Conservation*, 1:58, 112, 138, 141, 150.

camps in the Washington, D.C., vicinity. The "Cs" were here to stay.[52]

Hiring practices for CCC staff, as with other New Deal programs, rewarded Democrats. Bureau chiefs, Stuart included, received lists of party faithful from which to select new employees. Little pressure was exerted to hire incompetents; surely somewhere there was a qualified Democrat for any job. The administrator could reject list after list but eventually would find the Democrat he needed. Opposed by many—generally Republicans—this modified spoils system seemed to threaten the nonpartisan nature of the Forest Service. Along with the incredible demands to meet mobilization schedules, it no doubt contributed to Stuart's collapse and death.[53]

Statistically the average CCC enrollee was twenty years old and from a family of six. He had an eighth-grade education and had been unemployed for at least nine months, as had his father. After passing rigorous military tests for health and discipline, the enrollee was assigned to one of several thousand work camps. There he lived in wooden barracks, working and playing hard. Good food, laboring out of doors on conservation projects, and relaxation were part of the typical day. Self-improvement was an important duty. The young men had access to a range of training programs and a small library, from which the "subversive" magazines, *The New Republic* and *Nation*, were banned. All this and thirty dollars a month, most of which was obligated to dependents.[54]

The CCC became a popular success, rehabilitating land and men. The Forest Service administered nearly half the projects, the Soil Conservation Service, National Park Service, and other bureaus accounting for the remainder. By 1942 the CCC had spent well over 6 million hours combatting fire, with a loss of forty-seven men. Acres lost to fire were held to a record low during this period. Fire prevention as well as suppression was a responsibility, and thousands of acres were cleared of flammable materials. In California the Ponderosa Way, a fire break along the length of the Sierra, provided an 800-mile

52. John A. Salmond, *The Civilian Conservation Corps, 1933–1942* (Durham, N.C.: Duke University Press, 1967), pp. 26–47.

53. Elmo R. Richardson, "Was There Politics in the Civilian Conservation Corps?" *Forest History* 16 (July 1972): 13–14.

54. Salmond, *Civilian Conservation Corps*, pp. 135–44; Samuel T. Dana, *Forest and Range Policy* (New York: McGraw Hill, 1956), pp. 248–49.

tribute to CCC stamina. Of great importance to recreation was construction of a myriad of campgrounds and miles of trail. Nearly 8 million acres of national forests were purchased with CCC funds. Enrollment peaked in 1935 with over one-half million assigned to over twenty-six hundred camps. Three million men participated during the life of the program.[55]

Three high-ranking Forest Service retirees who watched the Civilian Conservation Corps in action remember its importance in different ways. Earl S. Peirce, retired assistant chief, points out that the CCC provided manpower in quantities that for the first time were adequate to tackle certain tasks. Large-scale forest thinnings and blister rust control both demanded large numbers of workers, as did road and trail building, campground construction, and fire suppression. Clare W. Hendee, retired deputy chief, sees the corps as having had a profound effect on professionalism and resource planning. The need for plans kept pace with increases in manpower and projects, and young foresters suddenly found themselves in charge of large crews and responsible for supervising construction of a reservoir or another equally awesome task. J. Herbert Stone, retired regional forester, remembers how plans were strained when the five CCC camps planned for on his national forest abruptly became ten instead.[56]

Popular and technical forestry literature contains many articles that support or expand on these observations. That some recollections may be larger than life in no way diminishes the importance of the Civilian Conservation Corps, either in deeds or in institutional pride. Undoubtedly, many executive decisions by mature men have been colored by their earlier CCC experiences, either as foresters who moved on to other responsibilities or as enrollees who had engaged in manual labor for the first and perhaps only time.

Land. During the interim between election and inauguration, Roosevelt considered combining land acquisition with relief for the unemployed. His staff prepared an 18-million-dollar proposal to purchase land east of the Mississippi River, with a like amount appropriated to pay for rehabilitation. Although this was only half the amount in dollars and acreage already proposed by the National Forest Reser-

55. Salmond, *Civilian Conservation Corps*, pp. 121–34; Dana, *Forest and Range Policy*, p. 249; "Various Major Accomplishments by the Forest Service, 1933–40," July 16, 1940, memo in box 13, RG 95-4, RFS.

56. Peirce, personal communication, July 1, 1975; Hendee, personal communication, June 9, 1975; Stone, personal communication, June 30, 1975.

vation Commission under the Weeks Law, Roosevelt's plan called for a more uniform distribution of purchase. Weeks Law acquisitions were limited to those states that had enacted enabling legislation.[57]

Pinchot, in a lengthy letter to "Dear Franklin," supported acquisition. According to Pinchot, there was no hope that industrial forestry could meet the needs of the time. "The solution of the private problem lies chiefly in large-scale public acquisitions of private forest lands." Pinchot was convinced that the Clarke–McNary Act ("private altruism, plus a government subsidy in the form of aid in fire protection, plus patting the lumberman on the back") had failed. Decisive federal intervention was crucial.[58]

Stuart wrote Wallace that recommendations for land acquisition were in "full accord" with established trends. He referred to Copeland Report proposals on land purchases—22 million acres from the public domain and over 130 million acres more to be acquired eventually from private holdings—and estimated that 134 million acres ought to be acquired, 8 million of them immediately. Stuart calculated that he needed 25 million dollars to purchase the 8 million acres in forty-two units in twenty states. The whole purchase scheme was closely linked to the Civilian Conservation Corps, for FDR wanted "plenty of land" for the corps to rehabilitate.[59]

Land prices were down, so it was a good time to buy. Since federal expenditures were already in excess of revenue, however, the Budget Bureau urged cautious consideration before buying. After much haggling within the administration, FDR authorized 20 million dollars by executive order. Further allocations brought the total to 76 percent more than all purchase funds available between 1911 and 1932. Nearly 8 million acres were added to eastern national forests, or two and one-half times more than had been purchased prior to the New Deal.[60]

In 1929 the National Lumber Manufacturers Association had advocated "greatly expanded" public land acquisition programs for purchasing primarily privately owned lands that were valuable for wa-

57. Nixon, ed., *Roosevelt and Conservation*, 1:126.

58. Pinchot to Roosevelt, Jan. 20, 1933, ibid., 1: 129–32.

59. Stuart to Wallace, Apr. 18, 1933, ibid., p. 143; Senate, *National Plan*, p. 1249; Stuart to Louis M. Howe, Apr. 29, 1933; Stuart to Wallace, Apr. 18, 1933; Roosevelt to Lewis W. Douglas, Apr. 24, 1933, all in Nixon, ed., *Roosevelt and Conservation*, 1:154–60 *passim*.

60. Lewis W. Douglas to Roosevelt, Apr. 26, 1933, in Nixon, ed., *Roosevelt and Conservation*, 1:160. See also pp. 182–98 *passim*. To keep the budget in better balance, FDR used NIRA funds instead of EWC. Dana, *Forest and Range Policy*, p. 250.

tershed or recreation. The lumbermen believed that "forests of this class logically belong in public ownership," since such chiefly public values should not have to be supported by private sources. President Hoover's Timber Conservation Board also had recommended public purchase of forest land. Industrial spokesmen balked a few years later, however, when they realized that the Forest Service and other federal agencies were acquiring large amounts of land of potential commercial value. One from the Far West wryly observed that the national debt stood as a high enough monument to FDR, he did not need land, too. Central to industrial concern was the possibility of local economies being seriously disrupted by a drastically altered tax base.[61] The fact that much of the acquired land was already tax-delinquent weakened this argument considerably. Regardless, corporate spokesmen remained uncomfortable about New Deal land purchase programs.

Shelterbelt. Another well-publicized forestry program of major proportions received help from a New Deal manpower pool, this time from the Works Progress Administration (WPA). The Prairie States Forestry Project, better known as shelterbelt, was to establish strips of trees at one-mile intervals on cultivated lands within a belt 100 miles thick to intercept the prevailing winds between Texas and North Dakota. When the project was transferred to the Soil Conservation Service in 1942, 18,000 miles of shelterbelts offered protection to 30,223 farms.[62] Of equal importance were the thousands of jobs made available through WPA to unemployed locals.

The concept of planting windbreaks was far from new in 1933. Forty years before, Fernow had sent questionnaires to "intelligent men" to find out "what effect, if any, timber belts or windbreaks exert upon the growth of various crops." John Warder, founder of the American Forestry Association, had written about shelterbelts in 1858, and Russian foresters had long been familiar with the technique. In any regard, Raphael Zon is generally credited with convincing

61. Wilson Compton, "Forestry Policy Statement," Aug. 8, 1929, box 2, WFCA Records, Oregon Historical Society, Portland; C. S. Cowen to H. H. Chapman, Jan. 9, 1937, box 50, SAF Records; John J. Haggerty, "Local Effects of Federal Land Acquisition and Consideration of Policy," Natural Resources Board, June 1935, "Confidential—not for public inspection," copy in drawer 241 B, Research Compilation File, 1897–1935, Record Group 95, RFS.

62. Dana, *Forest and Range Policy*, p. 251; "Various Major Accomplishments by the Forest Service," July 16, 1940, U.S. Forest Service, processed; Allan J. Soffar, "The Forest Shelterbelt Project, 1934–1944," *Journal of the West* 14 (July 1975): 95–107. For the most complete view of shelterbelt see Wilmon H. Droze, *Trees, Prairies, and People: Shelterbelt Planting in the Trans-Mississippi West* (Denton: Texas Woman's University Press, 1976).

FDR to back the project, arguing that the value of the idea's practical applications outweighed the consideration of its Russian precedents, which the president wanted to play down. Roosevelt, always enthusiastic about tree-planting projects, had already exchanged a series of letters on the topic with the secretary of agriculture and Chief Stuart when the 1934 Great Plains drought provided him with a logical justification for such a program. The president made general proposals, suggesting "expenditures of say $1,000,000." Using this figure as a starting point, the Forest Service developed a program with Paul H. Roberts as its director, working out of Lincoln, Nebraska. Foresters disagreed about whether the project would succeed, and many derided the idea of planting trees where God had neglected to; but in retrospect one can say that the belts thrived amazingly well.[63]

Political forces nearly ended the shelterbelt in 1936. Secretary Wallace warned the president that congressional cuts in the shelterbelt budget were serious. Countering, Wallace proposed a cooperative tree-planting program with the prairie states to "prevent wiping out the shelterbelt" and attached a draft bill for FDR's attention. The president jotted his answer at the bottom of Wallace's letter: "H.A.W. Yes—if *you* put it through. F.D.R." The secretary of agriculture agreed to follow with a bill broadened to include farm forestry on a nationwide basis, as such a proposal would fare better on Capitol Hill. Senator Norris and Congressman John M. Jones of Texas introduced companion bills to implement the farm forestry proposal. Norris' bill passed the Senate, but the House Committee on Agriculture failed to report Jones's bill out, and the effort died. Using the time before the next session of Congress, Forest Service staff conferred with state foresters, representatives from land grant colleges, and others who were interested in the shelterbelt concept. Norris reintroduced a modified version of the bill, and Wall Doxey of Mississippi introduced its companion in the House. Following negotiations between the Forest Service, Bureau of the Budget, state foresters, and members of Congress, the Norris–Doxey Cooperative Farm Forestry Act became law on May 18, 1937. It authorized 2.5 million dollars annually for promo-

63. Form letter dated Apr. 13, 1892, copy in FHS clipping file; John A. Warder, "Tree Planting in Shelterbelts," *Agricultural Review*, May 1882; Rexford G. Tugwell, *The Brain Trust* (New York: Viking Press, 1965), p. 71; FDR to Stuart, Aug. 17, 1933, Stuart to Wallace, Aug. 15, 1933, Roosevelt to Wallace, Sept. 13, 1933, and E. A. Sherman to Roosevelt, Sept. 8, 1933, all in Nixon, ed., *Roosevelt and Conservation*, 1: 198–206. One of the strongest forestry supporters of shelterbelts was Carlos G. Bates who, as one of Zon's staff at the Lake States Experiment Station, wrote a series of enthusiastic if technical articles about the project.

tion of cooperative farm forestry. The shelterbelt project continued, now complemented by a substantially broader program of federal-state cooperation. Portions of the Norris–Doxey Act were administered by the Soil Conservation Service, and the Forest Service was responsible for the remaining sections, including shelterbelt. In 1942 shelterbelt was transferred to SCS, with the remaining Norris–Doxey functions being passed to it in 1945.[64]

New England Salvage. If relentless prairie winds had suggested the need for shelterbelt, even stronger winds prompted another emergency program. In September 1938 a hurricane swept across New England, killing 682 and leaving 3 billion board feet of blown-down timber strewn in its wake. The immediate problem was economic. By summer, however, fire hazard would be paramount. Proclaiming that the Forest Service was responsible for public welfare in relation to forest lands "no matter who owns them," the agency launched relief efforts.[65] Earl Peirce, chief of the Division of State Cooperation, was responsible for overall direction.

Fifty CCC camps and 15,000 WPA workers that were in the area prior to the hurricane provided the manpower pool for cleanup operations, under Forest Service direction. The Northeastern Timber Salvage Administration of the Federal Surplus Commodities Corporation coordinated the salvage of downed timber with hazard reduction efforts. The Forest Service supervised debris removal from over ten thousand miles of roads and trails and over two thousand acres of roadside strips. Nearly 5 million man-days of labor were expended on fireproofing the disaster area. The salvage program cost 16 million dollars, 15 million dollars of which were retrieved through the sale of over 600 million board feet of salvaged logs and lumber.[66]

Storm salvage on private land, shelterbelts, and youth work corps were not typical of the Forest Service. Availability of public funds for social rehabilitation during the New Deal and coincidental weather extremes explain the existence of these programs more than does gen-

64. Wallace to Roosevelt, May 27, June 1, 1936, in Nixon, ed., *Roosevelt and Conservation*, 1: 527–30; "History of the Norris–Doxey Act," n.d., ms. in box 59A 1753/81, RFS, Washington, D.C., FRC; Dana, *Forest and Range Policy*, pp. 264–65. The Cooperative Forest Management Act of 1950 replaced the Norris–Doxey Act, which was repealed in June 1951.

65. Forest Service, *Annual Report* (1939), p. 19.

66. Ibid., pp. 19–22; Dana, *Forest and Range Policy*, pp. 251–53; Earl S. Peirce, *Salvage Programs Following the 1938 Hurricane*, 1968, Regional Oral History Office, Bancroft Library. Peirce feels that the salvage operation should not be viewed as a New Deal program. To him, it was only coincidental that storm salvage took place in 1939. Peirce, personal communication, July 1, 1975.

eral Forest Service policy. The New Deal encompassed unusual times for Americans, no less so than for their chief federal forestry agency. There was, however, another timber salvage program in Texas in 1944, and the 1960s and 1970s would see work corps again in the national forests. What is important about these more or less atypical additions to Forest Service history is the way the agency reacted to crisis situations and abrupt shifts of emphasis. Massive studies and careful plans had to be flexible enough to accommodate new demands which seemed to appear on an almost daily basis. Even then, distraction proved so severe that recommendations made with a sense of urgency were replaced by still more urgent needs. Substantial backlogs of data and plans enabled the Forest Service to be a major participant in FDR's New Deal.

CHAPTER IX

A Contest of Authority and Crisis of Identity

PUBLIC REGULATION of logging is a "quid pro quo in the public interest."[1] At least this opinion was so expressed by Earle Clapp when he wrote the introduction to the Copeland Report in 1933. The Clarke–McNary Act and other legislation had given lumbermen public protection from fire, support for their tax problems, and nurseries to aid reforestation. Pinchot had insisted more than twenty years earlier, when testifying in favor of a tariff on lumber, that if the public provided assistance to the private sector, then the public deserved something in return—Clapp's "quid pro quo" of logging regulation.

The 1924 Clarke–McNary Act may be viewed in context of the regulation issue. By authorizing federal matching funds for state forest fire programs, the act has contributed significantly to woodland protection. Because Clarke–McNary seemingly resolved the issue of regulation or cooperation in favor of the latter, it was also a key step in the evolution of forest policy. In the years immediately following 1924, the regulation issue simmered, while policy makers attended to other problems.

The crash of 1929 changed priorities, in the lumber industry as elsewhere. Sheer survival became the goal. Conservation concepts that required looking to the future had seemed realistic one day. Now suddenly such investments appeared impossible. The uncertain future at best was bleak.

Before the depression the lumber industry had been roundly crit-

1. U.S. Congress, Senate, *A National Plan For American Forestry*, Sen. Doc. 12, 73 Cong. 1, March 13, 1933, p. 78.

icized. George P. Ahern, a close associate of Pinchot, had just published his *Deforested America*, a scathing indictment of private practices. Others described industrial efforts as "show window forestry" and accused lumbermen of hiding behind the "taxation smokescreen." Forest Service economist W. N. Sparhawk, however, defended the private owner, insisting that forestry could be practiced only to the extent that it was profitable. If the public wanted increased forestry efforts in the private sector, he claimed, then additional public assistance in the form of fire protection and tax relief would be necessary.[2]

Wilson Compton was a moderate spokesman for the industrial position. In a policy statement for the National Lumber Manufacturers Association in late summer of 1929, Compton asserted that the industry had a great stake in "perpetuating the sources of its livelihood." He agreed that the lumber industry should adopt a leadership role but warned that "it cannot be forced beyond the rate dictated by prudent and responsible business management of timber properties. It must be economically self-sustaining." Privately, Associate Chief Sherman agreed with most of Compton's views. Chief Stuart advised Compton of Forest Service policies to favor established sawmills while protecting the public interest. Stuart apparently trusted Compton and asked others more skeptical than he to be patient. In a memorandum to the secretary of agriculture, the chief outlined the parallel goals and attitudes of the Forest Service and the lumber industry.[3]

Compton approached President Hoover directly with a proposal to ease the crushing economic situation that the industry faced. Compton claimed that overproduction of lumber contributed to destructive lumbering, waste, premature cutting, loss of land from tax rolls, and unemployment and was the key to its economic problems. Compton asked Hoover for a fact-finding board to study forest conditions and recommend policies.[4]

2. George P. Ahern, *Deforested America* (Washington, D.C.: Privately published, 1928); Minutes of Washington, D.C., Chapter, SAF, Dec. 28, 1928, box 144, SAF Records, FHS, Santa Cruz, Calif.; W. N. Sparhawk, "Problems in Determining the Economic Feasibility of Forest Use," *Journal of Farm Economics* (July 1929): 406.

3. [Wilson Compton], "Forest Policy Statement," Aug. 8, 1929, box 2, WFCA Records, Oregon Historical Society, Portland; E. A. Sherman to Compton, Aug. 6, 1929, ibid.; Stuart to Zon, March 26, 1929, Raphael Zon Papers, Minnesota Historical Society, Saint Paul; Stuart to Secretary of Agriculture, Dec. 18, 1930, drawer 208, Records of the Office of the Chief, 1908–47, Record Group 95 (hereafter cited as RG 95–4), RFS, National Archives.

4. "Concerning a Proposed National Timber Conservation Board," confidential memo to the president from NLMA, Apr. 19, 1930, copy in box 73, SAF Records.

The proposal received mixed response from forestry groups. Ovid Butler of the American Forestry Association supported Compton's plan, but Tom Gill of the American Tree Association opposed it. Gill's opposition stemmed partly from his belief that the industry wanted substantial public assistance, including modification of the Sherman Anti-Trust Act to allow for self-regulation, without offering to do enough in return. Butler's support stemmed from his acceptance of the industrial position on overproduction.[5]

Hoover responded favorably to Compton's proposal and on November 12, 1930, appointed the Timber Conservation Board. The president charged the group with the responsibility to develop "sound and workable programs of private and public effort, with a view to securing and maintaining an economic balance between production and consumption of forest products." A year and a half later, the board reported back to the president with eighteen proposals.[6]

Besides the time-worn pleas for equitable taxation and increased fire protection, the Timber Conservation Board principally recommended an expanded public forest land acquisition program, modification of antitrust restrictions, and increased forestry research. Three recommendations dealt with the need to coordinate public and private timber supplies to achieve a stable market. This proposal to coordinate supplies was labeled "sustained yield," a new use of the term that had meant a botanically oriented yield-equals-growth policy. Over a period of time, the volume logged would not exceed the volume grown. The new definition was market-oriented: supplies would be kept level with demand by holding public timber unavailable as long as private supplies were adequate. When private supplies were diminished, public forests would provide needed timber while the private lands regrew. Thus, yield would be sustained in coordination with demand.[7]

The idea was not particularly original. Pinchot, Graves, and Greeley had successively pledged to keep national forest timber noncompetitive. Professor Burt Kirkland of the University of Washington College of Forestry had advocated coordination of public and private

5. Ovid Butler to John C. Merriam, Apr. 19, 1930, box 7, John C. Merriam Papers, Library of Congress; Tom Gill to Arthur Newton Pack, Apr. 15, 1930, box 4, Tom Gill Papers, FHS; Gill, memo to file, n.d., box 8, Gill Papers.

6. Secretary of Commerce, Chairman of Timber Conservation Board, to President, June 18, 1932, copy in box 8, Gill Papers.

7. Timber Conservation Board, "Conclusions and Recommendations," n.d., copy in box 8, Gill Papers.

supplies as early as 1924. As is often the case, however, credit for an idea goes to the man who was able to be heard. In this case, the man turned out to be David T. Mason, ex–Forest Service field man in the West, ex–forestry professor, and, in 1930, a consulting forester. Mason insisted that sustained yield offered the most practical way to deal with overproduction, industry's greatest problem.[8]

The Forest Service was critical of Mason's proposal to the Timber Conservation Board. Stuart wrote the board that Mason had exaggerated the amount of federal timber suitably located for sustained yield units and also overestimated the potential impact on production, should agreements be entered into. Stuart did agree with one portion of Mason's plan, that cooperative agreements should be limited to one growing cycle. The chief forester further suggested that selective logging requirements, where only a portion of the trees were cut at any one time, ought to be imposed as a stipulation in any such agreement. Selective logging, Stuart believed, offered "possibilities for financial relief" to western lumbermen. All in all, Stuart opposed Mason's version of sustained yield because there appeared to be no practical way to enforce an agreement. If, for example, a company went bankrupt, there would be little chance of recovering the public investment.[9]

The Timber Conservation Board could only recommend to an administration beset by economic crisis. A New Deal program for industrial recovery—the National Industrial Recovery Act (NIRA) of

8. Burt Kirkland, "Development of Federal Forest Policies as Related to the Pacific Northwest," University of Washington *Forest Club Quarterly* 2 (June 1923): 5–12; Rodney C. Loehr, ed., *Forests for the Future: The Story of Sustained Yield as Told in the Diaries and Papers of David T. Mason, 1907–1950* (Saint Paul, Minn.: Forest Products History Foundation, Minnesota Historical Society, 1952), p. 82. The lumber industry has traditionally viewed controlling supply as the best means to influence price. Attempts to stimulate demand, through advertising or through codes prohibiting use of "wood substitutes," were generally unimaginative and heavy-handed. The decentralized nature of the industry made dealing with supply difficult and with demand almost impossible.

9. Stuart to Timber Conservation Board, Oct. 31, 1931. (Stuart's letter summarized an eighteen-page "Report on Possibilities of Cooperative Management of National Forest and Private Lands," by Fred Morrell.) Copies in drawer 50, Research Compilation File, 1897–1935, Record Group 95 (hereafter cited as RG 95–115), RFS; Sherman memo to Stuart, July 14, 1931, drawer 194, RG 95–4, RFS. Stuart's reference to selective logging concerns an extremely complex, technical issue that is still not well understood. Two Forest Service researchers proposed that selective logging in Douglas fir was an acceptable—perhaps preferable—alternative to clearcutting. Stuart's mention of financial relief referred to the lucrative nature of selective logging that allowed the taking of only the most valuable trees. For a variety of reasons the technique failed to live up to expectations and was soon abandoned. Ironically, selective logging techniques originally introduced to aid lumbermen have recently been discovered by environmentalists who advocate it as a means to control what they view as the destructive nature of clearcutting.

June 16, 1933—implemented many of the board's proposals. Early in 1933, Earle Clapp suggested to Stuart that the recovery act should be utilized to end destructive logging. Stuart in turn presented the idea to Secretary Wallace and Undersecretary Tugwell. It was Ward Shepard, however, Clapp's former assistant, who made the point to the president that very specific conservation goals could be achieved through NIRA.[10]

Shepard urged Roosevelt to "seize the initiative" and use the bait of production controls as means to end destructive logging practices. The Forest Service, through Wallace, strongly endorsed Shepard's proposal, telling the president that the National Recovery Administration of NIRA was a means to implement major recommendations of the Copeland Report. Regulation of logging would be a "quid pro quo to the public for the privilege of curtailing output." Franklin Reed, spokesman for the National Lumber Manufacturers Association before becoming executive secretary of the Society of American Foresters, wrote to Roosevelt that the SAF could be counted on to offer its full support to Shepard's proposal.[11]

Wallace asked Roosevelt to authorize the Forest Service to administer the conservation portions of the recovery act that dealt with the lumber industry. The president agreed, instructing Wallace to write to the National Lumber Manufacturers Association that "the President asks me to tell you that he trusts any code relating to the cutting of timber will contain some definite provision for the controlling of destructive exploitation."[12]

The industry responded quickly with a draft of Article X of the Lumber Code. Greeley and Mason served as principal authors of the article that would set forth minimum standards of logging performance. The Forest Service drafted its own version of Article X with much stricter regulatory features. Mason branded the Forest Service version "vicious." After a series of negotiating meetings, the two factions sent a compromise draft to Roosevelt, who promptly rejected it. The president wanted stiffer controls. On August 19, 1933, after

10. Clapp, memo to file, Feb. 2, 1946, box 11, Earle Clapp Papers, Record Group 200, National Archives (hereafter cited as RG 200).

11. Ward Shepard to FDR, May 29, 1933, in Edgar B. Nixon, ed., *Franklin D. Roosevelt and Conservation, 1911–1945*, 2 vols. (Hyde Park, N.Y.: General Services Administration, 1957), 1:165; Wallace to FDR, June 15, 1933, ibid., p. 171.

12. Franklin Reed to Roosevelt, June 10, 1933, Calendar to FDR Conservation Letters, microfilm copy at FHS; Wallace to Roosevelt, June 16, 1933, in Nixon, ed., *Roosevelt and Conservation*, 1:181; June 16, 1933, ibid., p. 182.

another drafting, Article X of the Lumber Code received presidential approval.[13]

Rules adopted under Article X became effective on June 1, 1934. Operators were to submit management plans to the code authority for approval. Mason, Greeley, and others worked long hours with their colleagues to bring the lumber industry under the code as quickly as possible. Their success was short-lived, however. The Supreme Court ruled on May 27, 1935, that the NIRA was an unconstitutional delegation of legislative authority.

During its brief life, some thought that the Lumber Code was successful. Clyde Martin of the Weyerhaeuser Timber Company reported an end to widespread forest devastation, an increased cooperative spirit, and a closer working relationship between the industry and the Forest Service. At least to members of the Western Pine Association, however, after the demise of the National Recovery Administration there was a general feeling of relief, because it was becoming apparent to them that the code with its regulatory features would never be really successful. But the concepts contained in Article X—forest conservation—would be retained on a voluntary basis. The National Lumber Manufacturers Association issued a statement affirming retention of Article X by the industry. Seventy-five percent of its affiliates agreed, and NLMA hired seven foresters to oversee the conservation program.[14]

The code, with its commitment of substantial federal cooperation, was dead. Even so, Mason continued to advocate his version of sustained yield as a means of stabilizing the industry. He urged Chief Silcox, Stuart's successor, to help inaugurate sustained yield in certain regions, especially in the Pacific Northwest. Greeley, too, thought sustained yield would allow the industry to maintain a leadership position rather than waiting to oppose ideas that the Forest Service might put forth. Congress gradually accepted Mason's sustained-yield proposal and authorized the secretary of the interior in 1937 to establish sustained-yield units in order to stabilize communities dependent

13. Wallace to Roosevelt, July 20, 1933, in Nixon, ed., *Roosevelt and Conservation*, 1:194; Loehr, ed., *Forests for the Future*, pp. 117–19; William B. Greeley, *Forests and Men* (Garden City, N.Y.: Doubleday, 1951), pp. 133–34; Samuel T. Dana, *Forest and Range Policy* (New York: McGraw-Hill, 1956), pp. 254–56.

14. Clyde Martin to Chapman, Dec. 3, 1934, box 65, SAF Records; Western Pine Association, *Annual Report* (1935); "Forestry Legislation Recommended by the Joint Committee of the National Forestry Conference," copy in box 73, SAF Records; Dana, *Forest and Range Policy*, pp. 256–57.

upon forest industries. The department eventually created twelve units on the revested Oregon and California Railroad lands. Timber cut from these units had to be given primary manufacture within specified market areas.[15] Congress followed suit eight years later for the national forests.

To the Forest Service, sustained-yield units to achieve a stable forest industry and therefore a stable local community, although in principle desirable, were not an effective means to end destructive logging practices. Unacceptable logging continued on private land, and voluntary compliance with Article X of the defunct NRA code seemed inadequate. Public regulation of private logging practices—a quiescent issue since enactment of the Clarke–McNary Act in 1924—again became a major thrust of forest policy.

Many foresters were skeptical about the effectiveness of Clarke–McNary's cooperative approach. Within five years of its enactment, impatience with the rate of improved logging practices caused Robert Marshall, soon to become a well-known advocate of wilderness preservation, to declare that "the period for voluntary private forestry is over." Federal law, he claimed, was necessary to force compliance with acceptable logging standards. If a landowner protested that he could not afford to change his methods, then his land should be purchased for public ownership. The Timber Conservation Board, Copeland Report, and NIRA had satisfied many that adequate effort was being made to solve the problem of destructive logging, but Marshall continued to be a major spokesman for those who were not. He and eleven other foresters strongly criticized the editorial policy of the *Journal of Forestry* for its "unsocial" qualities. To Marshall, the Society of American Foresters and its journal no longer advocated programs of social benefit but of industrial benefit instead.[16]

Past editor Emanuel Fritz was outraged by Marshall's accusations and claimed the article proved that Marshall was more interested in

15. Mason to Silcox, March 2, 1935; Greeley to E. T. Allen, Nov. 27, 1936, box 67, NFPA (formerly NLMA) Records, FHS. In 1916 the federal government took back lands granted to the Oregon and California Railroad, which had failed to honor requirements of the grant. The 2.5 million revested acres contain extensive stands of timber, which were managed by the Department of the Interior. Since its creation in 1946, the Bureau of Land Management has had jurisdiction over these lands. Nearly a half million acres of them are intermingled with national forest lands, causing an occasional jurisdictional dispute. In 1954 Congress assigned jurisdiction of the contested area to the Forest Service. Dana, *Forest and Range Policy*, p. 285.

16. Robert Marshall, "Forest Devastation Must Stop," *Nation* 129 (Aug. 28, 1929): 218–19; see also Marshall, *The Social Management of American Forests* (New York: League for Industrial Democracy, 1930); Marshall, "Should the Journal of Forestry Stand for Forestry?" *Journal of Forestry* 32 (Nov. 1934): 904–8.

malice than in "promoting forestry in the woods." Defending Marshall and themselves against the indignation of Fritz and others, two of Marshall's associates responded. One lightheartedly jibed at Fritz: "Apparently you—and others—have conceived the idea that a group of Bolsheviks are on the rampage with war whoops and knives . . . to take your scalp." More somberly, another explained that their goal was to get "forestry out of a world of dreams and into a world of realities, to motivate it not by a profit incentive but by a spirit of true social service." Those who supported Fritz and the general policies of the Society of American Foresters protested Marshall's "socialistic and trouble-making attitude." H. H. Chapman, SAF president, was warned to guard against having the organization turned into a "*Red Camp*." The writer quickly assured Chapman he did not think that all Forest Service men were "New Deal or Reds," but those that were not probably preferred anonymity.[17]

Although rather extreme, the two factions represent the issue of public regulation at the time of NIRA demise. Those advocating regulation claimed that social welfare demanded public intervention. Those opposed rejected regulation on the philosophical grounds that public regulation was incompatible with free enterprise. Since the Forest Service was the heir apparent to any federal regulation scheme, the agency was deeply embroiled in the dispute, usually as an outspoken proponent.

Chief Silcox prompted much adverse comment when he proposed a six-point program of public regulation to a 1935 meeting of the Society of American Foresters. Those who felt that the lumber industry had made great strides saw Silcox's proposal as a socialistic threat. The chief attempted to assuage industrial feelings by expanding his views before the Western Forestry and Conservation Association later that same year. Fears of federal domination, he said, were coming from an industry unwilling or unable to provide an alternative to public action. Which was better, Silcox asked, public programs or no programs at all? He asked for one example anywhere in the United States where a company had recently purchased cutover land and had reforested it. He reminded his audience of large-scale abandonment of cutover lands. In one record week, 8 million acres of such land had been offered to the national forest system by owners no longer able to

17. Fritz to Marshall, Sept. 13, 1934, box 65, SAF Records; E. Munns to Fritz, Oct. 10, 1934, ibid.; L. F. Kneipp to Fritz, Nov. 30, 1934, box 1a, ibid.; A. E. Wackerman to Fritz, Nov. 12, 1934, box 47, ibid.; E. D. Fletcher to Chapman, Jan. 19, 1935, box 1a, ibid.

afford the property. Countering claims that public ownership was leftist or socialist, Silcox argued that if rehabilitating abandoned, tax-delinquent land and creating jobs for the destitute was socialism, so be it. He challenged the industry to offer something in return for its repeated requests for reduced taxation, publicly guaranteed loans, and increased fire protection. Ending on a brief note of compromise, Silcox assured his audience that he favored cooperation and supported sustained-yield programs to assure community stability.[18]

The National Lumber Manufacturers Association responded to Silcox's challenge by endorsing specific forest practices by its members. "It shall be the aim and the purpose of industry that all lands hereafter to be cut over shall be left in restocking condition," read the first industrial goal. Implementation included improved fire protection and application of selective logging, "where and to the extent practicable." In return, the industry asked for lower taxes, increased public assistance in fire protection through expansion of the Clarke–McNary program, higher public funding for forestry research, and the pooling of public and private timber supplies through sustained-yield units.[19]

Not satisfied with industrial efforts, Silcox worked to attain presidential support for "control within a firmly democratic pattern." He proposed adoption of the Swedish system of decentralized administration of regulations, responsive to local needs. He would seek legal advice to avoid the constitutional questions that would surely arise. Agreeing in principle, Roosevelt asked for a draft message to Congress that would propose a joint committee to construct the necessary legislation. Later, Silcox explained to the SAF that even though more controls would be imposed, democracy would be safeguarded. Past court decisions upheld congressional authority to establish broad guidelines for forest management.[20]

Silcox, with Clapp's enthusiastic support, solicited advice from forest officers in the field. In a letter stamped "highly confidential," Clapp asked regional foresters and experiment station directors for recommendations on regulation. Repeatedly emphasizing the con-

18. Unidentified newspaper clipping, Jan. 29, 1935, box 143, ibid.; F. A. Silcox, transcript of remarks to WFCA, Dec. 13–15, 1935, copy in drawer 353, RG 95–115, RFS.

19. John B. Woods to C. S. Chapman, March 15, 1937, box 65, NFPA Records; Forest Conservation Conference, *Report*, Apr. 7–9, 1937, NLMA, pp. 42–48.

20. Silcox to Ding [Jay N.] Darling, Feb. 2, 1937, box 39, RG 95–4, RFS; Roosevelt to Secretary of Agriculture, Jan. 27, 1938, copy in box 12, Clapp Papers, RG 200; Silcox, "A Federal Plan for Forest Regulation within the Democratic Pattern," speech to SAF, Dec. 15, 1938, copy in box 26, NFPA Records.

fidential nature of the plans, he said regulation "was one of the most important decisions that the Forest Service has ever made."[21]

The industry, of course, was generally aware of what was happening. There was concern that Silcox was willing to "throw overboard the mutual confidence which has been developed between the industry and the Forest Service by years of cooperation." John B. Woods, once again an industrial spokesman after leaving his position with the SAF, feared a "show-down" with the Forest Service and thought that Silcox had allowed the agency to deteriorate while busying himself with "saving the world."[22]

George F. Jewett, a western lumberman who frequently wrote to Silcox, also registered disappointment. He told the chief that the choice between capitalism and socialism was at hand. Threats of regulation would scare capital away from forestry. Silcox rejected the lumberman's logic and accused him of using the term socialism to avoid the real issues. Silcox explained that it was not a question of trees but of human beings; regulation had become necessary because the industry was unwilling or unable to "redeem its responsibilities." He could not see how Jewett could disagree with the "basic philosophy" that forests must be managed to benefit society as well as the industry. But Silcox failed to convert his correspondent, who worried to his colleagues about "the tremendous impetus given socialism by the present administration," of vital concern "to all who believe in private forestry."[23]

Silcox was able to discuss the issues more effectively with others. Greeley agreed with the chief that public welfare should prevail over private interest and suggested that "education, enlightened self-interest, self-regulation and State regulation" would go far in achieving harmony of effort. Silcox wanted more. Seemingly questioning the historical basis of the American lumber industry, he told Greeley that the "present situation is largely due to the fact that forest industries (generally) have been living in a vacuum created by subsidies issued originally in the form of free—or nearly free—forest lands." As drastic as it seemed, the only long-range solution was to "wipe out"

21. Clapp to Regional Foresters and Directors, Aug. 16, 1938, box 52, RG 95–4, RFS.

22. John B. Woods, memo to file, Feb. 23, 1938, box 22, NFPA Records; Woods to Charles W. Bryce, Nov. 23, 1938, box 25, ibid.

23. G. F. Jewett to Silcox, Jan. 15, 1938, box 22, ibid.; Silcox to Jewett, [Feb. 1938], ibid.; Jewett to Members of NLMA Conservation Committee, Aug. 14, 1939, box 1, WFCA Records.

speculative values that had been capitalized into the cost of doing business.[24]

During these discussions, which lasted over several years, Congress responded to the presidential suggestion and created the Joint Congressional Committee on Forestry on June 14, 1938, with Senator John H. Bankhead of Alabama as chairman. The committee was charged with investigating the present and prospective forest situation. Two weeks later, Silcox called representatives from private conservation groups and industrial associations to a meeting in his office to prepare a presentation for the newly created committee. He wanted the ad hoc committee to reach a consensus on major problems and to set priorities for congressional action.[25]

Not all believed in consensus. The following month, at the National Lumber Manufacturers Association meeting in Chicago, a spokesman for the Southern Pine Association recommended the strategy of ridiculing the Forest Service to the Joint Congressional Committee on Forestry. The ever-moderate Wilson Compton, on the other hand, proposed asking the committee what the industry could do to help. Officially cooperative, the association eventually agreed only "that the industry should not undertake to attack the socialistic trends of the present Administration, although we should have material available if such attack becomes necessary." To Emanuel Fritz it seemed, too, that nothing productive could be achieved by battling with the Forest Service, although it was useful to "smoke out" Silcox from time to time. The Berkeley professor thought that Forest Service staff did not "give a damn for anything about forestry, except the aggrandizement of their bureau."[26]

Forest Service testimony to the congressional committee emphasized the need to establish a nationwide forest economy that would "help solve problems of rural poverty and unemployment and create added security and stability for families, communities, industries, and labor in every major forest region of the country." Criticizing the industry for exaggerating the deleterious effects of taxes on forest management, the Forest Service insisted that public regulation was needed. Claiming that the "liquidation philosophy is so deeply en-

24. Greeley to Silcox, March 9, 1939, box 11, RG 95-4, RFS; Silcox to Greeley, March 22, 1939, ibid.

25. U.S. Congress, Senate, Concurrent Resolution No. 31, 75 Cong., June 14, 1938; memo on JCC, dated June 28–29, 1938, in box 19, SAF Records.

26. NLMA, "Minutes," July 19–20, 1938, box 19, SAF Records; Emanuel Fritz to E. T. Allen, Sept. 15, 1938, box 4, WFCA Records.

trenched that it is practically impossible to change it by purely voluntary action," the agency defended controls. Otherwise, the public would be "left holding the sack." The public needed assurance that its increased support for cooperative fire protection actually yielded improved logging practices. Federal foresters suggested that another advantage of regulation would be protection of lumbermen who logged properly from unfair competition by those engaged in the more profitable destructive practices. Rather pompously, the agency told the committee that promoting the public welfare was a test of government: "In our own country to question the possibility is to question democracy."[27]

Lumbermen were displeased. John Woods of the National Lumber Manufacturers Association complained that the Forest Service, "due to Clapp's scheming," dominated committee hearings and left too little time for other views. George Jewett castigated the Forest Service for "dumb or vicious leadership." Remarking that Silcox approved of socialism, he compared him to Hitler, an epithet used with increasing ease by industrialists fearful of New Deal erosion of their freedom. On another occasion Jewett warned that the Forest Service was attempting to "sugar coat" the regulation issue by presenting it as an aspect of cooperative forestry.[28]

Neither the Forest Service nor the lumber industry received much comfort from the pair of influential forestry organizations, the American Forestry Association and the Society of American Foresters. The American Forestry Association endorsed regulation, but regulation by states, which concurred with some industrial views. The SAF, representing professional foresters, offered much weaker guidance, merely endorsing "in principle public regulation to the extent necessary in each local situation." The foresters would not commit themselves to which level of public intervention was preferable or even comment on whether conditions justified such an action.[29]

27. U.S. Forest Service, "Summary of Recommendations Presented by the Forest Service on February 16, 1940 to the Joint Congressional Committee on Forestry with Respect to a Forest Program for the United States."

28. John Woods to Wilson Compton, Dec. 16, 1935, box 36, NFPA Records; G. F. Jewett to Ward Shepard, Aug. 28, 1940, ibid.; Jewett to NLMA Conservation Committee Members, Feb. 21, 1940, box 1, WFCA Records. Frank H. Lamb's "Hitler in the Woods," a radio broadcast speech on Oct. 16, 1940, is representative of the most vicious attacks. Lamb referred to Henry Wallace as a "Nazi copyist" and claimed that Silcox "had all the traits of Naziism." Transcript in box 2, RG 95–4, RFS.

29. SAF Council Statement, approved Dec. 21, 1940, box 63, SAF Records; AFA, "Resolution on Public Control," adopted Oct. 20, 1939, AFA Records; Report of the Committee on Forestry and Land Use to AFA Directors, Sept. 22, 1941, Box B-5, AFA Records.

The industry and the Forest Service continued to hold polarized views, but with increasing support for seeking middle ground. Greeley believed that the best industrial defense against Forest Service domination was a positive plan of its own. Part of this program, he suggested, should include public regulation by state agencies—a view he had advocated nearly twenty years earlier when as chief he supported the Snell Bill. Descriptions of Greeley's attitude were contained in confidential reports from Regional Forester Lyle Watts to Clapp, acting chief since Silcox's death in 1939. Portraying Greeley as "anything but antagonistic" toward the principle of public regulation, Watts related that the ex-chief intended to promote industrial support for state legislation in the Pacific Northwest. Greeley hoped that the Forest Service would not be involved, Watts said, because many industrialists would bolt at the sight of federal officers. Clapp was satisfied that Greeley was sincere and agreed not to interfere lest his efforts be jeopardized. Clapp still envisioned, however, a broad federal program into which state efforts could be merged.[30]

A minor scandal bubbled briefly to the surface when a copy of one of Clapp's letters fell into the hands of H. H. Chapman. Clapp had written to regional foresters and experiment station directors to bolster their morale in the conflict with industry over regulation. "We should not allow ourselves," Clapp said, "to be deterred by fear of criticism by timberland owners or reactionaries whose point of view is frequently selfish and contrary to public interest." Clapp went on to chastise the field men for too often deciding for themselves whether or not they were going to follow orders; decentralization had limits. Chapman wanted to publish Clapp's letter in *SAF Affairs*, issued by the Society of American Foresters; but since he lacked permission from Clapp, publication was denied. As word spread about Clapp's "dictatorial" letter, some nonfederal foresters refused to appear on panels with Forest Service men. Years later a still-rankled Emanuel Fritz remembered the letter as "socialism reduced to dictatorship."[31]

30. Greeley to S. V. Fullaway, Jr., Apr. 15, 1940, box 64, NFPA Records; Lyle Watts to Clapp, March 15, 1940, box 3, RG 95–4, RFS; Clapp to Watts, Apr. 18, 1940, box 3, RG 95–4, RFS.

31. Chapman to Korstian, Sept. 9, 1940, box 32, SAF Records; Clapp to Regional Foresters and Directors, March 30, 1940, ibid.; Henry Clepper to Korstian, Sept. 14, 1940, ibid.; Walter Damtoft to Clepper, Nov. 2, 1940, box 62, ibid.; Clepper to Damtoft, Nov. 4, 1940, box 62, ibid.; Damtoft to Clepper, Nov. 8, 1940, box 62, ibid.; Clepper to Damtoft, Nov. 12, 1940, box 62, ibid.; Emanuel Fritz, *Emanuel Fritz: Teacher, Editor, and Forestry Consultant*, Oral History Interview by Elwood R. Maunder and Amelia R. Fry, 1972, Regional Oral History Office, Bancroft Library, p. 76.

This episode, too, died down, but nerves were raw and small provocations spawned severe repercussions.

Correspondence between forest officers and industrial spokesmen continued to carry charges and countercharges of the industry's ignoring public welfare and the Forest Service's seeking dictatorship. Clapp was philosophical about being criticized, believing that those who took strong stands had to expect controversy. He became skeptical, however, that well-meaning industrialists could achieve workable regulation at the state level. Industrial proposals then under discussion, he feared, would "accomplish nothing in the woods" and mislead the general public into thinking that destructive logging was at an end. Yet, Clapp was hesitant to oppose officially the forest practice bills Greeley and others supported in state legislatures. For the time being he was content to wait and see while still advocating federal regulatory legislation.[32]

On February 16, 1940, the Forest Service presented lengthy recommendations to the Joint Congressional Committee on Forestry. It began by showing the need for increased cooperative programs, including sustained-yield units in areas where public and private ownership was mixed. The agency then quickly brought up the subject of regulation. From the Forest Service point of view, regulation was appropriate "as part and parcel of cooperation" because public interest had to be protected. Supporting the Forest Service recommendations, Undersecretary of Agriculture Paul Appleby listed for the committee three alternative regulation proposals in order of priority. Most wanted was direct federal regulation administered by the Forest Service. If this proposal was too severe, then a system should be devised giving states an opportunity to regulate logging, backed up by federal intervention should the states falter in their responsibility. Least desirable would be encouragement to end destructive logging through increased funding for cooperative programs, with authorization to withhold funding should industrial practices not improve.[33] Clearly, the Forest Service had lost whatever confidence it once had that anything short of force could end destructive logging practices.

When in its report the Joint Committee recognized the need for public regulation, it seemed to be a victory for the Forest Service. But

32. Clapp to Watts, Feb. 5, 1941, box 3, RG 95-4, RFS.

33. U.S. Forest Service, "Summary of Recommendations Presented by the Forest Service . . . to the Joint Congressional Committee in Forestry . . . ," Feb. 16, 1940, pp. 3-30; Paul H. Appleby to John H. Bankhead, March 18, 1941, box 3, RG 95-4, RFS.

when President Roosevelt asked for an evaluation of the committee's report, he learned from Secretary of Agriculture Claude R. Wickard that the agency was disappointed because regulation by states—not the federal government—was proposed. The president commiserated and branded the committee recommendations as "pitifully weak." He was "inclined to think that there must be a wholly new approach to the general problem—an approach wholly Federal in its scope."[34]

Senator Bankhead introduced his "Forestry Omnibus Bill" in the fall of 1941. The official Forest Service position was that the bill did not go far enough. Conversely, the forest industry believed that it would result in "federal dictation to the States and their ultimate domination, for the regulation of privately owned forest lands." Industrial spokesmen feared that cooperative principles adopted under Clarke–McNary would be disrupted and discarded. They asked for an amendment that would offer state regulation without federal guidelines.[35]

Before the bill was introduced, Clapp skirmished briefly with the industry over whether public regulation was constitutional, not to mention desirable. Clapp's trump card was an opinion by the solicitor of the Department of Agriculture that the regulatory concepts then under study were indeed constitutional, since Congress had the right to promote general welfare.[36] The issue came to naught, however; lacking support from the Forest Service and facing opposition from the lumber industry, the omnibus bill died in committee.

Clapp was disgusted. In a personal memo several years later he described the members of the Joint Congressional Committee as disinterested, lazy, ill-informed, or in their dotage. The committee was "conservative or violently reactionary." Congressman Walter M. Pierce of Oregon was the only liberal, but he was too old to be effective. Even the committee clerk, according to Clapp, changed from being "utterly uninformed about forestry" to eventually falling under industrial influence. In a more positive vein, Clapp saw that the

34. Claude R. Wickard to Roosevelt, May 12, 1941, box 12, Clapp Papers, RG 200; Roosevelt to Wickard, memo dated May 14, 1941, ibid.

35. G. Harris Collingwood, "The Forestry Omnibus Bill: An Interpretive Analysis," NLMA, Dec. 1941, pp. i-ix. Assistant Chief Christopher Granger takes credit for drafting Bankhead's bill. See Granger, "*Forest Management in the United States Forest Service, 1907–1952*," Oral History Interview by Amelia R. Fry, 1965, Regional Oral History Office, Bancroft Library, pp. 90–91.

36. G. Harris Collingwood to Clapp, March 7, 1941, box 61, NFPA Records; Clapp to Collingwood, March 27 1941, ibid.; Mastin G. White, USDA Solicitor, to Secretary of Agriculture, memo dated March 10, 1941, ibid.

committee at least had publicized forestry problems and prepared the way for enactment of sustained-yield legislation.[37] The issue of regulation itself was still far from dead in 1941. Even World War II failed to distract those who believed controls were necessary. The next major confrontation, however, would come nearly a decade later.

Transfer Threatens

If regulation made the Forest Service and the lumber industry seem to be enemies, a concurrent event made them appear to be allies. In this instance, conservationists joined with industrialists to praise the Forest Service. At issue was the revived proposal, which was objectionable to many, to transfer the agency to the Department of the Interior. The threat to the agricultural identity of the Forest Service, occurring as it did during the regulation battle, caused the fascinating spectacle of industrial spokesmen both damning and praising the agency, with scarcely a pause for breath. Whatever shortcomings the Forest Service might have, these men claimed, they would only be accentuated by transfer to Interior.

Fernow had proposed a federal forestry bureau in the Department of the Interior as early as 1886. It had seemed a realistic step to him because Interior already held jurisdiction over the vast public domain, while Fernow's own department, Agriculture, offered only token support for forestry. When Pinchot succeeded Fernow in 1898, he began working almost immediately to have the forest reserves transferred to Agriculture. Pinchot opted for Agriculture because he realized that outside it he would have much less freedom and, moreover, he wanted to avoid what he believed to be Interior's ineptness and corruption.

After Pinchot's seven-year campaign to transfer the reserves ended successfully in 1905, he was able to concentrate his energies on building the Forest Service and clarifying its responsibilities. During Henry Graves's term as chief, bills to transfer the national forests to state ownership were repeatedly introduced in Congress but with no effect other than distracting staff from other tasks. More serious was Greeley's battle with Secretary Fall in the early 1920s over returning the national forests to Interior. Again, the Forest Service emerged

37. Earle Clapp, "The Joint Congressional Committee on Forestry," memo dated Feb. 3, 1946, box 12, Clapp Papers, RG 200. Clapp's evaluation of Pierce was probably unfair, as the ex-congressman lived until 1954.

intact when the Teapot Dome scandal ended for the time being Interior's expansionistic ideas.

When Herbert Hoover was president, the question of governmental reorganization, including resource agencies, rose again, only to be set aside for the more pressing problem of a collapsing economy. The New Deal, however, brought Harold L. Ickes, a long-time political associate of Pinchot, to Washington, D.C., to be secretary of the interior. Ickes would soon inaugurate a transfer program that nearly succeeded where Fall had failed. If the Forest Service were transferred to the Department of the Interior or to a proposed Department of Natural Resources, Ickes, of course, would be at its head. E. T. Allen had once predicted that if a strong conservationist became secretary of the interior, it would be difficult to prevent transfer of the Forest Service. Ickes was a strong conservationist, and chances are that if increasing American involvement in World War II had not drastically curtailed Roosevelt's ability to concentrate on any but the most pressing domestic problems, Ickes might have achieved the transfer. Not to be discounted, however, was the vigorous opposition by the Forest Service and its supporters to retain the agency's agricultural location.

To coordinate the many New Deal emergency programs, Roosevelt in the spring of 1936 appointed Louis Brownlow, Charles E. Merriam, and Luther Gulick as a committee on reorganization of the executive branch of the federal government. By fall, the committee had completed its assignment and presented recommendations on reorganization, which included transferring the Forest Service. Roosevelt was delighted and enthusiastically set about implementation, only to snag on an obstinate Congress. Not to be discouraged, the president drew upon his vast political skills and began maneuvering his program over the various political hurdles.[38]

Professional foresters had not been unanimous in opposing the proposed Forest Service transfer. In fact, when a similar reorganization seemed probable during the waning days of President Hoover's administration, some foresters were actively supportive, believing that Greeley had a good chance to be named "assistant secretary of conser-

38. For this portion of the chapter the author is heavily indebted to Richard Polenberg, *Reorganizing Roosevelt's Government: The Controversy over Executive Reorganization, 1936–39* (Cambridge, Mass.: Harvard University Press, 1966); for another version see also Polenberg, "The Great Conservation Contest," *Forest History* 10 (Jan. 1967): 13–23. Clarence L. Forsling, then assistant chief for Research, remembers Silcox reporting that FDR had offered him the post of assistant or associate secretary of the interior, if only he would support the transfer. Forsling to Frank Harmon, Oct. 27, 1975, U.S. Forest Service History Office files.

vation." Others favored reorganization, too, but most foresters and lumbermen were opposed. Emanuel Fritz found it amusing that opposition was usually voiced in terms of forestry's innate kinship to agriculture, for, as he correctly pointed out, most of the reluctance really stemmed from long lack of respect for the Department of the Interior.[39]

The issue of reorganization was not always clear-cut, because several proposals were often under consideration at the same time. Early in the New Deal, Wallace and Ickes apparently agreed to having the Forest Service moved to Interior, confirming the belief of some that Wallace knew or cared little about forestry. By late 1934, however, Wallace had changed his mind. The Taylor grazing bill again raised the question of which agency could best administer the public range. Ickes meanwhile suspected that Agriculture wished to gain control of grazing, a suspicion borne out by publication of *The Western Range*.[40]

Indeed, the Forest Service pushed vigorously to capture grazing. With a disdain typical of forest officers toward Interior, one regional forester justified the merger because stockmen were "fed up on unredeemed promise, and cock-sure statements [by Interior] that in due time prove to be pretty much hollow shells and delays." Earle Clapp moved to get the range report published without the usual outside review, and he asked staff for memoranda to justify merging the Grazing Service with the Forest Service. He ordered a careful review of the situation; to him, the Forest Service had once wrongly supported the placement of the Grazing Service in Interior, and he did not want another decision for which they would be sorry.[41]

Staff reports to Clapp on grazing administration not surprisingly favored the Forest Service over the Grazing Service, emphasizing the superiority of agricultural personnel and policy to administer the range. An exception was Paul Roberts, with many years in Forest Service grazing administration but currently directing the shelterbelt program, who wondered whether grazing on the public range was not too complicated for the Forest Service to handle properly. Perhaps, thought Roberts, a whole new agency was needed, in Agriculture, to take care of all federal grazing. Ickes correctly guessed what was going

39. Tom Gill to Arthur Newton Pack, May 28 1932, box 4, Gill Papers; Emanuel Fritz to H. H. Chapman, Dec. 8, 1930, box 51, SAF Records.

40. Polenberg, *Reorganizing Roosevelt's Government*, p. 102; Franklin Reed to H. H. Chapman, May 5, 1934, box 47, SAF Records.

41. Evan Kelley to Clapp, May 29, 1936, box 36, RG 95–4, RFS; Clapp to Show, Apr. 20, 1936, ibid.; Clapp to Pooler, May 8, 1936, ibid.

on and protested that not only was his department under attack but that he ought to be able to review any official report before publication that dealt with matters of concern to Interior. Wallace defended the apparent breach of protocol by saying that their differences were too irreconcilable to yield to negotiation.[42]

The interdepartmental battle heated up. Pinchot and Ickes had been friends, but by now they engaged in name-calling and unfettered abuse. Ickes accused Pinchot, the unyielding advocate of retaining the Forest Service in Agriculture, of joining "a motley crew of lumber barons" to further his personal aspirations for a cabinet appointment or even to run for president. Ickes assigned one of his staff to dig into Pinchot's past to uncover derogatory information. The aide reported back in a confidential memo: "I truly do not see anything that could be used to discredit the Governor—at least not without embarrassing you at the same time." For his part, Pinchot compared Ickes to Hitler, with the pronouncement that "grabbing for power is not well regarded in the world of today." Greeley, too, got carried away, referring to the transfer proposal as threatening the Forest Service with dismemberment, "like Poland, and its fragments strewn among other agencies."[43]

With the Brownlow Committee recommendations, threat of transfer seemed near. Then a new element was added—the so-called Forest Service lobby. Charles G. Dunwoody, a lobbyist for the California Chamber of Commerce, had been frequently in touch with Earle Clapp about research projects of interest to California businessmen. Clapp told him about the Brownlow Report and the proposed transfer, suggesting that he contact Pinchot. Soon Dunwoody was in Washington, D.C., setting up an office and directing a highly efficient propaganda effort. Congressional proponents of transfer were deluged with protests; opponents received praise and support. A deft achieve-

42. Roberts memo to Clapp, May 11, 1936, box 36, RG 95-4, RFS; Rachford memo to Clapp, May 18, 1936, ibid.; E. W. Loveridge memo to Clapp, Apr. 27, 1936, ibid.; Polenberg, *Reorganizing Roosevelt's Government*, p. 103.

43. Ickes speech to New York Rod and Gun Editors Association, Feb. 23, 1937, box 226, Harold Ickes Papers, Library of Congress; speech to Pennsylvania Democratic State Committee, Oct. 27, 1937, ibid.; Leona B. Graham to Ickes, Oct. 20, 1936, ibid.; Pinchot to Greeley, Sept. 21, 1939, William B. Greeley Papers, University of Oregon; Greeley testimony to Congress, n.d., copy in box 55, NFPA Records. The tone of Ickes' reference to Pinchot's presidential aspiration was misleading, for he had once advocated that the Republican party should dump Herbert Hoover and run Pinchot in his stead. Ickes to Mark E. Reed, March 12, 1932, Mark E. Reed Papers, University of Washington library. The author is grateful to Robert E. Ficken for calling this letter to his attention.

ment was Dunwoody's enlistment of the General Federation of Women's Clubs and its two million members.[44]

Pinchot turned to Greeley to gain industrial support for the lobby. Greeley was surprisingly cool, perhaps still bruised from Pinchot's earlier attacks, but agreed to help because the Forest Service had to stay in Agriculture. Dunwoody's vigorous efforts achieved the necessary support from the lumber industry as well as others, and Congress did defeat transfer attempts in 1938. Dunwoody took much of the credit—indeed, more than he deserved—for the victory.[45] He had coordinated the overall lobbying efforts, but to some measure the thwarting of Forest Service transfer was due to efforts by the Society of American Foresters. With leaders and general members overlapping with the Forest Service roster, vigorous SAF opposition to the transfer plan is not surprising.

Leading the Society of American Foresters in support of the Forest Service, on the issue of transfer at least, was its outspoken president, H. H. Chapman of Yale University. His involvement predated the Brownlow Committee and Dunwoody's efforts. In July 1935 Chapman issued a news release as SAF president in opposition to the reorganization bill. He referred the release to the SAF Council for permission to have it published in the *Journal of Forestry*, the official organ of the society. One of the council members, Christopher M. Granger, assistant chief of the Forest Service and past president of the SAF, said that he would vote for publication if Chapman thought a direct attack on the Department of the Interior was the best strategy. Other members of the council, including E. L. Demmon, director of a Forest Service experiment station, also voted yes. The single negative vote was cast by Emanuel Fritz, of the University of California forestry school, who thought that Chapman's release needed polishing. Abstaining from the council vote was Ward Shepard, who had worked for the Forest Service and Clapp but now was with Interior's Bureau of Indian Affairs.[46]

44. Polenberg, *Reorganizing Roosevelt's Government*, pp. 114–17; Charles O. Dunwoody, Oral History Interview by Amelia R. Fry, Nov. 26, 1966, Regional Oral History Office, Bancroft Library, p. 8.

45. Pinchot to Greeley, Sept. 21, 1939, Greeley Personal File, 1931–36, Greeley Papers; Greeley to Pinchot, Sept. 20, 1939, ibid.; Polenberg, *Reorganizing Roosevelt's Government*, pp. 120–21.

46. Samuel N. Spring to Chapman, Aug. 13, 1935, box 68, SAF Records; Emanuel Fritz to Chapman, Aug. 17, 1935, ibid.; Hugo Winkenwerder to Chapman, Aug. 7, 1935, ibid.; Austin Hawes to Chapman, Aug. 9, 1935, ibid.; C. E. Korstian to Chapman, Aug. 6, 1935, ibid.; C. M. Granger to Chapman, Aug. 6, 1935, ibid.; Ward Shepard to Chapman, n.d., ibid.

Ironically, Chapman began his statement by rebuking Ickes for claiming that the Forest Service dominated the SAF. Then he severely criticized past and present management practices followed by Interior's various resource bureaus and vowed to maintain his mistrust until the department could show "evidence of a grasp of . . . fundamental principles." Ickes dismissed Chapman's views and returned brickbats in kind, describing them "as so narrow in substance, so inaccurate in reasoning, and so arbitrary in their pronouncement that they are discredited on utterance." Ickes told Chapman that he could not take him seriously.[47]

The squabble did not rise to higher plateaus with the passage of time. Character assassination was commonly aimed at Ickes, who absorbed the best (or worst) that Forest Service supporters had to offer and could volley back with gusto. By 1937, when the Brownlow Committee recommendations prompted serious efforts in Congress to transfer the Forest Service to Interior, the factions were well drawn and the adversaries had become familiar veterans.

Chapman thought the Brownlow-inspired plan for reorganization to be "a most dangerous threat to the future progress of forestry and conservation." The Yale professor was joined by other educators. Samuel T. Dana of the University of Michigan thought consolidation of all resource agencies was neither "logical nor practical." Hugo Winkenwerder of the University of Washington predicted that transfer would impair the efficiency of the Forest Service. Walter Mulford at Berkeley voiced even stronger fears than his northern colleagues, predicting that the transfer "would probably be disastrous."[48]

All of this correspondence and a great deal more passed through the various offices of the SAF as the group of professional foresters worked to defeat the proposed transfer. Their strategy was aimed mainly at portraying Harold Ickes as power-crazed and the Department of the Interior as frequently corrupt and always inept. The attack reached its peak in late 1938 when Chapman (no longer SAF president) touched off still another fusillade at his favorite target. Charging that the general public had been misled about the desirability of national parks over national forests since creation of the Park

47. Herman H. Chapman, "The Case against the Ickes Bill Restated," *Journal of Forestry* 33 (Oct. 1935): 834–42; Ickes to Chapman, Aug. 19, 1935, quoted in Polenberg, "Great Conservation Contest," p. 19.

48. Chapman memo, Feb. 25, 1937, box 53, SAF Records; Dana to Joint Committee on Reorganization, Apr. 15, 1937, ibid.; Winkenwerder to Senator J. C. O'Mahoney, Apr. 16, 1937, ibid.; Mulford to Senator Joesph T. Robinson, Apr. 19, 1937, ibid.

Service in 1916, he concluded that extension of Interior influence had "no merit." This time he went too far. Emanuel Fritz, who generally supported retaining the Forest Service in Agriculture, protested to the SAF Council. Describing Chapman's article as an "outburst of vituperation," Fritz claimed that loss of the Forest Service to Interior was justified if it would end such attacks. The Berkeley professor insisted that the SAF should contribute to better forestry practices instead of serving as a platform for views that were unfair, biased, venomous, and "such a perversion of facts."[49]

Fritz was not alone. Professional foresters in Interior employ had suffered in silence for years while Chapman and others had a long field day lampooning not only their agencies but their personal achievements as well. Eighty-six foresters employed by the Department of the Interior wrote to the SAF Council decrying Chapman's published statements. Reminding the council that officers of the SAF had made many accusations against their department, they stated that "aspersions cast on the Department are reflections as well on the professional integrity of the Department's forestry trained personnel." Obviously wounded deeply by years of ridicule, they asserted that they were graduates of the same forestry schools and had the same ideals and ambitions as their antagonists. The eighty-six foresters resented all implications made by Chapman and others and would be "impelled to resign" from SAF if they continued.[50]

SAF Executive Secretary Henry Clepper warned Clarence F. Korstian, Chapman's successor as SAF president, about the delicacy of the situation, pointing out that it would be a terrific blow to the organization if mass resignations occurred. Korstian recognized the excesses and called for moderation and greater tolerance. Threat of transfer continued, but passions cooled and participants stuck closer to real issues.[51]

Other organizations, such as the Izaak Walton League and the American Forestry Association, likewise supported the Forest Service. Although colorful exchanges flew between their spokesmen and Secretary Ickes, none could surpass the Chapman–Ickes affair. Lumber associations supported the Forest Service, too. Apparently everyone outside of government concerned with natural resources—

49. [H. H. Chapman], "A Challenge," *SAF Affairs* 4 (Nov. 1938): 193–96; Fritz to SAF Council, Nov. 18, 1938, box 58, SAF Records.

50. Letters to SAF Council, Nov. 14, 1938, box 32, SAF Records.

51. Henry Clepper to Korstian, Dec. 6, 1938, ibid.

the livestock industry being the major exception—had little affection for Interior. Roosevelt made several more attempts to attain congressional adoption of his reorganization plans, but to no avail. Distracted by the impending World War II, the president turned to more pressing matters. The Forest Service once again was safe in Agriculture.

FDR, however, was most displeased that his program had been openly opposed by the Forest Service. It was traditional for agencies to support the administration.

When Silcox died of a heart attack in 1939 after a period of ill health, Roosevelt named Earle Clapp acting chief. Rumors began flying as to Silcox's replacement. John B. Woods of the Oregon Forest Fire Association thought that Clapp would be a "calamity" as chief. He preferred Richard H. Rutledge, regional forester in Utah. Tom Gill of the American Tree Association thought Clapp was the logical choice but feared he would be passed over. The SAF voiced concern that FDR might appoint a nonforester, thus breaking the chain of professionals guiding the Forest Service which began with Fernow's administration. The foresters formally endorsed Clapp's appointment. Greeley preferred Rutledge. His second and third choices were Earl Tinker, a recently resigned assistant chief who was now secretary of an industrial association, and Fred Morrell, career Forest Service assistant chief in charge of the Civilian Conservation Corps program.[52] But FDR did nothing more, leaving Clapp as acting chief for three years. It is not clear why Roosevelt chose not to name a chief, but it is generally agreed that presidential pique had denied Clapp the full-fledged appointment.

In 1940 Clapp had written to the regional foresters about the still-impending transfer and used the phrase "threat of reorganization." Somehow the letter came to the attention of President Roosevelt, who then wrote to Secretary of Agriculture Claude Wickard demanding an apology from Clapp for his insubordinate language. The president further ordered Wickard to reprimand Clapp in writing. Both Wickard and Clapp complied with the presidential directive, although both insisted that Clapp merely had the best interests of the Forest Service at heart and had intended no disrespect.[53]

52. John B. Woods to Wilson Compton, Dec. 22, 1939, box 54, NFPA Records; Tom Gill to Arthur N. Pack, Apr. 30, 1940, box 4, Gill Papers; Korstian telegram to Roosevelt, Jan. 4, 1940, box 62, SAF Records; Greeley to Ovid Butler, Jan. 8, 1940, box C–4, AFA Records, FHS.

53. Roosevelt to Wickard, Feb. 6, 1941, copy in box 13, Clapp Papers, RG 200; Wickard to Roosevelt, n.d., ibid.; Roosevelt memo to Wickard, n.d., ibid.; Wickard memo to Clapp, March 6, 1941, ibid.; Clapp to the Secretary of Agriculture, March 4, 1941, ibid.

Obviously aggravated by having to apologize for what he believed to have been a justifiable action, Clapp dwelt on the presidential reprimand. He wrote a memorandum for his files in his own defense first in 1941, again in 1943, and again in 1946. Twenty-two years after the incident, he was still smarting. Roosevelt had broken promises made to congressional leaders, he remarked, and that action explained the defeat of reorganization. Rather than accepting the blame, Roosevelt had lashed out at the Forest Service and at Clapp for his loss. The whole affair, Clapp believed, had "left FDR with a very bad opinion of the Forest Service." The victory of staying in the Department of Agriculture had actually been a defeat "for the Forest Service and forest conservation." The transfer conflict had prevented Roosevelt from adopting a comprehensive legislative program for conservation.[54]

Clapp remained acting chief until FDR appointed Lyle Watts, regional forester in Portland, to be chief on January 8, 1943. A year later Associate Chief Clapp confided to a friend that although he had retained a nominal position of authority, he had been "completely shelved."[55] He retired on December 18, 1944, but remained active in forestry affairs. His personal papers contain manuscripts of many articles, a proposed book on international conditions, and his unpublished memoirs. Earle Clapp died July 1, 1970. He had been unsuccessful in acquiring authority over private logging practices, but the Division of Research, an equal partner in Forest Service administration, remains his lasting monument.

54. Earle Clapp, memo to file, March 15, 1941, Feb. 12, 1943, Feb. 13, 1946, ibid.; Clapp to Paul Roberts, Apr. 4, 1963, ibid.

55. Clapp memo, Jan. 21, 1944, box 3, ibid.

CHAPTER X

The War and After

World War II ended in 1945, the fortieth anniversary of Pinchot's successful efforts to have the forest reserves transferred to the Department of Agriculture. As the name of the agency was forty years old, too, the formative years between 1876 and 1905 seemed literally to be pre–Forest Service. More significant than the age of the forestry bureau was the age of its staff. The last men who had begun when Pinchot was chief now were retiring. Pinchot died the following year. The postwar years truly marked the end of an era in Forest Service history and began a new one—anticipated, sometimes impatiently, since Fernow's time. The war was the last hurrah for many forestry pioneers and brought a change of direction for American forestry.

Wartime Forestry

A sensitive reader would have been disheartened by Earle Clapp's use of a wartime phrase. In his 1940 annual report, the acting chief described Forest Service protection efforts as "blitzkrieging forest fires," prompting visions of the Nazi tide pouring across Europe. Although the United States would not officially enter until late the following year, American thoughts already were fastened on war.

To the Forest Service, war above all meant matériel. Military machines had a near-insatiable appetite for wood-pontoon bridges and railroad ties, gunstocks, ships and docks, barracks and cantonments, crates, mess halls, hospitals, post exchanges, and aircraft. Wood cellulose provided a main constitutent of explosives. Glycerol made from wood produced nitroglycerin; the addition of sawdust produced dynamite. Shrapnel in warheads came packed in rosin, and turpentine fueled flame throwers. Wood plastic formed panels and knobs for the rapidly expanding electronics field. Fighters, light bombers, and PT boats consumed mountains of especially graded Douglas-fir plywood. The single biggest use of wood was for packing crates to ship military supplies to the front. In all it required three trees to equip and maintain each American soldier.[1] During the "war to end all wars," when Greeley and many other American foresters served in France in 1917–18, the British had been forced to log their game preserves and great estates. America's brief participation had cost 6 billion board feet. With the list of military uses greatly extended, this new war would cause an even larger drain on American forests than the last.

After December 7, 1941, the United States was deeply involved in what had been Europe's war, as well as in the Pacific theater. Prewar estimates of military requirements proved low, as the nation dedicated 50 percent of its productive capacity to the war. The Forest Service calculated that for the year 1942, military and related civilian activities would consume 8 billion board feet of timber, in contrast to the 6 billion board feet that had been our whole contribution to World War I. The Army used 1 million board feet per day for truck bodies alone. The Forest Products Laboratory greatly expanded its defense-oriented projects, as did other research divisions. The War Production Board quickly classified wood as a critical material and froze softwood lumber for civilian use. An added complication was the fact that most softwood lumber came from the Far West, further burdening an already overtaxed transportation industry. And that was not all: almost two thousand Forest Service personnel joined the Armed Forces.[2] The agency, as with all other organizations, would have to accomplish more with fewer in the ranks to do it. Vacations and the forty-hour week were among the war's first casualties.

War intensified the need for priorities. Waste, duplication, and inefficiency—war's frequent hallmarks—seemed intolerable. Easy to say, but where are the philosopher-kings to distinguish between what is necessary and what is desirable? Secretary of the Interior Harold Ickes complained to President Roosevelt that the War Department had invaded the Colonial National Historical Park in Virginia to build

1. Forest Service, *Annual Report* (1940), pp. 1–5; Richard G. Lillard, *The Great Forest* (New York: Alfred A. Knopf, 1947), pp. 312–13.
2. Forest Service, *Annual Report* (1942), pp. 1–3.

a recreation facility. Worse, the Army proposed locating a bombing range for its Air Corps in the nesting grounds of the trumpeter swan. Blasting with rhetoric so familiar to friends of the Forest Service, Ickes compared the American military with Hitler's European conquests. FDR agreed that the Army was out of step. "The verdict is for the Trumpeter Swan and against the Army. The Army must find a different nesting place."[3]

Enemy action had made the importation of many strategic materials difficult, if not impossible. What had been low-grade ore was now in great demand. Forest Service engineering staff worked feverishly to complete or rehabilitate mine-access roads in national forests. One of the largest mining projects was under a cooperative agreement with the Anaconda Company in Montana. Unable to get its own equipment because of wartime shortages, the company contracted the Forest Service to build a vital road. Also bedeviled by shortages, the Forest Service engineers scoured their own agency for needed skills and machines. Despite an early winter, the ore trucks were soon rumbling down the grade. Such achievements, multiplied many times over, were the source of much pride to those laboring on the "home-front."[4]

A project of lesser significance than the mundane access roads, as it turned out, but capable of generating greater excitement at the time (and for some still in retrospect) began in California. Machines of war run on rubber tires, and the United States depended upon imports of raw rubber. Japanese domination of the Pacific cut rubber imports to zero. New sources had to be found. It had been known for some time that latex material could be extracted from the guayule plant, a shrub native to the Southwest. Little was known, however, about growing guayule under cultivated conditions or how to render it into rubber on a large-scale basis to meet military needs. Evan W. Kelley, regional forester in Missoula, with J. J. Byrne of the Forest Service engineering staff and Paul Roberts, recently having headed the shelterbelt project, tackled the problem of growing guayule in massive amounts. Rubber company chemists and other technologists began to perfect manufacturing techniques. The War Production Board assigned the project

3. Ickes to Roosevelt, Nov. 27, 1941, and Roosevelt to Henry L. Stimson, Nov. 28, 1941, in Edgar B. Nixon, ed., *Franklin D. Roosevelt and Conservation, 1911–1945*, 2 vols. (Hyde Park, N.Y.: General Services Administration, 1957), 2:540–41.

4. Jack Hamblet, "Mineral Access Roads—Region One," in *Engineering in the Forest Service: A Compilation of History and Memoirs, 1905–1969*, U.S. Forest Service, processed, n.d.; Hollis G. Stritch, "Defense Access Roads—Region One," ibid.

the highest nonmilitary priority, giving Kelley ready access to equipment and supplies.

Guayule grew well in a 100-mile belt running from central California through Texas. Salinas, California, was the site chosen for the first pilot project. Drawing heavily on the priorities assigned by the War Production Board, the Salinas project within three months had a 540-acre nursery to grow seed, a 1,500-man labor camp, a seed extractory, a pilot rubber extraction plant, 2 laboratories, and an equipment design and repair shop. Congress was impressed, upping authorization from an original 75,000 acres to be planted with guayule seed to 500,000 acres. New nurseries were sprinkled across the Southwest and up to 9,000 men were employed. By the spring of 1944, more than 200,000 acres of guayule were under cultivation.

As the war progressed, rubber supplies became even more critical. The two guayule extraction plants produced thirty tons each day. Guayule rubber was used for the inner plies of tires and mixed with natural rubber to make bullet-proof fuel tanks for aircraft. Three million pounds of guayule rubber were produced before the project was abandoned in 1945, when traditional sources became available once again. The mere 1,500 tons of rubber could not have altered the course of the war to a significant degree. If hostilities had continued, however, an ever-increasing production capability would have undoubtedly made guayule rubber live up to its much-heralded potential.

The two most sensitive problems solved along the way had been justifying the sacrifice of 200,000 acres of lettuce production to rubber and handling complaints from the California landowners that living conditions in the guayule labor camps were so luxurious they were causing dissatisfaction among the traditional migrant workers. A congressional investigating committee offered the owners little sympathy for their hardships caused by improving living standards. All in all, this was a challenging venture by the Forest Service into an area where it lacked prior experience.[5]

Priorities—lettuce or rubber—were difficult to assign. With "don't you know there is a war going on?" ringing in his ears, FDR in mid-1943 opposed opening the scenic Skagit River of Washington to logging. Chief Lyle Watts appealed the presidential decision on

5. J. J. Byrne, "Engineering on the Guayule Project," ibid. Another synthetic rubber project in twenty northern states focused on *koksaghyz*, a Russian dandelion.

grounds that the area under study contained 24 million board feet of prime Douglas fir needed for veneer, pontoon lumber, ship decking, aircraft construction, and other military-related processes. He assured Roosevelt that the area supported little recreational use and that it would be logged selectively where possible. Watts assumed full responsibility to protect "scenic and recreational values." The president reconsidered and authorized the sale and logging of the national forest timber.[6]

The War Production Board looked hungrily at Sitka spruce in Olympic National Park. The light elastic spruce was ideal for aircraft skeletons. Ickes forwarded an analysis of the situation by conservationist Irving Brant to Roosevelt. Brant was outraged, charging that the only justification for taking timber from the park was a shortage of loggers, which brought pressures to log the most accessible timber, even if it was in a national park. Why the shortage of loggers? They had either been drafted into the Armed Forces or lured away by high-paying defense jobs. Calling the misallocation of manpower "a monstrous failure of democracy," Brant asked for presidential reassignment of loggers. Rejecting in advance the cry that the lives of fighting men would be jeopardized by shortages, Brant charged that "if any soldier dies" because of protecting the scenic and scientific values of Olympic Park, the "fault will be upon those who drafted the lumberjacks out of the woods." Yet, the pressure was intense, and departmental opposition was withdrawn; Undersecretary Oscar Chapman told Ickes that the spruce must be yielded to the war. Only a last-minute shift by the War Production Board to another supply detoured the loggers from the park.[7]

Wartime Legislation. At the war's peak in 1944, Congress found time to enact four important forestry laws, three of which had been recommended ten years earlier during the drafting of Article X of the NIRA conservation code. Two of the new laws authorized funding increases under existing legislation: now three-quarters of a million dollars could be spent on forest survey, and the Clarke–McNary authorization was tripled to a maximum of 9 million dollars per year. The other two laws dealt with sustained yield and federal income tax.

6. Roosevelt to Marvin H. McIntyre, June 10, 24, 1943, and Lyle Watts to Roosevelt, June 18, 1943, in Nixon, ed., *Roosevelt and Conservation*, 2:576–77.

7. Ickes to Roosevelt, July 21, 1943, and Irving Brant to Ickes, July 7, 1943, ibid., pp. 578–79; Chapman to Ickes, July 5, 1943, box 67, Oscar Chapman Papers, Truman Library, Independence, Mo.; Elmo R. Richardson, *Dams, Parks and Politics: Resource Development and Preservation in the Truman–Eisenhower Era* (Lexington: University Press of Kentucky, 1973), p. 13.

Originally, American interpretation of sustained yield had been simplistic and biologically oriented: we would never run out of trees if we planted to replace each one logged. During the twenties and even more so during the thirties, largely on account of the tenacity of consulting forester David T. Mason, sustained yield took on another meaning. By shifting emphasis from the forest to the forest industry, sustained yield came to mean the continuous production of lumber rather than forests. Conservation meant conserving the industry and therefore the supply of raw material, in particular the raw material within economic transport range. To any individual sawmill owner, national timber inventories meant little; local supplies were his prime concern. Mason proposed to sustain these local supplies by tapping national forest timber.

In many areas of the West, railroad land grants and other nineteenth-century land disposal laws had left national forest acreages intermingled with or at least adjacent to private holdings. Why not pool the timber on both ownerships, Mason argued, and underwrite the needs of the local mill? Congress agreed and in 1944 enacted the Sustained-Yield Forest Management Act, which had been sponsored by Senator Charles McNary. The Forest Service and a lumber company could now enter into a long-term agreement promising a constant supply of public timber to feed the company's mill at or above appraised value, without competitive bids. This guaranteed supply was authorized only when community stability required federal timber not available through conventional sales.

The law recognized two types of sustained-yield units. One, which has received the most publicity, merged for management purposes national forest and private timber lands. The other type reserved national forest timber in a geographic region for use by operators in that area, thus eliminating outside competition and stabilizing the local economy. Supporters of the Sustained-Yield Act asked which was the greater; welfare payments to whole communities if the mill shut down, or revenues lost from noncompetitive timber sales. Totting up the costs of rehabilitating impoverished communities cinched the congressional decision in favor of sustained yield.

Promptly, two days after Roosevelt signed the sustained-yield bill into law, the Simpson Logging Company of Shelton, Washington, applied for a cooperative agreement. Simpson's mills provided jobs for over six thousand employees. With only ten years' supply of timber on its own lands, the company would soon have to lay off perhaps five thousand workers, an economic catastrophe for the community. The argument was persuasive, and on December 12, 1946, Simpson entered into a 100-year cooperative agreement with the Forest Service. Two hundred seventy thousand acres of forest lands were merged. Nearly 60 percent was Simpson's, but, significantly, 82 percent of the standing timber in the unit was federal, as Simpson's contribution was mostly cutover. That, after all, was the point of the cooperative agreement. Simpson was guaranteed access to Olympic National Forest timber while the young stand on its own lands matured.[8] There were many controls to safeguard public interest. Shelton did not become a ghost town.

It is not surprising that the Shelton unit is the only example of cooperative sustained yield. Companies holding little or no timber acreage feared that they would not qualify for an agreement, so although there were other applications, industrial pressures to stop this formal allocation of the national forests prevented the Forest Service from approving additional units. The Forest Service did, however, locate and approve five federal sustained-yield units that utilize only national forest land. These units, in New Mexico, Arizona, California, Oregon, and Washington, have reserved timber on 1.7 million national forest acres to stabilize five communities. Since 1953 communities seeking to stabilize their forest economy have been required to initiate the process, but the final approval rests with the secretary of agriculture.[9]

Not directly related to the Forest Service, but of enormous consequence to the lumber industry, was a 1944 amendment to federal income tax law which authorized lumbermen to report income from timber sales as capital gains instead of as conventional income. Since the maximum capital gains tax levy was 25 percent while regular income could at that time be taxed up to 92 percent, the amendment provided substantial savings. Before, capital gains could not be re-

8. Rodney C. Loehr, ed., *Forests for the Future: The Story of Sustained Yield as Told in the Diaries and Papers of David T. Mason, 1907–1950* (Saint Paul: Minnesota Historical Society, Forest History Foundation, 1952), pp. 232–48; Thomas C. Adams, "Cooperative and Federal Sustained-Yield Forest Units—A Problem in Resource Management," Ph.D. diss., University of Michigan, 1952, pp. 56–57; Samuel T. Dana, *Forest and Range Policy* (New York: McGraw-Hill, 1956), pp. 284–85. McNary died a month before FDR signed the bill into law. In 1937 Congress had enacted a similar law for the revested Oregon and California Railroad lands in western Oregon. These lands are administered by the Department of the Interior.

9. Adams, "Cooperative and Federal Sustained-Yield Units," *passim*; Dana, *Forest and Range Policy*, p. 285; for a study of one unit, see Paul W. Bedard and Paul N. Ylvisaker, "The Flagstaff Federal Sustained Yield Unit," Inter-University Case Program no. 37 (New York: Bobbs-Merrill, n.d.).

ported by an operator cutting his own timber, which obviously discouraged landowners from practicing forestry. The amendment thus encouraged that stable ownership vital to the practice of forestry. The U.S. Chamber of Commerce saw the tax measure as congressional recognition of federal responsibility in maintaining a supply of wood. FDR disagreed and vetoed the measure. Congress overrode the veto, 299 to 95 in the House and 72 to 14 in the Senate. David Mason jotted in his diary: "great news." His mood was dampened when he learned that Senator Charles McNary had died the same day.[10]

The war continued, as did shortages and the need for priorities. The War Production Board, through its Lumber Products Division, set up the Timber Production War Project (known to insiders as Teepee Weepee) to call on Forest Service expertise in solving production bottlenecks. Howard Hopkins, chief of the Division of Private Forestry, was named director of the project, which employed 633 full-time and part-time men—the largest organization of technical foresters in North America working in production problems.[11]

One of the first problems Hopkins tackled was the rapid closure of sawmills despite peak demand for lumber. These closures resulted largely from wartime labor shortages. Hopkins' project petitioned local draft boards for deferments to stabilize the labor base and won 75 percent of its appeals. Five thousand millworkers stayed on the job instead of being inducted. An active recruitment program brought thousands more to the mills, and after January 1, 1944, nine thousand prisoners of war worked on some phase of timber production. By easing labor and other shortages, the Timber Production War Project, according to its own reports, had added 1.75 billion board feet of lumber output by mid-1944.[12]

Postwar Planning. Demanding though a war might be, governments must be optimistic for the long run and plan ahead. Six days after the attack on Pearl Harbor, Richard E. McArdle, director of the Appalachian Forest Experiment Station in Asheville, North Carolina, was appointed Forest Service representative for the Appalachians to the Committee on Post-War Planning. The next week, Assistant

10. Dana, *Forest and Range Policy*, pp. 285–86; U.S. Chamber of Commerce, Resolution on Forestry, 1944, copy in box 16, NFPA (formerly NLMA) Records, FHS, Santa Cruz, Calif.; Loehr, ed., *Forests for the Future*, p. 248.

11. Howard Hopkins, "The Timber Production War Project," *Journal of Forestry* 42 (Nov. 1944):790; M. H. Collet, "An Opportunity and a Public Trust," in "The Timber Production War Project," ibid. (May 1944):338.

12. Hopkins, "Timber Production War Project," pp. 790–95.

Chief Raymond Marsh sent a memo to Washington office staff which, while acknowledging the difficulty of doing so, insisted that planning be maintained or intensified. The goal of postwar planning was to minimize unemployment of those returning to civilian life and the dislocations of converting to a peacetime economy.[13]

Within the Department of Agriculture, the Interbureau Committee on Post-War Programs coordinated Forest Service planning with other agencies. The committee had no administrative authority but served as a sounding board and idea exchange. From the reports, it is obvious that Forest Service views were accepted by the committee in toto. Timber, range, water, recreation, and wildlife programs were proposed and discussed. Most urgent was the need to stop destructive logging brought about by wartime demands. Within the Forest Service, however, there was dissatisfaction with planning at the departmental level. McArdle, at least, protested spending time making plans that were too vague or too general to be implemented. Realizing the impropriety of cutting free from the department, he recommended that the Forest Service begin a parallel effort and produce detailed plans of its own. Too much time, he claimed, was spent coordinating generally meaningless plans between agencies. He doubted the work to date was "worth very much."[14]

The main burden of administering postwar planning for the Forest Service fell to Assistant Chief Marsh. Proposals were to include programs extending twenty-five years or more beyond the war. His critiques of a seemingly endless stream of reports are heavy with fatigue. Night after night Marsh relentlessly pored over pages of recommendations by lower-echelon staff, tightening a statement here, requesting more information there about how to utilize surplus men and equipment to meet the needs of the Forest Service after the war. Reflecting upon his chores half a decade later, Marsh most vividly remembered the colossal amount of work involved. Apparently agreeing with McArdle's evaluation, Marsh decided that perhaps the

13. Roy I. Kimmel to Regional Chairmen, Dec. 13, 1941, box 1, Records of the Office of the Chief, 1908–47, Record Group 95 (hereafter cited as RG 95-4), RFS, National Archives; R. E. Marsh to Staff, Dec. 21, 1941, ibid.; Chief Agricultural Economics to Chairmen State Agricultural Planning Committees, Aug. 9, 1941, ibid.

14. Interbureau Committee on Post-War Progress, "Agriculture When the War Ends," Oct. 15, 1943, 59 pp.; "Topical Outline for State Plan for Land-Resource Conservation and Development," March 18, 1942, 28 pp.; "Work on Post-War Programs," May 27, 1944, 7 pp.; "Report of the Interbureau Committee on Post-War Programs at War's End," Sept. 27, 1945, 94 pp., all in boxes 2 and 6, RG 95-4, RFS; McArdle to Clarence L. Forsling, Feb. 12, 1944, box 28, ibid.

greatest benefit to the Forest Service had been increased understanding of the agency's relationship to other bureaus in the Department of Agriculture.[15]

There were postwar plans of the broadest scope. In the summer of 1944, Pinchot, now seventy-eight, lunched with Roosevelt at the White House. Not surprisingly, their talk included conservation needs following the end of hostilities. The president asked for a memorandum of Pinchot's ideas. Pinchot responded by proposing an international conference on conservation to bring together the heads of state of American allies. To Pinchot, conservation clearly was the key to a lasting peace. Harking back forty years, the ex-chief explained that conservation was use of resources for "the greatest good of the greatest number for the longest time. It is the basic material problem of mankind." He suggested that the conference agenda include a worldwide inventory of natural resources, and means of fair distribution.[16]

Roosevelt was pleased and sent Pinchot's proposal to the State Department, with the suggestion that the conference be held even before the United Nations convened. What did State think of this? The State Department favored waiting and making the conference part of the newly formed Food and Agriculture Organization of the UN. Roosevelt accepted the use of FAO but scolded Edward R. Stettinius, undersecretary of state, for signing a memo that revealed the department's "lack of understanding" about the need for a conservation conference. FDR wanted higher priorities. Stettinius, backpedaling, acknowledged the weakness of his earlier proposal and offered a new plan showing quicker action. After further dickering that included Pinchot, the conservation congress proposal suddenly died with Roosevelt in the spring of 1945. It would remain for various UN organizations, such as FAO, to deal with peacetime need for natural resources.[17]

15. Marsh to Fred Morrell, Aug. 30, 1944, box 1, RG 95-4, RFS. Marsh's critiques are found throughout the Records of the Office of the Chief. His memorandum on postwar planning of Jan. 12, 1951, is in box 63A 4027/22, RFS, Washington, D.C., FRC.

16. Pinchot to Roosevelt, Aug. 29, 1944, and [Gifford Pinchot], "Proposal for an International Conference on Conservation," Aug. 29, 1944, in Nixon, ed., *Roosevelt and Conservation*, 2:591–94. Pinchot, *Breaking New Ground* (1947; reprint ed., Seattle: University of Washington Press, 1972), pp. 368–372, summarizes his part in creating a postwar forestry program.

17. Roosevelt to Pinchot, Oct. 24, 1941; Roosevelt to Cordell Hull, Oct. 24, 1944; Edward R. Stettinius to Roosevelt, Nov. 10, 1944; Roosevelt to Stettinius, Nov. 22, 1944; Stettinius to Roosevelt, Dec. 16, 1944; Pinchot to Roosevelt, Apr. 10, 1945; William L. Clayton to Stettinius, Apr. 17, 1945, all in Nixon, ed., *Roosevelt and Conservation*, 2:599–644 *passim*. For a summary of FAO forestry activities, see Henry Clepper, "Forestry in FAO," *American Forests* 64 (Jan. 1958):10–13, 50–51.

Postwar Forestry

Within several months of FDR's death and the new presidency of Harry S. Truman came the surrenders of Germany and Japan. In August 1945 the world returned to an uneasy peace. Now that the war was over, Chief Lyle Watts told his staff, the "first-order of priority all-service job" was to complete the reappraisal and inform Americans of the "true" forestry situation and the steps needed for correction. Watts believed that the reappraisal was essential to all Forest Service policy formulation. At a staff meeting in March 1946, agreement was reached to give Watts the priority he sought. The American Forestry Association, a group of professional and lay conservationists, was also studying the forestry situation, and maximum agreement was imperative prior to releasing the report to the public.[18]

Relations between the Forest Service and AFA had been shaky for some time. During the battle with the lumber industry over regulation, Pinchot had fulminated against the conservation group, which had supported regulation by states. Acting Chief Earle Clapp controlled his hostility but confided to his files that the AFA had become a spokesman for the lumber industry.[19] When Watts became chief, he shared Clapp's and Pinchot's reservations.

He readily agreed that the AFA ought to study the forestry situation but strenuously denied the right of the association to pose as the "ultimate authority on the factual situation and as the impartial interpreter and advisor on policy and needed action." Watts pointedly brought up the need for "disinterested financing" to support such a study and wondered aloud whether the AFA was "in a favorable position to assume such a responsibility." AFA's Ovid Butler protested to Watts that he misunderstood. The AFA was not assuming the self-aggrandizing role Watts claimed but only wanted to help the cause of forestry. Samuel T. Dana, forestry dean at the University of Michigan, counseled Butler to be tolerant of Watts's suspicions, as the chief had been subjected to much criticism recently and was probably on edge.[20] Both groups continued their studies.

18. Watts to Assistant Chiefs, Division Chiefs, Regional Foresters and Directors, Jan. 15, 1946, March 28, 1946, box 1502, General Correspondence, 1905–52, Records of the Division of Timber Management, 1896–1952, Record Group 95, RFS.

19. Clapp, memo to files, June 9, 1942, box 11, Earle Clapp Papers, Record Group 200, National Archives (hereafter cited as RG 200).

20. Watts to Butler, Apr. 2, 1943, box B-5, AFA Records, FHS; Butler to Watts, Apr. 2, 1943, ibid.; Dana to Butler, Apr. 8, 1943, box 21, ibid.

"Timber Shortage or Timber Abundance?" was Watts's title for his 1946 annual report. The Forest Service reappraisal supported the agency's contention that the timber supply was dwindling at a rate of 18.6 billion board feet per year. Poor cutting practices on private land were a major contributor to the drain. Public control of logging was essential.[21]

The AFA completed its Forest Resource Appraisal the same year. John B. Woods, who had been and would again become an industrial spokesman, was director of the two-year, 225,000-dollar project. Agreeing with the Forest Service that depletion had occurred, the AFA laid substantial blame on public agencies. Forty percent of the national inventory was publicly owned but contributed only 9 percent of the yield. National forest production of forty-six board feet per acre, the AFA charged, could scarcely be rated as high productivity. Acknowledging that the Forest Service had traditionally held timber off the market at the request of and especially to benefit the lumber industry, the AFA thought the time was well past for opening public lands for logging. Countering Forest Service criticism of poor private practices, the AFA asserted that industrial efforts had "forged ahead." The report asked for more public aid to industry but regulation only at state—not federal—level.[22]

Although retired, Earle Clapp still championed the Forest Service cause, and now he was furious. To him, the AFA report was "anti-Federal, anti-Forest Service, anti-labor, by implication at least asocial, and pro-forest industry." Strong language, and more to come. In Clapp's view, the AFA wanted to remove the Forest Service "almost entirely from the picture, glorifying private efforts and condemning public programs." He called on the AFA's board of directors to overrule the Forest Resource Appraisal.[23]

Clapp was not alone. Chief Watts told the AFA-sponsored American Forest Congress in 1946 that he took issue with their Forest Resource Appraisal. First he criticized the so-called Higgins Lake Committee, appointed by the AFA to refine the appraisal, for not represent-

21. Forest Service, *Annual Report* (1946), pp. 1–10. See also "Forests and National Prosperity: A Reappraisal of the Forest Situation in the United States," USDA Misc. Pub. no. 688, 1948. Clapp had hoped that a "social approach" would have been included in the reappraisal. Apparently original plans had included such an approach. Clapp memo on Reappraisal, March 3, 1950, box 7, Clapp Papers, RG 200.

22. John B. Woods, "Report of the Forest Resource Appraisal," *American Forests* 52 (Sept. 1946):414–28.

23. Clapp memo, n.d., "The Forest Resource Appraisal of the American Forestry Association," box 2, Clapp Papers, RG 200.

ing the full spectrum of society keenly interested in the forest situation. Watts hoped that the forest congress would allow wider discussion. Defending past Forest Service policies as farsighted and bold, Watts described AFA proposals on timber management as "studded with allegations and innuendos." He protested "misleading statements" that exaggerated the productive capacity of national forests and might cause some of the public to think that there was plenty of timber. Watts spoke heatedly about the "Higgins Lake Proposal No. 12" for a thorough study of relationships between all agencies and categories of landowners. Impossible for a committee to achieve, according to Watts; besides, the suggestions seemed to include a moratorium on public land acquisition. Going slowly when problems were so pressing was out of the question. "When are we going to get down to brass tacks?" Watts demanded, "and put [land] into public ownership and start acceptable forest management . . . where experience keeps on shouting to us that public ownership is needed?" Watts then accused the writers of the AFA Forest Resource Appraisal of having "interlarded their pages so heavily with editorial matters on other subjects" that the "principal forest problem" of the United States—improving forest management on private lands—was obscured. Two days later, over NBC radio, the chief told an even wider audience that he was "shocked" by some of the Forest Resource Appraisal proposals.[24]

Watts must have gritted his teeth when Richard A. Colgan of the National Lumber Manufacturers Association spoke to the assembled forest congress. "The lumber industry subscribes to . . . the program recommended by the American Forestry Association." Colgan acknowledged a few differences, which he saw as divergences rather than disagreements. In his "humble opinion," private achievements in some situations surpassed those of public agencies—in fire protection, for instance. The pressing problem of private forestry was the 4 million owners of small parcels. A significant proportion of productive forest land was improperly managed, Colgan maintained, because forestry was not the main interest of the owner. Agreeing that some public ownership was desirable, Colgan asserted that the limit had been reached. In fact, reductions were in order.[25]

In a formal statement, attached as an appendix to the proceedings of

24. AFC, *Proceedings* (1946), pp. 29–32, 307. In sharp departure from objectivity, FRA Director John B. Woods was allowed to argue his position in footnotes to the proceedings whenever a speaker criticized his efforts.

25. Ibid., pp. 63–65.

the forest congress, the NLMA position on public regulation was expanded. In a mild tone, surprising considering the fevered emotions at the time, the association called for regulation by state agencies without specifically rejecting federal intervention.[26] This statement stands out as perhaps one of the few times that moderation could be used to describe attitudes toward regulation.

Regulation of Logging. Regulation is a tantalizing topic. Ample documentation makes it attractive to the historian always bedeviled by inadequate or inaccessible records. Overemphasis is tempting, as many readers will revel in sensational statements. Such attractions aside, regulation provides an extraordinarily valuable vantage point from which to view the history of the Forest Service and American forestry. No single issue better shows the relationship between public and private forestry sectors. No single issue better shows that American forestry is distinctly American. One must understand the degree to which personal convictions have tempered or even displaced technical consideration. When foresters argued about regulation, it was not in terms of silviculture or other technical topics. Rather, discussions revolved around personal political philosophies that favored or abhorred additional public participation and control of American forestry. It was as opinionated laymen, then, not as technical experts, that foresters debated property rights and whether the Forest Service was promoting socialism. Purely technical debates rarely caused such a drain of emotional energy. The regulation debate concerned not just the management of forests, but the defense of a way of life.

Regulation had been left in limbo at the beginning of the war. Recommendations of the Joint Committee on Forestry and the Omnibus Forestry Bill of 1941 failed to resolve the problem. In fact, it seemed to the Forest Service that war made the need to regulate the lumber industry even more important than before. Central to the controversy was whether industrial publicity programs were lulling the public into thinking everything was satisfactory. The Forest Service charged that the industry was twisting statistics and exaggerating its accomplishments. Further, lumbermen were using the war as a justification to log without adequate safeguards.[27]

The industrial publicity campaign that so concerned the Forest Service was launched on a new level of sophistication in 1941. Largely to blunt the Forest Service's advocacy of regulation during the 1930s,

26. Ibid., p. 320.
27. Forest Service, *Annual Report* (1942), pp. 5–7; ibid. (1943), pp. 1, 12–13.

the National Lumber Manufacturers Association established the American Forest Products Industries. The NLMA would continue to be the national lobby while AFPI would function strictly on the public relations front. In 1941 AFPI received substantial independence from the NLMA and went on its educational way. The Keep Green programs, tree farms, 4-H, and other groups received support or sponsorship from AFPI.

AFPI began with a three-year budget of three hundred thousand dollars. The public relations program was to "foster public good-will toward the industry" in order to counteract many years of unfavorable publicity. In particular, the term timber baron had to be eliminated. In examining 493 textbooks used in American secondary schools, AFPI staff found that 35 percent mentioned forest products in some way. Of that fraction, 80 percent of the texts said that forests had been exploited and that much forest land had been acquired through fraudulent means. The industry was of course irritated to learn that 92 percent of the books rated the Forest Service as doing a good job. AFPI had its work cut out. Judging by the howls of protest, effects of the educational campaign were quickly felt.[28]

The Society of American Foresters, representing the forestry profession, was called upon to judge AFPI. C. Edward Behre, a career Forest Service researcher, introduced a resolution describing the American forestry situation as "far from satisfactory" and charging that AFPI "is engaged in a campaign of publicity, cleverly distorting the facts as to forest depletion and forest productivity and otherwise misrepresenting the Nation's timber resource situation in a manner obviously calculated to cultivate complacency." Behre asked for SAF censure, declaring AFPI's publications, especially *Speak Up* and *John Q. Citizen*, to be "contrary to the public interest."[29]

The SAF Council rejected Behre's resolution by an eight to two vote. One dissenter said that the Forest Service was distorting the situation, too, so how could he vote against AFPI? SAF President Henry Schmitz submitted an alternative resolution, asking the industry to stand on its record, thus ending the propaganda. Having no success with the alternate, Schmitz tried a third version, which the

28. Wilson Compton to NLMA Board and Affiliates, Dec. 3, 1940, AFI (formerly AFPI) Records, FHS; Budget dated May 14–15, 1941, ibid.; Memorandum on Need for Public Relations, Dec. 1, 1940, ibid.; "What Is Said about the Forests?" (n.d.), ibid.

29. Henry Schmitz to SAF Council, Oct. 19, 1943, box 63, SAF Records, FHS.

council finally approved, seven to three: the SAF president would send a letter to AFPI expressing concern over false optimism.[30]

Wilson Compton, long-time industrial spokesman, responded. The industry realized that unwarranted optimism was as bad for itself as for the public. Obviously, Compton explained, the industry did not share the Forest Service's gloomy view. Emphasis on devastation and deterioration was wrong. The goal of AFPI was to make the public aware that timber was a crop and that the industry was working for "continuous forest production."[31]

The regulation issue had rubbed nerves raw, and every action was suspect. Early in 1945, Assistant Chief Christopher Granger wrote over Watts's signature (a common bureaucratic practice) a letter that went out to regional foresters and station directors. Granger probably later wished he had never been involved. Pointing out that the Forest Service was frequently called upon to recommend or endorse candidates for positions outside the agency, Granger asked that to avoid embarrassment, all such recommendations be forwarded through the chief's office. "It is important that all of us deliberate thoroughly before allowing ourselves to become party to unfortunate selection for positions of influence."[32]

As Theodore Roosevelt once said, "The opponents turned handsprings in their wrath." The Southern Group of State Foresters considered charging Watts with unethical conduct. If proven, the chief could be expelled from the Society of American Foresters. President Schmitz formally notified Watts of the accusations, adding that he agreed with the protestors that the letter had sounded "bureaucratic and coercive." Siding with Watts, one forester passed the affair off as an example of the special ability of the Forest Service to do things wrong. Another agreed that Watts's letter was within his right, noting that the "storm of protest" showed a lack of confidence in the federal agency. H. H. Chapman felt that the Forest Service had no concept of the impact Watts's letter had made on SAF members. Watts, however, kept calm, explaining to Schmitz that lower-echelon staff often lacked experience necessary to judge a potential forestry school dean or candidates for similar positions. It was for that reason recommenda-

30. Schmitz to SAF Council, Nov. 10, 22, 1943, Jan. 18, 1944, ibid.
31. Compton to Schmitz, March 7, 1944, ibid.
32. Watts to Regional Foresters and Station Directors, Apr. 27, 1945, copy in box 106, ibid.

tions needed review. The explanation satisfied Schmitz and he considered the matter closed.[33]

The industry mounted its campaign on another front. AFPI handled its publicity and education assignment well, but still needed was a high-level policy body. On May 11, 1943, in the AFPI offices, representatives from the National Lumber Manufacturers Association, American Paper and Pulp Association (APPA), and the American Pulpwood Association (APA) met to form the Forest Industries Council (FIC). Obviously, New Deal agencies had no patent on acronyms. The policy committee of the FIC faced a big task. Since the Forest Service had always assumed the lead in the past, "the forest industries cannot survive without a constructive counter policy that represents more than good intentions."[34]

Policy and education needed stronger support, it was decided. "Our industry has a great deal of political strength that has never been organized." It would be well, industrial strategists proposed, to list each member of the House and Senate Committees on Agriculture and Forestry and assign a lumberman to him. Each politician was to be contacted, establishing at least a friendly relationship. At war's end the industry's main concern was that the Forest Service would assert control by attaching strings to Clarke–McNary funds. Lumbermen should be leery of technical guidance that would "permit the long-arm of the federal government to reach in and exert influence, if not control." Expansion of Clarke–McNary would give the Forest Service a "toe-hold." The industry would completely reject any Forest Service effort to make fire protection contingent upon good logging practice. Outspoken Stuart Moir of the Western Forestry and Conservation Association suggested an industry-supported filibuster of the Forest Service budget.[35]

The industry was torn on the matter of the Forest Service budget. Philosophically it supported balanced federal budgets and local initiative. Federal funds for protection via Clarke–McNary was another

33. Southern Congress of State Foresters to Henry Clepper, Aug. 17, 1945, box 106, SAF Records; Schmitz to Watts, Sept. 5, 1945, ibid.; Harris Reynolds to H. H. Chapman, Aug. 31, 1945, ibid.; S. T. Dana to Chapman, Sept. 22, 1945, ibid.; Chapman to Schmitz, Sept. 6, 1945, ibid.; Watts to Schmitz, Sept. 21, 1945, ibid.; Schmitz to Watts, Oct. 3, 1945, ibid.

34. FIC, Minutes, Feb. 17, 1945, box 18, NFPA Records; E. A. Sterling to F. K. Weyerhaeuser, Jan. 29, 1944, ibid.

35. S. R. Black to George T. Gerlinger, June 18, 1945, box 20, ibid.; FIC, Minutes by G. H. Collingwood, Nov. 29, 1945, box 18, ibid.; A. G. T. Moore to Collingwood, Aug. 24, 1945, ibid.; Moir to David J. Guy, Feb. 9, 1946 (filibuster recommendations glossed on carbon copy to Collingwood), box 68, ibid.

situation, as was that portion of the Forest Service budget which supported sale of national forest timber. Money for land acquisition or regulation, on the other hand, met with stiff opposition from the private sector.[36]

The Clarke–McNary Act, long popular with the industry, began to appear ominous. Moir wanted to know if the Forest Service was deviating from its long-held policy of matching state funds. Assistant Chief McArdle answered Moir's query to Chief Watts. The formula was to have been 50 percent federal, 25 percent state, and 25 percent private. Instead, McArdle claimed, by 1947 the industrial contribution to its protection was only 11 percent. Since it was obvious that the private sector could not pay its full share, then the public agencies must fill the gap. In 1944 Congress had increased the Clarke–McNary authorization from 2.5 million to 9 million dollars. A substantial increase, McArdle agreed, but still not enough. In 1947 the Forest Service failed to match state efforts by 4 million dollars. He advocated expanding the total protection program to 40 million dollars.[37]

All well and good—protection was crucial. But to the philosophically conservative, federal spending had to stop increasing. The staggering war debts and inflation must be dealt with. S. V. Fullaway of the Western Pine Association thought that there were too many "good causes." More federal protection money eventually, perhaps, but "we should postpone full cooperation from the federal government until the federal treasury is in better position than now to do its share." The stereotyped image of civil servants rankled him, too. It seemed to Fullaway that most progress would come from foresters in private employ who could work without "bureaucratic restraint" and did not have to follow "swivel-chair developed wood practices and instructions."[38]

Some in the industry feared that large federal sums would entice state foresters to go along with Forest Service plans. "It would be well for us to know the individual reasons for acceptance" by the states. Industrial representatives met with the state foresters and elicited an agreement that minimum federal intervention was the best. All to the good, thought Clyde Martin of the Weyerhaeuser Timber Com-

36. Moir to George M. Fuller, March 25, 1948, box 68, ibid.; Colgan testimony to House Subcommittee on Agriculture of the Appropriations Committee, Feb. 27, 1948, copy ibid.

37. Moir to Watts, July 21, 1948, box 1, ibid.; McArdle to Moir, n.d., ibid.

38. S. V. Fullaway, Jr., to R. A. Colgan, March 5, 1949, ibid.

pany, because the states had "little conception of the hidden motives of the Forest Service." Martin and others thought that the Forest Service was capitalizing on the popularity of Clarke–McNary and attaching regulation features to it.[39]

It was hard to hold industrial ranks together. In the West, the Forest Service was dominant via the vast national forest system. In the South, even with decades of purchase, the national forests constituted only a small fraction of timberland ownership. State protection agencies were more influential in southern policy decisions, and when it seemed that the South was beginning to side with the Forest Service to realize an increase in Clarke–McNary funds, westerners reacted. Southern cooperation would break down the intent of Clarke–McNary and "set the pattern for federal control of private forestry," said Clyde Martin. As much as federal dollars were needed, westerners were distinctly separating fire protection from all other federal activities. Fearing that salaries paid by the Forest Service would inevitably lead to allegiance to the agency, Martin thought it would be a "tragedy" if the states "abandoned" cooperation with industry in return for federal subsidies. Potentially, Clarke–McNary would become the forestry pork barrel.[40]

Assistant Chief McArdle, in a memo to the files, believed that state foresters had to choose between the Forest Service and the NLMA. To him, the industry was trying to sell the states a bill of goods. He also saw the industry infiltrating the American Forestry Association and predicted that the lumbermen would turn on the conservationists as soon as their goals were achieved.[41] Both sides were very suspicious.

Lumbermen were unanimous in their opposition to federal regulation, but at least one felt uneasy. Southerner Frank Heywood confided to William Greeley, chairman of AFPI, that Greeley's organization perhaps was painting too rosy a picture about industrial forestry. The Southern Pine Association never had adopted an action program and it simply was untrue that the industry was focusing on small woodlots. He feared that AFPI's program was actually harmful: "You and I know that AFPI and SPCA were organized purely as

39. R. A. Colgan to Selected NLMA Members, Feb. 28, 1949, ibid.; Henry Bahr to Clyde Martin, Dec. 16, 1949, ibid.; Martin to Bahr, Dec. 19, 1949, ibid.

40. Clyde Martin to Henry Bahr, Nov. 25, 1949, ibid.; Martin to H. C. Berckes, Nov. 29, 1949, ibid.

41. McArdle memo, Dec. 27, 1949, box 59A1753/81, RFS, Washington, D.C., FRC.

defensive action against the Forest Service." Now AFPI was so successful that many companies were using it as "a screen to hide their lack of activity in the field." He called on Greeley to break the "logjam of inertia." Greeley tried to pacify his friend, calling him industry's "most distinguished pessimist." But Heywood was not soothed, telling Greeley that he was most troubled by the "steadfast indifference" toward logging quality by large companies working on small holdings. The industry, he insisted, should use the same standards on the other fellow's property as on their own. The only "clear solution" seemed to be federal regulation, and he was philosophically opposed to that.[42]

Regulation became an incredibly complex issue, much of it semantic, most of it philosophical, almost none of it technical. When Earle Clapp resumed his quest to regulate the lumber industry in 1942, after scoring a near miss in 1941, he tried to get FDR to issue an executive order under the Second War Powers Act. The USDA solicitor supported this tactic as valid because wood was a critical material, and the president could act to safeguard supply. In fact, Roosevelt could dictate stocking requirements after logging, prohibit cutting of young timber, regulate logging techniques, and prohibit clearcutting. Clapp was sorely tempted, as the president's emergency powers brought regulation as close as the flourish of a pen.

The opportunity might never appear again. Still, he had to consider the long run, and Clapp decided that legislation was preferable to a temporary, wartime executive order. He felt obliged, however, to go through the motions of supporting the order that Secretary of Agriculture Claude Wickard and Pinchot wanted so much. Wickard passed Clapp's recommendation for an executive order along to FDR, adding that permanent legislation was the best way to defeat the "unholy entente of lumbermen and State Foresters."[43]

Clapp's underlying reluctance to use an executive order was supported by actions of the War Production Board. Seeing that the lumber industry was "beset with numerous difficulties" in maintaining production, Donald M. Nelson of the board reported to Wickard

42. Frank Heywood to Greeley, Nov. 22, 1950, and Greeley to Heywood, Dec. 5, 1950, AFI Records. Greeley must have been satisfied with industrial progress, for seven years earlier he had shown similar concern for exaggerated claims. Greeley to Charles R. French, Jan. 15, 1943, ibid.

43. Solicitor to Clapp, Aug. 29, 1942, box 12, Clapp Papers, RG 200; Clapp to Secretary of Agriculture, July 25, 1942, ibid.; Wickard to Roosevelt, Aug. 27, 1942, ibid.

that he was opposed to any additional controls. After all, if the nation was sacrificing its youth overseas, it could tolerate sacrificing some forests to logging back home.[44]

Clapp's attitude toward the executive order tactic might call into question the depth of his dedication to regulation. Several personal memoranda in his papers offer a true measure of his commitment. At one point he had even tried to develop a "board of strategy" to serve as his "general staff for a major forest war." Clapp felt that he was never able to wage this war to full effect because most conservation organizations and individuals had lost their militancy. He saw the Forest Service as complacent and lethargic and becoming more and more politically conservative. He believed, too, that Lyle Watts had not been aggressive enough.[45]

When Lyle Watts became chief in 1943, he continued regulation efforts. He soon became convinced that the only workable method was to apply federal standards to state controls. He thought that tying regulation to Clarke–McNary, so feared by the industry, would be unsatisfactory. That method would allow enforcement only by withdrawing protection funds, causing much ill will. New federal legislation was needed, as was state legislation for on-the-ground implementation of regulation.[46]

Existing state regulatory laws were clearly inadequate. Before World War II, only five states had enacted legislation to curb destructive logging practices: Nevada, New Hampshire, Idaho, Louisiana, and New Mexico. By 1945, Oregon, Washington, California, Mississippi, Maryland, Massachusetts, and Vermont had passed forest practice acts. Other states followed in due course. The emphasis of state laws was on fire protection and logging in ways to allow for natural regeneration.[47] They did not require successful regeneration, just encouraged it with provisions such as for seed trees. In Washington, Oregon, and California, at least, the state forest practice acts were drafted by the industry which lobbied successfully for their passage. There is no question that to the lumber industry state regulation was preferable as the lesser of two evils.

Many in the Forest Service were not satisfied. The distrust of the states was exceeded only by distrust of the industry, and the feelings

44. Donald M. Nelson to Secretary of Agriculture, Sept. 5, 1942, ibid.
45. Clapp memos, Feb. 5, 6, 9, 1946, boxes 12, 13, 14, ibid.
46. Watts to Regional Foresters and Directors, Oct. 5, 1943, box 46, RG 95-4, RFS.
47. Dana, *Forest and Range Policy*, pp. 302–7.

were returned. As the Forest Service stepped up the regulation tempo following the war, the industry mounted its aggressive defensive, utilizing the AFPI, FIC, NLMA, and other outlets. The industrial effort would proceed on two fronts: promoting the image of good logging practices and a "vigorous and uncompromising" frontal attack "full blast against any and all attempts to promote federal regulation." The long-feared threat was introduced in 1949 by ex-Secretary of Agriculture Clinton P. Anderson, now a senator. "A Bill to Provide for Establishment of Forest Practices for the Conservation and Proper Use of Privately Owned Forest Lands, and for Other Purposes" was the title.[48] Industrial reaction was predictable and intense.

Following Watts's preference of five years before, the bill proposed federal guidelines for state administration. "One of the longest steps in recent years along the road toward complete Federal control," anguished a report for the California Chamber of Commerce, while Earl Tinker of the American Paper and Pulp Association saw "collectivism or socialism" inherent in the proposal. The industry prepared for an all-out battle, believing that after decades of threats, the Forest Service was ready to go all the way. By late 1949 the lumber industry had organized its forces against federal domination.[49]

AFPI led the way. Materials flowed to newspapers, professional and trade journals, and popular magazines. The frontal assault on Forest Service propaganda blanketed the nation. Emphasized were two "facts": the industry was doing a good job, and private enterprise was more efficient than the federal government. Less obvious were industrial efforts to line up support from state foresters and sympathetic Forest Service employees.[50]

Congress was the target. Every congressional district was to have an industrial contact. Each man was to be fully instructed in the industrial position and the best methods of approaching congressmen and senators with this information. Witnesses at hearings needed careful grooming to assure optimum presentation. Little was left to chance. A significant factor was Bureau of the Budget opposition, for Budget could not support the measure until existing cooperative programs proved inadequate, until proposed regulation was by states without

48. NLMA, "Action Program" confidential, n.d., copy in box 152, NFPA Records; S1820, 81 Cong. 1, May 10, 1949.

49. California Chamber of Commerce, May 27, 1949, box 147, SAF Records; R. A. Tinker, speech at University of Michigan, July 28, 1949, box 2, NFPA Records; R. A. Colgan to A. J. Glasson, June 27, 1949, box 2, NFPA Records.

50. "Suggested Program and Outline of Specific Activities in Opportunities to Proposals for Federal Forest Regulation," Nov. 1, 1949, copy in box 2, NFPA Records.

federal intervention, and until the political winds could be better tested. The political factor was the most crucial of the three. Opposition, including a two to one referendum vote by the SAF against federal regulation, was so intense that Congress would not act. Anderson's bill died in committee.[51]

Down but not out, the Forest Service kept up the pressure. In December 1951, Assistant Chief Edward C. Crafts spoke at Yale University on the need for federal logging rules. Crafts saw that the regulation issue was philosophical, not technical, evidenced by opponents using terms such as totalitarianism or socialism. The aim of regulation, Crafts asserted, was to make private land so productive that public ownership (socialism) would not be required. Crafts insisted that the public had rights, too—in this case, rights to protect its interests. Regulation was necessary because the nation needed forest products, because private land was in bad condition, and because nonregulatory measures had failed to improve the situation materially. Crafts warned against complacency brought about by the apparent closing of the gap between forest growth and cut. Since earlier inventories were of questionable accuracy, we really did not know how badly our standing inventories had been depleted. Crafts charged his opponents with lacking faith in the democratic process. He tried to sweeten the bitter pill by stating that the Forest Service would not tell a man "when, where, or how much he could cut." Instead, state agencies would regulate under federal guidelines. Only if a state was ineffective would the Forest Service step in. Federal agents would be withdrawn "in due course," after attaining acceptable standards.[52]

Crafts's assurances had little effect. Thinly veiled threats of public ownership unless conditions improved did little to calm lumbermen. Members of the NLMA received copies of the speech with a cover letter explaining: "It deserves your careful study." Another attachment caused even more concern—a copy of a letter from Watts to regional foresters and station directors, referring to Crafts's speech as "a clear and persuasive statement of Forest Service position."[53] As

51. Charles F. Brannon to Frederick J. Lawton, Apr. 24, 1951, copy in James C. Rettie Papers, University of Oregon; C. F. Evans to members of Senate Committee on Agriculture and Forestry, July 12, 1950, box 109, SAF Records. Senatorial committee records in the National Archives on S1820 contain only letters of protest from many groups.

52. Edward C. Crafts, "The Case for Federal Participation in Forest Regulation," speech at Yale Forestry Club and Yale Conservation Club, Dec. 5, 1951.

53. NLMA memo to Members, March 5, 1952, box 2, NFPA Records; Watts to Regional Foresters and Station Directors, Jan. 10, 1952, ibid.

with the introduction of Anderson's bill two years before, the reaction was predictable.

Walter S. Johnson of AFPI thought that Crafts's speech was an "exposé" of Forest Service attitudes on regulation. He had a "deep conviction that men who think as Crafts are leading us straight into a totalitarian government and ultimately socialism." The assistant chief was a "dangerous man in the position he occupies." Harry T. Kendall of the NLMA agreed, saying he had been "jarred" by Crafts's statements because he believed regulation had been a dead issue.[54]

Larger, presidential considerations abruptly settled the matter. On the campaign trail in Oregon, Dwight D. Eisenhower stated flatly that he did not want "federal domination of the people through federal domination of their natural resources." Eisenhower would select Governor Sherman Adams of New Hampshire to be presidential assistant. Adams, also a lumberman, had written in the November 1951 issue of *American Forests*: "The great issue, it seems to me, is whether we are smart enough to conserve and distribute the natural products of our land without succumbing either to dictatorship or national socialism. . . ." He hoped that self-discipline would preserve the "autonomy of control," but if regulation proved necessary in the public interest, it should be "exercised as close to the grass roots of government as possible." The election of November 1952, which swept Eisenhower to the presidency, marked the end of federal regulation as a Forest Service goal.[55]

Eisenhower was following the sense of his party's platform, which called for "restoration of the traditional Republican public land policy." Ownership by citizens and "elmination of arbitrary bureaucratic practices" were high on this 1952 list of traditions. Democrats were scarcely more imaginative, pledging wise use of natural resources. Eisenhower's impressive victory seemed to place Chief Richard E. McArdle, Watts's successor, in a difficult position. Crafts remembers that it was "touch and go" whether McArdle would be replaced by the incoming Republican administration. Rumors flew. Fred E. Hornaday of AFA and a long-time observer of the Washington forestry scene heard from a Forest Service associate that McArdle and all USDA bureau chiefs had been called to Secretary Ezra Taft Benson's

54. W. S. Johnson to H. T. Kendall, March 21, 1952, box 8, ibid.; Kendall to Johnson, Apr. 2, 1952, ibid.

55. Eisenhower speech reprinted in *New York Times*, Oct. 8, 1952, clipping in box 62, ibid.; Sherman Adams, "People and Conservation," *American Forests* 57 (Nov. 1951):25–26.

office. The secretary reportedly announced that he did not anticipate any immediate dismissals "provided the department heads did their jobs and followed the policies set up by the new adminstration and as outlined in the Republican party platform."[56]

Two principal participants recall the affair differently. Crafts believes that to avoid McArdle's removal for political reasons, the Forest Service turned away from regulation and concentrated on administering the national forests and research. McArdle disagrees, not with the facts but with Crafts's interpretation. When McArdle became chief on July 1, 1952, he decided to let the regulation issue "ride." He wanted to wait until after the upcoming election to see who would be president and what stance the new administration would take on regulation. To the new chief, there were other issues of equal or greater importance that needed attending to. McArdle believed that the extremely long battle over regulation had lost the Forest Service much important support for fire control and similar measures. He had decided to drop regulation in favor of other problems before Eisenhower's election, and by the time of inauguration he "wasn't much interested in what my new bosses thought of regulation." McArdle attests that he "never discussed regulation" with any member of the new administration.[57] One can only wonder about McArdle's decision had Adlai Stevenson become president instead.

McArdle's recollection that regulation was abandoned before Eisenhower's election is supported by Bureau of the Budget opposition to Anderson's 1949 bill. Budget Bureau positions reflect White House attitudes. Although the Forest Service is traditionally closer to Congress than to the Executive, Budget opposition might well have been effective. The record is incomplete and far from clear.

In support of Crafts's view, there is no doubt that industrial representatives met with Secretary of Agriculture Benson and obtained support for ending regulation pressures. Introduced by Presidential Assistant Sherman Adams, an FIC delegation met with Benson on January 27, 1953. Clyde Martin as chairman of FIC; Corydon Wagner, NLMA; James Madden, APPA; E. O. Ehrhart, APA; and

56. Edward C. Crafts with Susan R. Schrepfer, "Congressional Liaison in the Forest Service," *Forest History* 16 (Oct. 1972):13; Hornaday to Robert W. Sawyer, Jan. 24, 1953, box C-2, AFA Records. A news item on page 9 of the January 1953 *Country Gentlemen*, for over a century a farm news magazine, reported that McArdle was " 'iffy' to stay on his job."

57. Richard E. McArdle, *Dr. Richard E. McArdle: An Interview with the Former Chief, U.S. Forest Service, 1952–1962*, Oral History Interview by Elwood R. Maunder, 1975, FHS, pp. 58–59.

Greeley made up the group. Their strategy was to support McArdle as a chief they could work with (at the same time McArdle had agreed to accept FIC as spokesman for the forest products industry, unless there were protests from unaffiliated industrialists) and to advance cooperation — not regulation — as the best means to serve the public interest. At their meeting, Benson, who had already assured McArdle that he could stay on, agreed with the lumbermen that the chief would be retained, at least for the time being, and that cooperation would be given priority over regulation.[58]

Regulation thus dropped abruptly from sight and has not returned as part of Forest Service advocacy. The uneasy detente inspired industrial efforts to keep the peace. Kenneth R. Walker, a California lumberman, was nervous because he understood that McArdle had endorsed Crafts's Yale speech. Leo Bodine of the NLMA soothed his western associate by pointing out that there was little else McArdle could have done. Bodine was convinced that the chief did not support regulation. Greeley scolded AFPI leadership about complaints that they appeared "antagonistic and uncooperative" toward public forest agencies. He reminded them that since the Benson/McArdle agreements, the industry should "bury all of the old hatchets." Leonard Carpenter, another AFPI trustee, felt that the industry should make every effort to avoid conflicting with the Forest Service. He urged that great care be taken that industrial appraisals and predictions of the forestry situation would not conflict with those produced by the Forest Service.[59]

Considering the open antagonism of the forties, this was an incredible about face. Even so, bruises were still evident nearly two decades later. "The likes of Ed Crafts in this nation and in the government are increasing," said Wilson B. Sayers of AFPI. "Those of us who favor the cause of free enterprise and capitalism will have to be firm and in many cases unyielding in our opposition to their creed."[60]

58. W. S. Bromley telegram to Benson, n.d., Corydon Wagner Papers, University of Washington; Clyde Martin to E. D. Erhart, Jan. 7, 1953, ibid.; McArdle to E. W. Tinker, Jan. 8, 1953, ibid.; E. W. Tinker to FIC, Jan. 19, 1953, ibid.; W. S. Bromley to Karl Butler, Jan. 15, 1953, ibid.; Clyde Martin to Benson, Jan. 27, 1953, ibid.; confidential memo, Jan. 27, 1953, "Meeting with Members of FIC Committee and the Honorable Ezra Taft Benson, Secretary of Agriculture, January 27, 1953," ibid.; McArdle to Frank Harmon, May 29, 1975, U.S. Forest Service History Office files.

59. Walker to Bodine, Feb. 9, 1953, box 8, NFPA Records; Bodine to Walker, Feb. 17, 1953, ibid.; Greeley to C. A. Gillett, Apr. 16, 1954, Research Reading File, AFI Records; Leonard Carpenter to Gillett, Nov. 29, 1955, AFI Records.

60. Wilson B. Sayers to John Benneth, Aug. 2, 1967, Research Reading File, AFI Records.

Regulation of Range. Timber on private land was not the only difficult management problem for the Forest Service to resolve. Different from timber in several key aspects, range use constantly demanded attention. Creation of the Division of Grazing in 1935 (which became the Grazing Service in 1939) in the Interior Department to administer the Taylor Grazing Act, and its merger with the General Land Office a decade later to become the Bureau of Land Management, had worked to stem deteriorating conditions on range land in the public domain. National forests, too, contained vast acreages of range, and the Forest Service continued to develop grazing policies begun at the turn of the century.

Grazing was an important use in 100 of the 152 national forests, regulated by 10,000 allotments. Nearly 12 percent of all cattle and 25 percent of all sheep grazed on national forest ranges. Forest Service policies were crucial to the livestock industry. As with timber, World War II had demanded full use of American ranges. During the post-war readjustment period, stockmen pressed for significant changes in range policy.

Vital to the stockman was his grazing permit. Wanting more than the stability offered by a ten-year permit, he sought instead outright ownership of the range. In 1946 cattlemen and woolgrowers had jointly resolved to work for the "ultimate goal" of transferring public land to private ownership. Then with a play on Pinchot's words, the Joint Livestock Committee on Public Lands stated: "The greatest good for the greatest number comes through private initiative and private ownership." During what promised to be a long interim, stockmen, through Senator Patrick A. McCarran of Nevada, moved to have existing permit holders given preference. The Forest Service opposed McCarran's bid, as it seemed to grant vested rights for what was really a privilege. Not only must new applicants have equal opportunity, but since the range was overstocked, substantial herd reduction was in order. F. E. Mollin of the American National Live Stock Association accused the Forest Service of trying to pit the little rancher against the "terrible large operator."[61]

The efforts to reduce stock in national forests were of even greater

61. American National Live Stock Association to Members, Aug. 26, 1946, box 25, ANCA (formerly American National Live Stock Association) Records, University of Wyoming, Laramie; F. E. Mollin to Watts, Oct. 31, 1946, box F-11, AFA Records; Watts to Mollin, Nov. 14, 1946, box F-11, AFA Records; W. L. Dutton to Mollin, Apr. 8, 1946, box 558, General Correspondence, 1905–52, Records of the Division of Range Management, 1905–52 (hereafter cited as RG 95-63), RFS.

Manufacturing wooden propellers for World War I aircraft at the Forest Products Laboratory, 1918. Courtesy of the U.S. Forest Service, Forest Products Laboratory

The Forestry Regiment, 10th Engineers, World War I. Courtesy of the U.S. Forest Service

B-17 propeller packed in a crate specially designed by the Forest Products Laboratory, World War II. Courtesy of the U.S. Forest Service, Forest Products Laboratory

Wartime inspection of the Tongass National Forest, Alaska, 1944. Left to right: Chief Lyle F. Watts, Assistant Chief Christopher M. Granger, Assistant Regional Forester Charles G. Burdick. Courtesy of the U.S. Forest Service

concern to stockmen than the matter of ownership. In 1947 Wyoming Congressman Frank A. Barrett began an investigation of federal grazing policies, and stockmen asked for a three-year moratorium on reduction while the investigation was in progress. The Forest Service was confident that Barrett's study would support contentions that reductions were necessary to allow for range rehabilitation. There were two factors to keep in mind. The national forests were managed on a multiple-use basis, and grazing was only one consideration. Moreover, stock reduction was an ongoing range rehabilitation program, which would benefit the grower. The average reduction rate of 2 percent per year since 1908 was not all that drastic as far as the Forest Service was concerned. As proof of good faith, however, the Forest Service agreed to suspend reduction for one year.[62]

Legislatures in states where the livestock industry was important adopted strongly worded resolutions protesting Forest Service policy. In 1947 New Mexico claimed that not only had the Forest Service made stock reduction its policy, but forest officers were guilty of abusive and "near libelous" statements about stockmen. The same year the Colorado legislature charged that the Forest Service was dictatorial and had inadequate congressional authorization to administer grazing. The agency should limit itself to forests, and the Grazing Service should have responsibility for all federal ranges, as it administered more practical applications and offered fairer "adjudication of grievances."[63] Stockmen's preference for agencies in the Department of the Interior had been fairly consistent throughout the twentieth century.

By 1949 the range industry began advocating federal legislation to specify clearly Forest Service objectives. They wanted local grazing advisory boards to be legalized, as well as the ten-year permit, concise definition of fees, and a determination of how much of the fee should be earmarked for range improvement. Attorneys had warned them that the Forest Service had lengthy legal precedent in its favor; the stockgrowers ought to settle for new legislation because chances of winning in court were slim.[64]

62. National Woolgrowers Association memo, Jan. 28, 1948, box 43322, RFS, Denver FRC; Brannon speech to Colorado Stockgrowers, June 21, 1946, ibid.; Dutton to Alan Rogers, Oct. 18, 1948, box 25, ANCA Records; Clinton P. Anderson to Welch, Feb. 19, 1947, box 557, RG 95-63, RFS.

63. New Mexico, Senate Joint Memorial no. 2, 1947; Colorado, Senate Joint Memorial no. 2, 1947.

64. Mollin to National Forest Advisory Committee, July 30, 1949, box 27, ANCA Records; Frank A. Barrett to J. Byron Wilson, Aug. 22, 1949, ibid.

As it turned out, the legislation that showed up on the floor of Congress was not what the graziers had bargained for. Assistant Chief Raymond Marsh believed that Congressman Walter K. Granger of Utah would be the logical sponsor of the so-called Fiscal Omnibus Bill, "to facilitate and simplify the work of the Forest Service." Among other features, the bill simplified management procedures for private range intermingled with national forests, authorized ten-year grazing permits, and recognized local advisory boards. Edward J. Thye of Minnesota agreed to sponsor the bill in the Senate because he believed that the Forest Service surpassed all federal agencies in ability and public service. The Granger–Thye Act was signed into law on April 24, 1950.[65] The industry was far from satisfied, however, and continued to work for correction of what it considered to be deficiencies in public land management.

About the same time, the stockmen won a victory. For decades the Forest Service had taken Pinchot's "greatest good of the greatest number" advice to heart and had favored reducing the size of larger grazing permits in order to increase the size of the smaller permits or to allow new settlers an opportunity to run stock on national forest ranges. Stockmen opposed the policy because it was arbitrary and discriminated against the large owner. Many Forest Service field men opposed the distribution policy because it drained off energy better spent rehabilitating deteriorated range. Walt Dutton, Lyle Watts's chief of grazing, remembers clearly how one day Secretary of Agriculture Brannan summoned Watts to a meeting about range matters. As was customary, Dutton attended, too. Brannan was upset by a large number of complaints to his office about Forest Service grazing policy, including distribution. After a preliminary discussion, Brannan said to Watts: "Lyle, several times in the past I have asked you to take distribution out of your grazing policy; now I am telling you to do it." Watts moved quickly to comply with the secretarial order.[66]

Watts retired soon thereafter, and his successor, Richard E. McArdle, commented on the change of policy in his 1953 annual report. McArdle referred to distribution as a seldom-used government "right" that had caused resentment. He offered data which showed fewer than 1 percent of permits to have been affected by distribution. He acknowledged that the most onerous application of distribution had

65. Marsh memo to file, Feb. 2, 1949, box 62A481/3, RFS, Washington, D.C., FRC; Howard Hopkins memo to file, Sept. 22, 1949, ibid.

66. Dutton to Frank Harmon, Aug. 30, 1975, U.S. Forest Service History Office files.

been to reduce the size of the permit at the time grazing privileges were transferred to a new applicant. The so-called transfer adjustments, McArdle promised, would cease. Distribution would continue, however, but not be timed to transfer.[67] Having won few battles with the Forest Service, the range industry would have to settle for this victory of principle.

"It appears," William Voigt of the Izaak Walton League wrote in an open letter to conservationists in January 1953, "that the western livestock people are pushing to try to win a quick victory over the Forest Service in the first few months of this new administration." Stockmen had approached Secretary Benson of the new Eisenhower administration in a move to end the stock-reduction policy, just as the lumbermen had approached him on the regulation issue. McArdle claims that the February 1953 meeting between Benson and stockmen began with the secretary saying to Assistant Secretary Coke, "These are our friends. Give them what they want."[68]

Voigt further charged that Forest Service personnel who had resisted industrial pressures had been transferred, causing a serious loss of morale. Reacting to Voigt's plea for support, Fred M. Packard of the National Parks Association asked Benson not to be "stampeded" by industrial pressure. Packard acknowledged that there were ways in which the Forest Service could be improved but pleaded that "the entire conservation program of the country" was based upon the Forest Service. Anything that reduced the agency's effectiveness would be a serious setback for the conservation cause.[69]

Never ones to mince words, stockmen had earlier threatened to have the Forest Service budget slashed if the agency continued to insist upon stock reduction.[70] Since powerful congressional allies had severely cut Bureau of Land Management budgets in retaliation for recommending grazing fee increases, this was no idle threat. The Forest Service had stood firm and weathered the protest; then, shortly after Eisenhower's inauguration, Montana Congressman Wesley D'Ewart introduced his Uniform Federal Grazing Act. The bill applied to all federal range lands the general principles of continuity of grazing privileges, the transferability of permits, and uniform fees

67. Forest Service, *Annual Report* (1953), pp. 10–12.

68. Voigt to Conservationists, Jan. 6, 1953, McArdle to Harmon, May 26, 1975, U.S. Forest Service History Office files.

69. Fred M. Packard to Benson, Jan. 12, 1953, box 3, NRCA Records, FHS.

70. New Mexico Woolgrowers' Association, Resolution, Feb. 3, 4, 1948, box 37703, RFS, Denver FRC.

for all bureaus. To stockmen, the bill seemed similar in philosophy to cooperative agreements reached under the 1944 Sustained-Yield Forest Management Act.

In addition to protesting to Benson, conservationists sought aid from Sherman Adams, Eisenhower's assistant. Ralph C. Carpenter, director of the New Hampshire Fish and Game Commission, told Adams that D'Ewart's bill was unacceptable because it eventually would secure vested rights for grazing permittees. Carpenter believed that stockmen had long resented the Forest Service because they had been unable to dominate the agency the way they had the Bureau of Land Management. The executive committee of the American Forestry Association also protested: "[The bill] contains provisions that would weaken the authority of the Forest Service and the Bureau of Land Management to properly manage the grazing use of the public lands. . . ." The protests were effective. In response to Benson's reports of heavy mail concerning the grazing act and anticipated Forest Service opposition, Adams gave his assurances that the administration would do nothing to endanger the national forests.[71]

Lumbermen were appalled by the clumsy tactics of their stockmen associates. Leo Bodine of the NLMA and Irvin H. Luiten of Weyerhaeuser agreed that the stockmen were politically "inept." They also noted the strong and effective manner in which conservationists had rallied in support of the Forest Service against the livestock industry. It was obvious to Bodine that lumbermen would have to cultivate those groups and individuals that "conceive of the USFS as their knights in shining armor."[72] Fresh from their own meetings with Benson about Forest Service advocacy on regulation, it seemed wise for the NLMA to not associate publicly with the heavy-handed stockmen.

Even more appalled was the waspish and influential author, Bernard DeVoto. Since the war, DeVoto had been as outspoken in his affection for the Forest Service as he was in his disdain for lumbermen and especially stockmen, whom he accused of attempting a huge "landgrab." From his perch in the "Easy Chair" column of *Harper's* and in more than three dozen lengthier "conservation polemics" in

71. Ralph C. Carpenter to Adams, March 30, 1953, box 1088, White House General File 140-D, Eisenhower Library, Abilene, Kansas; Benson to Adams, Apr. 20, 1953, ibid.; Adams to Carpenter, Apr. 29, 1953, ibid.; *American Forests* 59 (May 1953):6. To place the issue in a broader context see Elmo R. Richardson, *Dams, Parks and Politics*, pp. 100–6.

72. Irvin H. Luiten interoffice [NLMA] memo to W. J. DeLong, Oct. 9, 1953, box 96, NFPA Records.

that and other journals, DeVoto zestfully pursued the range industry with razor sharp and at times outrageous prose. A master at turning a caustic phrase, DeVoto portrayed the stockmen's position as one of demanding that the federal government "get out and give us more money." He personally assumed credit for stopping the graziers' legislation program dead in its tracks. Undoubtedly this was an overstatement—one among many—but DeVoto's name crops up in the highest levels of industry and government, attesting to his influence.[73]

Undoubtedly, too, DeVoto's salvos were among the first as public opinion began its ever-accelerating shift from conservation and use toward preservation. What seemed like excessive demands from the industry were merely a continuation of decades of effort; what was new was the public awareness. During the 1950s the Forest Service, seemingly the defender of natural resources, began to lose some of its luster. The "white-hatted" ranger was becoming suspect. It was truly a new era in American forestry.

73. Bernard DeVoto, *The Easy Chair* (Boston: Houghton Mifflin, 1955), pp. 231–347 *passim*; Wallace Stegner, *The Uneasy Chair: A Biography of Bernard DeVoto* (Garden City, N.Y.: Doubleday, 1974), pp. 298–322. Richardson's *Dams, Parks and Politics* gives a fuller account of the issues and draws frequently on DeVoto correspondence. W. J. DeLong of Weyerhaeuser's public relations branch sent dossiers to Corydon Wagner on DeVoto, Senator Richard Neuberger of Oregon, and conservationist Robert W. Sawyer with the admonition to consider them as opposition. Oct. 8, 1953, box 3, Wagner Papers.

CHAPTER XI

Multiple Use: The Greatest Good of the Greatest Number

To THE PUBLIC—hiker or logger—the Forest Service is the ranger district. The chief might speak for the agency to Congress and basic policies may emanate from the Washington office, but the district ranger and his staff provide on-the-ground management of the national forests. Before World War II, the ranger had performed largely the custodial duties of protecting and inventorying his domain against that day when demands were certain to increase. His responsibilities have always been heavy because of Pinchot's 1905 edict that local needs would receive first consideration and subsequent decentralization of administration. More and more these responsibilities were articulated in terms of multiple use, a balanced allocation of timber, water, range, recreation, wildlife, and other resources found in the national forests. As the tempo of activities and demands on the district increased, the ranger felt the growing need for an official balancing mechanism with which he could proceed evenly amid accumulating pressures. The balancing mechanism would be the Multiple Use–Sustained Yield Act of 1960. Typically, it had been the Washington office staff that promoted the legislation, but the needs of the district ranger were well in mind. The decade of the 1950s had set the stage.

THE MCARDLE ADMINISTRATION

Richard E. McArdle became chief of the Forest Service on July 1, 1952, replacing Lyle Watts, who was retiring. McArdle first came to work for the Forest Service in 1924 and since then had been given a variety of assignments that assured him of broad familiarity with the agency's operations. Most of his time had been spent in Research and State and Private Forestry. McArdle also had been dean of forestry at the University of Idaho during 1934–35; his Ph.D. from the University of Michigan at a time when holders of the doctorate were not common gave him high standing in the academic world.

McArdle was well liked by Forest Service personnel. His gruff but affable manner fit well with an extremely informal style. To men in the field he was "Mac" or chief, never Dr. McArdle, a lofty title reserved for use by secretaries in the Washington office. He supplemented his famed ability to remember people with a list that eventually contained ten thousand names, which he would consult to refresh his memory before dropping in on a national forest headquarters or other field station.

McArdle further endeared himself to the rank and file by demonstrating his realization that a chief on a field inspection would be steered into seeing projects that a regional forester or forest supervisor wanted to show off; those that had been less than successful could be detoured around. Dodging shrewdly, McArdle, to the delight of everyone except a fidgeting forest supervisor, would on occasion jettison a careful itinerary, plop down into the chair next to a startled clerk, and ask, "I'm McArdle, who are you?"[1] Loyalty to such a man was immediate.

Even those outside the Forest Service found McArdle's natural charm difficult to guard against. A year after McArdle became chief, William D. Hagenstein, a western industrial forester who in the mid-1960s would rise to national forestry prominence by being elected president of the Society of American Foresters, complained to an associate about a recent encounter. He had told the chief that the Forest Service was not doing as good a job as possible—it was not achieving the allowable cut and it was busy at tasks it had not been specifically authorized to do. Also, the agency was spending too much time "poisoning the public mind against private ownership." But Hagenstein admitted that he had met his match: "You know him. He gave me little satisfaction except to smile and be his usual disarming self."[2]

1. Richard E. McArdle, *Dr. Richard E. McArdle: An Interview with the Former Chief, 1952–1962*, Oral History Interview by Elwood R. Maunder, 1975, FHS, Santa Cruz, Calif., pp. 23–26.

2. William Hagenstein to Corydon Wagner, Aug. 28, 1953, box 3, Corydon Wagner Papers, University of Washington.

All of McArdle's qualifications—his education, experience, and charm—would be put to the test during the 1950s. Ascendancy of Republican politics within the federal government coincided with a burgeoning of public interest in the national forests. Timber management, land acquisition, mining claims, and wilderness preservation were typical of the issues that he faced.

Timber Protection and Management. When McArdle became chief in 1952, he believed that the regulation issue had distracted many, particularly outsiders, from forestry matters of equal or greater importance. At least partly for that reason, he favored turning away from regulation in order to consolidate gains on other fronts. Since the lumber industry was an integral part of national forest policy, a good working relationship was vital. The goal of both the Forest Service and the lumbermen was to supply wood to meet society's demands. Only through the industry could this goal be realized, unless, of course, the forest industries were nationalized, and no serious proposals to do this were heard. Greeley's cooperative philosophy of the 1920s, severely questioned during the 1930s and 1940s, again dominated during the 1950s and beyond.

Wartime demands for national forest timber had been matched by peacetime demands at war's end in 1945. During the war, civilians had accumulated purchasing power and the desire for homes. Now they would have them. Rehabilitating wartorn Europe also consumed American lumber. From Pinchot's time the Forest Service had held the vast national forest system in reserve to be used when private supplies were depleted. Now was the time. Since the war, timber had been logged in the national forests at a faster and faster rate. Timber management had ceased to be largely custodial, and good logging practices were no longer something to be advocated for private lands alone.

The industry, for its part, was pursuing a self-improvement program. The short-lived National Industrial Recovery Act of the New Deal had caused industrial associations to formulate forest practice rules under Article X of the Lumber Code. Following the court-ordered demise of the National Recovery Administration, the lumber industry, through the National Lumber Manufacturers Association, agreed to implement the conservation code on a voluntary basis. In 1941 the tree farm system began on Weyerhaeuser Timber Company land in western Washington. Other forest-land owners who were able to meet the standards for reforestation, protection, and general continuity of management followed suit. By July 1, 1975, over thirty-one thousand certified tree farms encompassed over 77 million acres. Although an occasional cynic would sneer that a tree farm was where you would put up one sign and cut a thousand trees, the industry was proud of this voluntary conservation program for private land.[3]

Getting their conservation house in order, lumbermen realized, would have to include convincing members of the public and the Congress of their progress. The "educational" focus of the American Forest Products Industries emphasized the significance of the tree farm movement and introduced and supported keep-green programs in state after state. The literary talents of Stewart Holbrook and James Stevens, on AFPI retainers, added substantially to an improved public image of the industry. A series of radio programs beamed at the young listener—*The Adventures of Peter Pine*—explained how after a long life in the forest, Peter Pine was lonely and wanted to go to town where all the people were. A forester, after due deliberation, marked Peter, and a logger sent the tree on its way. Then Peter was converted into many useful products.

Joint or parallel ventures with the Forest Service in public education caused occasional uneasiness. When the National Advertising Council adopted Smokey the Bear for its public service fire-prevention symbol, at least one lumberman resented the implication that the Forest Service deserved all of the credit for protection. Other industrial feathers were ruffled when Johnny, the bellhop symbol of the Philip Morris tobacco company, lost out to Smokey. It had been planned that Johnny would be displayed showing care with cigarettes on tree farms. When the Forest Service moved in with the indomitable Smokey, the Johnny program was scrapped, wasting a lot of industrial public relations energy.[4] Despite these minor scrapes, however, the Forest Service and the lumber industry were able to work together in public education.

Fire prevention, of course, was much more than distributing Smokey the Bear bookmarks through public libraries across the nation. The Weeks Law of 1911 had marked the beginning of federal-state cooperative fire control programs. The 1924 Clarke–McNary

3. AFI, "Tree Farm Progress Report," n.d. In 1964 AFI retained Opinion Research Corporation to measure public familiarity with the tree farm system. Fifty-nine percent of those polled had heard of tree farms, but only 12 percent realized that it was an industrial program. Blue Paper no. 7, Editorial Reading File, AFI (formerly AFPI) Records, FHS.

4. Sidney Ferguson to Charles Gillett, Dec. 9, 1952, AFI Records; Gillett to Ferguson, Dec. 18, 1952, ibid.

Act had increased the amount of federal funds and extended protection to privately owned forest lands lying outside watersheds of navigable rivers. In 1944 Congress had tripled the fire prevention authorization under Clarke–McNary. Research efforts of the Forest Service and college faculties meanwhile had probed into the means of preventing and combatting fires.

Fire damaged or destroyed timber, wildlife, and, more importantly, watersheds, thereby exposing fragile soils to wind and water. Fire was a spectacular enemy of forests—easy to identify and easy to explain to the public. Of even greater danger to the forests, however, were invasions of disease, insects, and other pests. The older, more valuable stands of timber were the most susceptible to attack (although a young forest might have problems of its own, such as rodent damage and browsing deer). When it came to such disasters as fire, insects, and disease, property lines meant nothing; both public and private lands were subject to damage. In 1947 Congress unified scattered efforts to deal with individual forest enemies by enacting the Forest Pest Control Act. Regardless of ownership, the federal government would work both independently and cooperatively to prevent, control, or eradicate insects and disease.

Protection against all disasters was the sort of mutual venture that McArdle was reluctant to jeopardize by continuing the regulation campaign. Lumbermen had long advocated increased funding for protection, without regulatory strings, and the use of national forest timber to augment private supplies. Many in the Forest Service shared this view and supported McArdle's policy shift away from ever-disruptive regulation proposals.

Decades of custodial management had maintained the national forests relatively intact until logging would commence. Following the war there was broad agreement for the need to step up the tempo of logging on national forests. In 1952 Assistant Chief Christopher Granger wrote an enthusiastic article for the readers of *American Forests*, reporting that national forest receipts had topped expenditures. The goal of operating "in the black" had been pursued, then abandoned, by earlier chiefs; but according to Granger's arithmetic, revenues mainly from timber sales had increased rapidly enough to provide a modest budgetary surplus. Never mind, he insisted, that 25 percent of receipts were returned to local jurisdictions for roads and schools, which some would say should be treated as an expense not as revenue. Granger's zeal to see the national forests operate in a so-called businesslike way prompted him to overlook the fact that the 25 percent payment to counties was traditionally considered compensation for federal lands not appearing on local tax rolls. Few businessmen would list taxes in their profit column.

Granger told his readers that they could count on a continued increase in timber sale revenues. He added that the increases were by no means due only to increased demands for national forest timber by the industry; much of the increase was "due to the initiative of Forest Service men going out and getting business."[5]

The forest industries, of course, shared Granger's enthusiasm for "upping the cut" and took advantage of every opportunity to promote sales. Getting at the timber, as it turned out, was a serious problem. Postwar lumbermen and foresters looked at millions upon millions of acres of virgin forests and realized that no suitable access existed. Over the years, Congress had been niggardly in appropriating funds to the Forest Service or the Bureau of Public Roads to provide an extensive transportation system in the national forests. What funds there were had to be assigned to fire protection needs or salvaging diseased or insect-infested trees. If logging was to increase substantially, timber-oriented roads were too few and far between.

Federal funding for forest road systems began during the late teens and early twenties, the latter years of Graves's tenure as chief and the early portion of Greeley's. Following World War II, Greeley, then in his job with the West Coast Lumberman's Association, thought that much of the Douglas fir timber in the Pacific Northwest was deteriorating because it was inaccessible. Perhaps building a straw man, Greeley claimed that the industry was operating at only 60 percent of capacity because of this limited access. The Forest Service agreed that more roads were needed; in fact, the agency for decades had continuously pressed for more roads to increase management capability. The outbreak of war in Korea in the summer of 1950 intensified demands for lumber and for roads.[6]

5. Christopher M. Granger, "The National Forests Are in the Black," *American Forests* 58 (July 1952):6–9. The Forest Service continues to pursue the goal of operating "in the black." In 1975 Chief John R. McGuire reported that receipts had exceeded expenditures by over 72 million dollars. See "To Our Shareholders," ibid. 81 (Sept. 1975):16.

6. Jenks Cameron, *The Development of Governmental Forest Control in the United States* (Baltimore, Md.: Johns Hopkins Press, 1928), pp. 322–31; Greeley to Mathias Niewenhous, May 20, 1946, William B. Greeley Papers, University of Oregon; [Greeley] "Memorandum on the Need for the Rapid Construction of Access Logging Roads in the National Forests of the Northwest," March 2, 1946, Greeley Papers; West Coast Lumbermen's Association, "Forest Conservation Committee of PNW Statement on Access Roads," July 18, 1950, Greeley Papers.

Timber access received strong support. In 1953 the American Forestry Association revised proposals included in its 1946 Forest Resource Appraisal. These earlier proposals had recommended a substantial increase in federal timber sales. Now the AFA urged the Forest Service and other public forestry agencies to "provide for construction of access roads to permit utilization of timber, not now accessible, either to meet emergencies or to carry out management plans." Three years later, in 1956, Chief McArdle, during a routine report, explained to a senatorial oversight subcommittee the long-held view that proper forest management depended upon transportation. The extent of the forest road system influenced not only the amount of timber cut but also the size and duration of sales, spatial distribution of logging, and all aspects of protection. National forest logging roads had been traditionally amortized out of stumpage, which the chief commended as still a good system. Nonetheless, to finance entire road systems in totally undeveloped regions would require sales of such massive proportions that few if any individual companies could handle them. Without appropriated public funds to build the transportation system, many commercial areas of the national forests would remain undeveloped.[7]

McArdle recounted to the subcommittee how in 1952 inadequate access had caused the national forest cut to remain nearly 70 percent below its potential. The Forest Service at that time had asked for 112 million dollars to construct primary access systems. He reminded the senators that only 41.4 million dollars, far less than half that amount, had been appropriated between 1952 and 1956. Trying to clear up any doubts that these funds might constitute a subsidy to the private sector, the chief explained that when the operator built a road, the cost was deducted from the gross value paid for the timber. Therefore, the public indirectly paid for operator-constructed roads. Timber sold in areas opened by access funds could be sold at accordingly higher prices, and the final return to the treasury would be the same. The important difference was that by using appropriated funds, forest management plans would be less susceptible to short-run market fluctuations and logging operations of all sizes could be accommodated. Access funds were vital to good management.[8]

7. "A Proposed Program for American Forestry," *American Forests* 59 (Aug. 1953):30; Richard E. McArdle, "National Forest Timber Sale Problems," Statement to Subcommittee on Legislative Oversight Function of the Senate Committee on Interior and Insular Affairs, Feb. 21, 1956, copy in box 88, NFPA (formerly NLMA) Records, FHS.

8. McArdle, "National Forest Timber Sale Problems."

McArdle had other points to make about Forest Service timber policies. The terms "sustained yield" and "allowable cut," he noticed, were frequently used as synonyms. The time seemed appropriate to make a clear distinction. Allowable cut was the upper limit of logging permitted during short periods; sustained yield set quotas over longer time spans. The difference was significant. For example, eastern national forests acquired through Weeks Law purchases were frequently in a cutover condition. In such a situation, the allowable cut would be low until the forest was re-established and reached maturity. Then allowable cut would be increased to a much higher figure approximating sustained yield, because the site would be managed at full biological potential. Having made his point, McArdle then explained to the perhaps confused senators the difference between sustained yield and sustained-yield units as provided for by the 1944 law: the first was biological and the other economic and social.[9]

Although it was implicit in his overall statement to the subcommittee, McArdle did not delve into an important effect of establishing allowable cuts and sustained-yield targets. For generations the general public had been fed a steady diet of timber famine fare. Time and time again, the cry had been heard that the United States was going to run out of timber. The simplistic notion of sustained yield—plant a tree for each one logged—would do away with fears of future scarcity. Since relatively little logging had occurred on national forests, however (less than 2 percent of the prewar cut), postwar implementation of allowable cut/sustained yield meant an increase in logging, not a decrease, because the national forests were capable of producing much more than they had been required to. The distinction between acceptable logging practices applied to public land and "devastation" on private ownership was too subtle for the nonspecialist. The feeling of relief on the part of conservationists that sustained yield would save the national forests from the same fate private timberlands had suffered now turned to dismay as logging tempos increased, more than doubling during the first half of the 1950s. Decades of public-relations rhetoric was catching up with the Forest Service; more and more it seemed that the agency was adopting the very methods it had held up for so many years as examples of bad land management.

Timber Resources Review. Another aspect of forest management, which McArdle also discussed in his statement to the senators, was a

9. Ibid.

national inventory of timber resources. To him, this issue was every bit as vital as improved access. Periodically the Forest Service had estimated the amount of standing timber and compared it to the rate of cut. Predictions of future rates of growth and consumption were generally a part of the inventories as well. The postwar reappraisal had been useful, but to McArdle a full-fledged examination and evaluation of the timber situation now seemed in order.

In 1952 the Forest Service announced plans for a major inventory—the Timber Resources Review. Industrial participation was sought, even though the regulation issue had yet to be resolved. Assistant Chief Edward Crafts, thought of by many as the major regulation spokesman within the Forest Service, was assigned general responsibility for the inventory. Members of the Forest Industries Council pondered three tactics toward the proposed Timber Resources Review: (1) oppose the project because it would distract the Forest Service from more important work, (2) cooperate in order to channel the end product along lines favorable to the industrial position, and (3) ignore the invitation altogether in order to be in a "better position to attack the results." The industrial representatives in FIC decided on the second option.[10]

The decision to cooperate failed to end industrial ill will fostered by the regulation controversy. Alf Z. Nelson of the National Lumber Manufacturers Association reviewed a draft of the Forest Service proposal, jotting comments in the margin: "Dangerous . . . this section will deal with the place of timber resources in the national economy." When the Forest Service proposed to establish social, political, and economic forms of reference, Nelson thought it "pure and simple hogwash." Volume estimates of unutilized wood following logging and the manufacturing process Nelson also branded as "dangerous."[11]

Greeley, too, was anxious. Crafts's Yale speech on regulation in 1951 had contained disparaging references to earlier inventories as having been too inaccurate to be used to substantiate any claim that cut versus growth ratios were improving. According to Greeley, "the Left-Wing group in the Forest Service has taken the position that AFPI was in error in comparing the 1945 Survey figures with earlier surveys and thereby drawing simple mathematical deductions that rates of growth in all Categories have substantially increased." Greeley feared that Forest Service strategy would "by one process or another" show a

10. FIC, Agenda and Minutes, Feb. 18, 1952, box 19, NFPA Records.
11. Nelson's glossed copy of the TRR proposal in box 24, ibid.

deteriorating forest situation. He would personally urge McArdle to insist that his staff utilize all the facts. It might also be beneficial, Greeley thought, for the industry to bring in its own consultant to examine Forest Service data.[12]

Compilation and interpretation began to take form, drawing heavily on data provided by Forest Survey, a branch of Forest Economics Research created to carry out the inventory provisions of the 1928 McSweeney–McNary Act. Drafts were sent out for review. Suspicions continued to mount, not only of Forest Service motives but that certain industrial groups might be "playing ball" with the Forest Service in order to win favors over other sectors of the industry. In the spring of 1954, a McArdle-led contingent of Forest Service staff met with western industrial representatives in San Francisco. In the lumbermen's view, McArdle's behavior validated their worst fears. The chief announced that he had a tight schedule and could allot only two hours for the conference. He also said that the Forest Service would need most of that time to make its presentation.[13]

Vigorous protests won more time for industrial views, but the whole episode left a bad taste in industrial mouths. Among themselves, lumbermen saw McArdle trying to split the western lumber industry from its eastern counterpart. To them the chief was portraying the westerners as obstructionists by comparing them to the more cooperative easterners. Even more unsettling was the thought that McArdle's perceptions might be right. Since he had made confidential overtures for western support at the expense of the East, might he not be playing the opposite game as well? "The West is really aroused over this situation," reported Weyerhaeuser's Clyde Martin, "and needs some reassurance that the APPA [American Paper and Pulp Association] is not encouraging some of its members to make a favorable deal for themselves." Later when Alf Nelson asked for closer contact with the Forest Service, Crafts replied that he had no recollection of any agreement to work further with the industry on the TRR. In Crafts's opinion, the Forest Service had taken all reasonable steps to discuss and explain.[14]

The Forest Service released a preliminary report in 1955, asking for

12. Greeley to James L. Madden, July 25, 1952, AFI Records.

13. Clyde S. Martin to E. W. Tinker, March 22, 1954, box 23, NFPA Records.

14. Martin to Tinker, March 22, 1954, box 23, NFPA Records; Alf Nelson to McArdle, May 14, 1954, ibid.; Crafts to Nelson, June 24, 1954, ibid. The Forest Service had formed an advisory committee representing many interest groups to provide a forum for outside views on the TRR.

comments from all interested parties. The Forest Industries Council retained John A. Zivnuska, a specialist in forest economics who took leave from his job at the University of California at Berkeley, to analyze Forest Service data and interpretations. Zivnuska expressed reservations over the preliminary findings. Some of the data, he thought, were actually only estimates which needed to be used with caution. Moreover, he charged that convention was not always heeded when using economic terminology, such as "demand." The report was most vulnerable, Zivnuska believed, in its estimates of American needs for wood in the future. He agreed that the TRR was extremely valuable, but he hoped that the final version would take the industrial point of view into account.[15]

Not to be outdone, the Forest Service brought in James C. Rettie of the Division of Forest Economics Research to analyze Zivnuska's evaluation. Rettie stoutly justified the predictions included in the preliminary report. His style of defense was exemplified in his explanation of Forest Service predictions about probable new uses for wood. Rettie agreed that no one could anticipate technological advances and consumer preferences with certainty, but a government agency had an obligation to make an attempt. He offered several examples of previously unforeseen uses: 1 billion board feet of lumber used annually to construct pallets; 700,000 cords of pulpwood for paper milk containers; and increased exports to assist developing nations as a Cold War tactic. A dozen years earlier, all three uses had been unforeseen, and Rettie argued that certainly the future would hold similar surprises.[16]

Zivnuska continued to take the Forest Service to task for its efforts to predict the future. One favorite analogy was to show how futile it would have been for an observer in 1920 to predict technological advances or market conditions in 1960. How then could the Forest Service now look ahead to the year 2000 and make realistic plans?

15. John A. Zivnuska, "More Timber Today—and Tomorrow" (New York: FIC, n.d.).

16. James C. Rettie, "Rejoinder to Some of Zivnuska's Criticism of the Timber Resource Review." In his 59-page report to Crafts, an overly defensive Rettie belittled Zivnuska's comments as silly, naïve, dogmatic, and biased, hoping that "some charitable person would do well to send our forest economist friend an invitation to come into the Twentieth Century." Zivnuska's reference to the misuse of "demand" Rettie rejected as rigid adherence to elementary textbooks. When he discounted other criticism as "surface manifestation of a point of view that has much deeper ramification," he unwittingly showed that differing personal philosophies—his and Zivnuska's—were as much the issue as predictions of future needs for wood. Similar to the regulation issue, preferences again colored professional judgment on the national timber inventory. A copy of Rettie's critique may be found in the Rettie Papers, University of Oregon.

Surely, Zivnuska reasoned, the year 2000 will hold as many surprises to them as 1960 would have to Chief Henry Graves.[17]

Industrial leaders feared that the Forest Service might release the preliminary findings to the general public, for they hoped that the final TRR would reflect all of Zivnuska's comments. As far as the industry was concerned, the gloomy mood of the preliminary text was not supported by statistics contained in the report.[18]

The TRR failed to include an action plan. Logically, the Forest Service could have been expected to explain in detail how the nation's future demands for wood as described in the report could best be met. The already nervous lumber industry managed to enlist help from a sympathetic Eisenhower administration to force the deletion of such plans that might include regulation. Technical findings of the TRR were not suppressed, however, and *Timber Resources for America's Future* appeared in 1958. McArdle introduced the report with cautious optimism: "There is no 'timber famine' in the offing although shortages of varying kinds and degrees may be expected." The 715-page report stated that there was no surplus of commercial forest land, so productive potential would have to be utilized more intensively. Whether supply would remain equal to demand depended upon whether "forestry would continue to intensify and accelerate as indicated by recent trends."[19]

A number of other assumptions were offered about population growth, rate of economic increase, and the American participation in international affairs. Specifically, the forestry projections for 1975 and 2000 were based upon assumptions of continued growth of population, prosperity, standard of living, and an unending need for military preparedness. Demand for timber products would increase, and most of this demand would have to be supplied from domestic sources. The basic assumption that demand could be met was tempered by the realization that 52 million acres needed planting, and that fire, insects, disease, and weather had killed 44 billion board feet of timber in 1952 alone, approximately nullifying new growth for that year. The losses that year had also exceeded the amount of timber taken for commer-

17. John A. Zivnuska, personal communication, Aug. 1975.

18. Alf Nelson to McArdle, March 14, 1956, box 25, NFPA Records.

19. Edward C. Crafts, "The Saga of a Law," *American Forests* 76 (June 1970):15; USDA, Forest Service, *Timber Resources for America's Future*, Forest Resource Report no. 14, Jan. 1958, pp. iii, 102.

cial purposes. Obviously, only through more intensive forestry practices could domestic supply be maintained.[20]

Protecting a Land Base. The report on timber stated that no longer was there a surplus of forest land in America. Urban and agricultural expansion and other nontimber uses had whittled down what at one time had seemed a more-than-ample timberland base to an absolute minimum. If potential wood shortages were to be avoided, this minimum would have to be used with maximum effectiveness.[21]

McArdle reported to his regional foresters and experiment station directors the impressions that he had gained from talking with Secretary of Agriculture Benson and Assistant Secretary Coke about land acquisitions. Both had told McArdle that they were "strongly—very strongly—opposed to further enlargement of Federal land holdings." The departmental policy was consistent with Benson's speech to the 1953 American Forest Congress, where he had said that the national forest system ought to remain at its present gross acreage.[22]

Greeley was troubled, telling McArdle that Eisenhower and Benson seemed to be looking to Congress for guidance on national policy. He was alarmed that this approach ignored grass-roots sentiment; apparently the AFA/FIC proposal for a land study would not receive serious consideration. McArdle tried to reassure Greeley by explaining that the Forest Service was planning a study of its own, confiding that he himself generally supported the AFA proposal. He agreed with Greeley, however, that under Eisenhower there was a "tendency to look to Congress for policy guidance on matters which are sometimes substantially administrative in nature."[23]

The second Higgins Lake Program, sponsored by the AFA in 1953, had proposed federal and state committees "to consider a desirable pattern for ownership of federal, state and private forest, range and other conservation lands."[24] Ownership had been an intermittent issue since 1891, when Congress had authorized creation of federal forest reservations. In the 1930s the issue burned hot, then flared again during the 1950s, when some feared and others hoped that the new Republican administration would curb federal acquisition of commercial forest land.

20. Ibid., pp. 102–6.
21. Ibid., p. 104.
22. McArdle to RF & Directors, July 2, 1953, box 63A 4027, RFS, Washington, D.C., FRC.
23. Greeley to McArdle, Aug. 3, 1953, Greeley Papers; McArdle to Greeley, Aug. 24, 1953, ibid.
24. AFC, *Proceedings*, Oct. 29–31, 1953, p. 347.

During a panel discussion on forest land ownership at the 1953 American Forest Congress, the issue was quickly clarified. Christopher Granger, former assistant chief of the Forest Service, abandoned the polite conversation usual to such sessions. Granger believed that an editorial in the July 1953 *American Forests* represented AFA policy against expansion of the national forest system. He felt that an earlier editorial had taken a view more supportive of the Forest Service. Granger challenged the AFA to be "counseled by discretion" and to be consistent in its utterances. He was deeply upset by the rather moderate editorials and, after accusing the organization of "trying to ride its horse in several directions," closed his remarks with a folksy story, which concluded with: "Get the hell off that horse or I will smack the hell out of you."[25]

Granger threw a punch at the lumber industry as well by claiming that although most federal forests were under good management, most private forests were not. For good measure he said that state programs could stand substantial improvement, too. Edward Rettig, an Idaho lumberman, passed off Granger's disparaging comments with an obscure joke and then outlined the industrial position as he saw it. He reminded his audience of Secretary of Agriculture Benson's remarks earlier in the Congress, apparently alluding to Benson's view that the national forests should not be increased in total area. Rettig then moved on to criticism of the Forest Service for not authorizing logging of dead and dying timber in wilderness areas. To allow this timber to go to waste, he said, was contrary to the tenets of conservation through wise use. The lumberman cited instances of federal ownership hampering local development and insisted that the policy of acquiring additional land "should be stopped and, in fact, reversed." Rettig offered as solid evidence of responsible private ownership the 4,622 industrial tree farms in thirty-six states comprising almost 29 million acres. He closed his remarks by endorsing AFA's second Higgins Lake program and its recommendation for a study of forest land ownership.[26]

Speaking with a sharpness that made Granger's statements seem mild, Anthony W. Smith of the CIO next voiced his concerns about forest conservation. Organized labor had pretty much ignored conservation until after World War II. Then its interest focused on protecting sources of raw materials needed to assure jobs. Quality of life for

25. Ibid., pp. 296–300.
26. Ibid., pp. 304–8 (Benson's comment on p. 48).

workers during leisure hours had become a factor, too. Now Smith charged that management of large corporate holdings was bad and that small holdings were in even worse shape on account of specific actions of the large corporations that prevented federal assistance to the small operations. He claimed that the American public was being misled into believing that corporate forestry was doing a good job. Smith then zeroed in on the Higgins Lake proposal to examine land ownership. To him there was only one reason to advocate such a study: "Some people think that the political climate is favorable to the dissolution of the national forest system by transfer to the states and then to private interests, or by direct transfer to private ownership." As proof, he offered concurrent efforts by the grazing industry to acquire permanent rights to public range.

The American Forestry Association was not to be spared Smith's broadside. The AFA had "berated" Granger's defense of the forests; to Smith, the episode was "deplorable." He then accused the AFA of being conspicuously absent from a conservation coalition trying to stave off any reduction of the national forest system. Smith predicted that the raid on the national forests would be unsuccessful, with or without the AFA's opposition.[27]

Following the AFA congress, Granger wrote an open letter to Secretary Benson, which was published in the *Journal of Forestry*. He branded AFA and FIC proposals to study land ownership as potentially "gravely contrary to the public interest." Granger feared that authorizing a study that even considered reducing the national forests would open the floodgates to powerful interests. He ended his letter by reminding Benson that the Republican party had occupied the White House when the national forests came into existence; surely the modern Republicans would not wish to participate in any program to dismember "these great national institutions."[28]

In making reference to Greeley in his open letter to Benson, Granger upset the ex-chief. He now wrote to Granger that the FIC had no plans to disrupt the national forest system, but he noted that the Forest Service proposed to acquire 34 million acres of private land, a possible disruption of industrial forestry. Greeley insisted that as well as contemplating additions, the Forest Service should also consider deleting parcels from national forests that were inappropriate for public ownership. Granger responded by repeating his fears that the

27. Ibid., pp. 316–18.

28. Granger to Benson, March 10, 1953, *Journal of Forestry* 51 (June 1953):448–49.

FIC was playing into the hands of those who advocated wholesale subtractions from the national forests.[29]

The issue of acquisition swirled within the rapidly forming conservationist/preservationist/environmentalist movement. While Bernard DeVoto agitated the public to be alarmed about stockmen's efforts to gain control of public range, the Natural Resources Council of America circulated an information sheet warning of a "land grab."[30] The council had been formed by organizations mutually concerned about natural resources as a loose coalition to keep abreast of issues in the many related resource fields. When Congressman Harris Ellsworth of Oregon introduced a bill in 1953 "To Prevent Federal Land Acquisition from Interfering with Sustained Yield Timber Operations," the council invited ex-Forest Service Chief Lyle Watts to analyze it for the entire membership.

The bill would allow any timber company operating on sustained yield (the cut equals growth version) to be allowed a choice of federal timberland as replacement for any land lost to the government. Watts noted that under Ellsworth's bill, timberlands within national parks, wilderness areas, and other noncommercial stands could be selected for private ownership. Pointedly, Watts charged that the "failure to except such areas from selection was deliberate." He suspected that proponents of the bill had anticipated loud objections, and by eventually dropping this obviously unacceptable clause, they would demonstrate their "willingness" to compromise. Such a compromise would require the conservationists to compromise in turn as a demonstration of their own good faith and willingness to negotiate. Watts listed a whole host of other defects and summed up the proposal as "bad legislation." He advocated that the bill be rejected in its entirety. The Ellsworth measure failed to receive adequate support and died in committee.[31]

The lumber industry kept up its pressure for a study of land ownership. As so often happens, the dispute polarized opinions. When al-

29. Greeley to Granger, June 11, 1953, and Granger to Greeley, June 29, 1953, Greeley Papers.

30. William Voigt, Jr., and C. R. Gutermuth to NRCA Members, June 12, 1952, copy in box F-4, AFA Records, FHS.

31. Watts's memo of Sept. 10, 1953, is included in NRCA Circular of Nov. 23, 1953, to all members. Ellsworth had been a character reference for Al Sarena, the principal figure in an Oregon land scandal in the late 1950s. The scandal broke into public view, largely on account of Drew Pearson's muckraking, at a time when the Republican party was trying to combat its "giveaway" image. Clearly no proposal to dispose of public land to private ownership would fare well in Congress.

most one hundred thousand acres of land within national forest boundaries became available for purchase in Arizona, the American Forestry Association urged public acquisition. Some felt that the AFA had been in the industrial camp because of the two Higgins Lake proposals for a land study. Now by advocating public purchase of timberland, to some the opposite seemed true. Alf Nelson of the NLMA wrote to James Craig, editor of *American Forests*, to see if the AFA was taking a different stance on land policy. It was too bad, Nelson wrote, that the AFA could not support the NLMA. Privately, Nelson confided to an associate that the AFA "may jeopardize its relations with the lumber industry if it continues to persist in taking actions on national legislation. . . ." The association needed industrial support, especially through sale of advertising space in *American Forests*. "[AFA] Board actions which run counter to policies of industry will do nothing to increase industry support. These are a few facts of life."[32]

The industry's friends in Congress helped to keep up the campaign to study what they viewed as the problem of encroaching public acquisition. In 1956 and 1957, bills were introduced to establish federal–state land study commissions. Such commissions, the industry believed, would finally provide means to resolve land ownership disputes. They anticipated major opposition from "organized wildlife-recreation-conservation groups who seem to resist any measure which might reduce, however slightly, the Federal land estate."[33]

In 1956 the American Forestry Association began "a few exploratory studies" of its own to guide the more extensive public land investigations it had proposed via the 1953 Higgins Lake study. Land in California, Minnesota, and North Carolina was the subject of three AFA-published books, which in turn showed the need for state commissions to provide overall guidance for land policy. A decade later, Congress created the Public Land Law Review Commission, which ultimately produced a multivolume analysis of the land situation. In the meantime, the industry had ordered its own study of public land acquisition, state by state. A summary appearing in 1963 showed 39 percent of the United States to be exempt from local taxation—all in public ownership, except Indian reservations. Since 1911, Weeks Law purchases and other federal acquisitions had added nearly 52 million acres to the public estate.

32. Nelson to Craig, Nov. 9, 1955, box 98, NFPA Records; Nelson to H. B. Shepard, Nov. 3, 1955, ibid.

33. Mortimer P. Doyle to Arch N. Booth, Oct. 3, 1957, box 2, ibid.

By the end of the 1950s, industry's biggest worry was loss of commercial forest land to recreational use. Power transmission lines, airports, and reservoirs had once seemed to be the biggest threat, but now recreational demands more and more appeared insatiable. Gradually it began to dawn on land managers, public as well as private, that the public at large would measure their programs with recreation or aesthetics as a yardstick. In 1956 Alf Nelson of the NLMA was advised "to get the views of the Izaak Walton and other outdoor and wildlife groups. . . . They appear to be powerful groups politically and seem to have pretty definite views." Seven years later, AFI's Charles A. Gillett claimed in a speech entitled, "Is More Federal Recreation Land Needed?" that "never before in the history of the United States . . . has the government had on its payroll so many persons whose job it is to see that the people have fun."[34] That recreation should receive the same priority as commodity production was a concept difficult for many to discuss in a calm voice. Although by this time the land manager had to begin to take the recreationist seriously, it would still be some time before he could muster respectful tones.

Timber Claims. "Miners were stealing timber on the basis of their mining claims," remembered Edward Crafts, as he recounted events of the 1950s when he had been assistant chief.[35] There were many who agreed that something had to be done to curb the flagrant abuse of public lands by miners whose purpose seemed to be other than mining.

As authorized by the mining law of 1872, miners could enter public land and stake a claim based on their discovery. The law did not require the mineral deposit to merit economic development or even to be mined. Abuses occurred when claims of little or no mineral value gave the miner surface rights—rights to valuable timber or even to control access by tying up the only logical road site. In actuality these became timber claims, not mineral claims. These abusive practices had been apparent for a long time, but priorities had tended to shuffle this problem down the list. Finally, during the 1950s, curbing these abuses on national forests received high priority, and the Forest Service gathered support for new legislation.

34. Daniel B. Noble to Nelson, Apr. 12, 1956, ibid.; Charles A. Gillett, "Is More Federal Recreation Land Needed?" speech to American Farm Bureau Federation, Chicago, Dec. 9, 1963, AFI Records.

35. Edward C. Crafts, *Forest Service Researcher and Congressional Liaison: An Eye to Multiple Use*, Oral History Interview by Susan R. Schrepfer, 1972, FHS, p. 47.

In 1951 the Forest Service released its summary of the mining situation. First of all, the miner was not required to file his claim with any but the appropriate county agency. Therefore, the Forest Service received no consistent reports on either the existence of claims in the national forests or whether they were being worked or had even been abandoned. Of the nearly 1 million acres of patented mining claims in the national forests, only 14.7 percent had ever operated on a commercial basis. All claims, including those remaining unpatented, occupied nearly 2 million acres of national forest land. Of the total, less than 3 percent were commercially active in 1950. The timber standing on all claims was appraised at over 57 million dollars. Despite these statistics, the report emphasized Forest Service support for mineral exploitation as an appropriate use among other national forest uses. Yet, something had to be done to stem abuse.[36]

During the winter of 1952, Forest Service staff told the "harrowing story" to a group of men representing public and private forestry. The thrust of the Forest Service proposal was to separate surface rights from subsurface.[37] The mining industry, as represented by the American Mining Congress, agreed that the time was at hand to face the problem.

The U.S. Chamber of Commerce, perennial opponent of any "threat" to the rights of the range industries, immediately accused the Forest Service of carrying on a propaganda program against miners. The problems were greatly exaggerated, according to the chamber. A spokesman for the Western Mining Council followed up by charging that the Forest Service was trying to pit lumbermen and stockmen against miners.[38]

The American Forestry Association offered substantial support to the Forest Service. During 1952–53 Cleveland Van Dresser authored a six-part article in *American Forests* under the main title, "Abuses under the Mining Laws." Even more effective than making the public aware that a problem existed was the AFA's provision of its good offices to expedite remedial legislation. On February 10, 1955, representatives of the mining industry and the Departments of Agriculture and Interior met in AFA headquarters to resolve their differences. After general discussions, a committee was appointed to draft legisla-

36. "Forest Service Report on Mining Claims in National Forests," processed, revised Oct. 18, 1951.

37. Henry Bahr to S. V. Fullaway, Jr., Feb. 11, 1952, box 104, NFPA Records.

38. Wendell T. Robie to Ernest Kolbe, March 25, 1952, ibid.; Richard W. Smith to George Fuller, March 11, 1952, ibid.

tion acceptable to the three main participants. The committee met the following month in Salt Lake City. Mining and Interior representatives quickly agreed on the issues at hand, but Reynolds Florance, assistant general counsel of the Department of Agriculture, doggedly insisted on retaining or inserting sections vital to the interests of his department. Persistence was rewarded, and the draft received approval at the departmental level.[39]

With combined support from the mining industry and the public agencies, swift congressional approval was assured. Strangely, the only conservation groups testifying at the Senate hearings were the National Wildlife Federation and the Izaak Walton League, both of which offered support. Conservation groups in general apparently felt that the issue of mining abuses was too narrow for their concern. The bill received final approval on July 23, 1955.

Legitimate exploration and mine operation were not to be affected by the new law. Mining claims would no longer pertain to extraction of sand, stone, gravel, common pumice, and cinders, however. Uses unrelated to mining would not be permitted on a claim, nor could the claimant remove timber, except as needed to operate the mine. One of the most important features of the law, from an administrative point of view, was the so-called *in rem* procedure. After proper notice, the claimant was required to demonstrate the validity of his claim. This process quickly eliminated countless thousands of abandoned and dormant claims and brought under scrutiny those of questionable validity. In the vast majority of cases, the Forest Service was soon able to manage the surface resources of mining claims. Over the years, the American Forestry Association and the Forest Service had been at the opposite ends of many issues, but in this instance, certainly, the AFA's quiet leadership had cleared the way for a resolution of the "number one unsolved problem with respect to administration of the national forests."[40]

A Multiple-Use Directive

Mining, timbering, recreation, and other legitimate demands on the national forests required the administrator to achieve a balance of uses. A rising population and expanding economy—but a fixed land

39. Lowell Besley to DeWitt Nelson, March 25, 1955, box 25, AFA Records. Florance transferred to the Forest Service shortly thereafter.

40. Besley to Ervin L. Peterson, Jan. 26, 1955, ibid. Edward Crafts believes that the AFA has never received proper credit for its key role. Crafts, *Forest Service Researcher*, p. 48.

base—made the balancing act more and more difficult. By the 1950s it was obvious that the national forests no longer held enough resources to meet burgeoning demands.

McArdle saw that careful planning was needed. Planning had long been central to Forest Service procedures (the 1933 Copeland Report, for instance), but the chief wanted such plans to reflect the desired balance of uses. His National Forest Development Program was designed in 1959 to take "a fresh, new look" at each Forest Service activity on the national forests "and to do more precise and more integrated planning." Long and short-term goals were studied and set.[41] Plans would not achieve a balanced operation by themselves, however, and the Forest Service turned to Congress for a solution.

In early June 1960, McArdle, while on a field inspection trip, was able to telegraph Assistant Chief Edward Crafts, "CONGRATULATIONS. A FINE JOB WELL DONE," with the signature, "McArdle plus Region 2 from Rapid City."[42] The chief was praising Crafts for skillfully steering the Multiple Use–Sustained Yield Act through Congress. President Eisenhower signed the bill four days later, on June 12. The Forest Service had since Pinchot's time operated within broad authorities. Now for the first time the agency had a specific congressional directive that established priorities for resource use. The distinction between a permissive authority and a specific directive was not subtle.[43] The Multiple Use Act stated that "the national forests are established and shall be administered for outdoor recreation, range, timber, watershed, and wildlife and fish purposes." The law defined multiple use as utilization of resources in combination to meet needs and stipulated that economic return was not in all cases to be the limiting factor. Policies for specific uses or resources had appeared on an ad hoc basis over the years as the need arose. A unifying statement had been missing. To McArdle and supporters of the new law, the Forest Service had long practiced multiple use. Now it was the law of the land.

The Sundry Civil Act of 1897, which had authorized a degree of management and protection for the forest reserves, referred specifically to timber and water. The Forest Service never felt constrained

41. McArdle, *Dr. Richard E. McArdle*, pp. 75–83. In 1974 Congress enacted the Forest and Rangeland Renewable Resources Planning Act (88 Stat. 476), which mandated much more comprehensive planning than McArdle had devised fifteen years earlier.

42. McArdle to Crafts, June 9, 1960, box 69A 5902, RFS, Washington, D.C., FRC.

43. McArdle points out in retrospect that in order to gain a congressional directive, the Forest Service had to name the uses, a stipulation that was not required to operate under broad authorities. McArdle to Frank Harmon, Aug. 7, 1975, Forest Service History Office files.

to interpret the law in its narrowest sense, that is, timber and water as the only legitimate resources. Rather, the agency since 1897 has to varying degrees recognized range, wildlife, recreation, and minerals as appropriate and desirable uses, along with timber and water.

Although it is always difficult to look into a man's mind, Pinchot as the first chief to be responsible for national forests seemed to be thinking more of users than of uses. Politically astute, he realized that a successful federal forestry agency needed support—or at least not opposition—from all the groups that made use of the national forests. Programs he supported tended to have something for everyone, be they farmers living within national forest boundaries or lumbermen seeking a source of logs. But little thought was given to equality of uses as is implicitly included in the Multiple-Use Act and clearly some uses, such as timber, received prime consideration. At any rate, the concept of multiple use seems not to have been a conscious policy in the Pinchot era, having instead evolved along logical and pragmatic lines through the years.

The 1933 Copeland Report, *A National Plan for American Forestry*, has been the only Forest Service attempt to study American forest policy and practices in their entirety. Specialists reported on all uses. S. B. Show, regional forester in California, was senior author of the section on future forest land ownership. Part of Show's report referred specifically to multiple use and its practicability. He thought it unnecessary to segregate timber production from watershed management and recreational demands. "A sikllfully [*sic*] managed forest can serve all these purposes at the same time." He acknowledged, however, situations that required some designation of one use dominant over others.[44] Show's portion of the Copeland Report was not the first time the phrase was uttered, but in the context of the report itself, his discussion gave modern form and substance to multiple use.

Inspection reports, those administrative linchpins for maintaining control of a decentralized organization, clearly show that the concept of multiple use has been an integral part of Forest Service operation. An example may be seen in the 1937 general integrating inspection report of the Southern Region by Earl W. Loveridge, assistant chief for administrative management and information, and J. A. Fitzwater, assistant chief of the Division of Timber Management. Their report

44. The Forest and Rangeland Renewable Natural Resources Planning Act of 1974 mandates a comprehensive study and projection of Forest Service and other resource agency operations; *A National Plan*, p. 1294.

described unsatisfactory relationships with some state foresters and a reduction in antagonism between National Forest Administration and Research. Loveridge and Fitzwater gave recreational efforts a "good" rating, but saw too little effort. They suggested giving recreation a higher priority in CCC projects. The southern wildlife program, they noted, showed a good cooperative relationship with the various states involved. Timber and other uses received Washington office scrutiny, too.[45] Follow-up inspections provided a measure of how regional programs and processes were bolstered or modified to meet criticisms found in the integrating inspection report. The integration of uses stressed in such evaluations was in fact implementation of multiple use.

At first, multiple use was not a controversial topic, because it did not connote any of the customarily inflammatory questions—regulation of logging, raising grazing fees, or keeping the forestry agency in the Department of Agriculture. There seemed to be enough land to provide a diversity of uses to meet the demands of all special-interest groups. During World War II, however, some predicted that postwar adjustments would bring a substantial increase in competition among users. Accordingly, the Society of American Foresters surveyed its membership on the wisdom of creating a mutliple-use committee. Comments were favorable, ranging from "It's about time" to "We've been practicing multiple use all along." SAF President Henry Schmitz appointed representatives from a spectrum of public and private forestry to serve on the committee. Deciding at the last minute that multiple use should include recreation as well, he asked J. D. Coffman of the National Park Service to help draft the policy statement.[46]

It took two years, but the July 1947 issue of the *Journal of Forestry* carried a referendum to the SAF membership. The policy basis of multiple use as adopted by foresters required "adequate recognition of all resources and benefits." When the referendum did not distinguish between public and private land, Greeley was quick to point out

45. "General Integrating Inspection Report: A Report on Forest, Watershed, Range, and Related Resource Conditions and Management—Southern Region," 1937, copy in General Integrating Inspection Reports, 1937–55, Record Group 95-41, RFS.

46. Bernard Frank to Samuel T. Dana, Oct. 23, 1943, box 20, SAF Records, FHS; Frank to Schmitz, Oct. 25, 1943, ibid.; Schmitz to Frank, Nov. 3, 1943, ibid.; Schmitz to SAF Council, Jan. 18, 1944, ibid.; Schmitz to Charles A. Connaughton, May 27, 1944, ibid. Despite foresters' apparent familiarity with multiple use concepts, an astute observer of American forestry, Richard G. Lillard, as late as 1947 described multiple use as a new and desirable Forest Service program. See *The Great Forest* (New York: Alfred A. Knopf, 1947), p. 328.

that "private forest owners should not be expected to sacrifice the essential interests or purposes of their ownership in order to conserve public interests as in recreation, scenery, etc. Where value of such public interests warrants, public agencies should acquire the land."[47] As multiple use became more than an abstraction, it was growing apparent that controversy was possible.

At its 1953 congress, the American Forestry Association assigned the topic of multiple use a half-day on the program. Assistant Chief Edward P. Cliff represented the Forest Service on the panel. The audience chuckled when he defined multiple use as causing multiple abuse on those who practiced it. The issue was heating up. Cliff explained that national forests were managed under sustained yield on the basis of multiple use. Every acre did not have to sustain every use, and watershed protection "ordinarily" received first priority. Multiple use was integrated management of all resources and values "consistent with the public interest." He agreed that talking about multiple use was much easier than practicing it.[48]

Although multiple use by definition included all uses, recreational use was unquestionably the most delicate. Conservation organizations had been supportive of Forest Service policies over the years, but by the 1950s increasing population and increasing demands for all resources steered conservationists and the Forest Service on a collision course. The semantic game of stamping the preservationist brand on this new militant conservationist was of little value in actually resolving issues. But under its ground rules, everyone could be a conservationist, while the preservationists could be scorned as extremists.

The most militant conservation group was the Sierra Club. Founded in 1892 by John Muir, the club had been mostly concerned with conservation issues in the Sierra of California, or at least the Far West. For much of its history the Sierra Club had been primarily an outing organization. When the Forest Service or the lumber industry talked of conservationists, it was usually in terms of the Izaak Walton League or some other unspecified wildlife organization. But David Brower's ascendancy to the executive directorship of the Sierra Club in the 1950s meant that its long-held support of the Forest Service would be carefully scrutinized. McArdle and Brower assumed the top positions in their respective organizations at about the same time.

47. Policy statement and Greeley's comment in box 60, SAF Records.

48. AFC, *Proceedings* (1953), pp. 161–64. The American Forest Congress in 1946 also had a session which explored multiple use concepts. See ibid. (1946), pp. 217–43.

While McArdle worked to achieve a congressional mandate to give equal consideration for all uses, Brower labored to guarantee protection for recreational use, especially wilderness recreation. Recreation and wilderness became the most controversial aspects of multiple use, to the extent that it is impossible to unravel the history of multiple-use legislation without examining concurrent attitudes about recreation.[49]

Several events coincided during the 1950s which cost the Forest Service the support of many conservationists. Since Pinchot's time, the national forests had been held in reserve to supply lumbermen when privately owned timberlands had been cutover and were in the regrowth cycle. Year after year and chief after chief reaffirmed the policy to log a substantial portion of the national forest system—at the appropriate time. The appropriate time followed World War II, when increases in population and affluence concurrently multiplied recreational pressures on all public lands. Couched in terms of timber famine and devastation, decades of propaganda by the Forest Service to justify federal regulation of logging had created the popular image of rangers protecting forests against rampaging, greedy lumbermen. Now the postwar public came to the forests—camping, fishing, hiking—and "discovered" logging in the national forests. To many of the public, national forests and national parks were the same, and logging either was bad. Implementation of long-term timber management plans collided with the results of an extremely effective public relations program against "destructive" logging. As a result, the public felt deceived.

Logging in recreation areas stood out as an obvious source of friction. Foresters saw certain trees as hazards to the public, as disease carriers, or at least as sources of economic waste if not utilized. Within the Sierra Club, doubts emerged concerning the justification for logging all diseased trees. Conservationist philosophies were beginning to differ "radically in this regard from that of the Forest Service."[50] Some disease and even some fire could be allowed, in the Sierra Club view, if aesthetic considerations were to receive consideration equal to commodity production.

In his presentation to the 1953 American Forest Congress, Assistant

49. The author is grateful to Susan R. Schrepfer for many of the following interpretations of Sierra Club policy, and also for the gracious loan of much research material on the topic. For a biographical view of David Brower see John A. McPhee, *Encounters with the Archdruid* (New York: Farrar, Straus, and Giroux, 1971).

50. Sierra Club, Conservation Committee Minutes, June 13, 1959, Sierra Club, San Francisco.

Chief Cliff had used sanitation salvage—removal of dead and dying trees—within recreation areas as an example of timber management being compatible with recreation. Multiple use, Cliff said, could accommodate even those supposedly contrary uses. While it may have been clear to the assistant chief and others in the Forest Service that timber harvest and recreation were compatible, to many recreationists it was not clear at all.

The Deadman Creek area in the California Sierra became the focus of one of a series of confrontations between multiple use and recreation. The episode is particularly significant because events at Deadman Creek ended any question in Brower's mind that the Forest Service no longer could be trusted. The Sierra Club was beginning its phenomenal growth in membership and recreation leadership, due more to a battle-to-the-death with the Eisenhower administration over damming the Dinosaur National Monument than to public suspicion of Forest Service policies. Although the Forest Service was not involved with the Dinosaur issue, it became identified with the whole "giveaway-landgrab" image that Bernard DeVoto and Brower had fastened onto public resource agencies.

Deadman Creek contained three thousand acres of pine. The area was too small for wilderness designation, but Brower thought it "a jewel." Additionally, the area contained "one of the few remaining" virgin Jeffrey pine forests in the United States. The American Museum of Natural History had studied the area and planned to make a diorama of it. When the Forest Service announced plans to allow logging—sanitation salvage—in Deadman Creek, Sierra Club reaction was swift. To log in unique stands of low-quality commercial timber for the purpose of providing logs to local sawmills already doomed to shut down caused Brower to doubt Forest Service judgment. Similar situations throughout the length of western mountain ranges that required choices between multiple use and recreational use prompted Brower to advocate legislation specifically protecting wilderness from the dangers of administrative decision. For its part, the Forest Service sought legislation to make official the policy of multiple use.

Wilderness Protection and Multiple Use. To the mind that accepts multiple use as an appropriate management policy for public lands, intense emphasis on wilderness seems out of balance. To the avid recreationist, however, wilderness was the issue, and compromise was rapidly becoming unthinkable. But American political process and administrative tradition accept compromise as fundamental to decision-making. Fully aware that compromise would be proposed, staunch wilderness defenders were no longer satisfied by the administrative protection that the Forest Service afforded wilderness.

On April 10, 1956, Senator Hubert H. Humphrey of Minnesota introduced a bill to protect and preserve the multiple resources of the national forests. The bill listed range, timber, water, mineral, wildlife, and recreation as resources. Humphrey proposed that citizens' advisory councils would make policy recommendations to the secretary of agriculture. The secretary would retain final authority, but the public could formally propose policy through the council system. Humphrey's proposal had been adopted a year earlier at a conference of conservation groups that had included the AFA and the Sierra Club.[51]

During the next two years, joint resolutions offered congressional support for a fifty-member commission to determine how best to use the many public resources "for the maximum benefit of the entire Nation now and in the future." As backing grew for multiple-use legislation, however, so did support to include the concept of wilderness. Don P. Johnston, AFA president, told AFA's Executive Vice-president Fred Hornaday that in order for their organization to legitimize its opposition to wilderness proposals, it must first demonstrate the adequacy of multiple use. Senator Humphrey had also introduced a bill to protect wilderness in 1956, the same year of his multiple-use bill. The AFA had quickly opposed the wilderness bill because wilderness constituted single use, and single use was contrary to the AFA's long policy of supporting multiple use. The National Park Service also opposed the wilderness bill, but for different reasons. Anticipating the compromise that would invariably occur, the Park Service did not want to jeopardize its domain by becoming involved in the discussions.[52]

Within the Forest Service itself, unanimous support for the multiple-use bill had been lacking. Some of McArdle's staff feared that if Congress failed to enact the bill, then a precedent would be established that could be dangerous to existing policies. It seemed

51. Humphrey news release, Apr. 10, 1956, copy in box 4, NRCA Records, FHS; news release from Outdoor Writers of America, Feb. 1, 1955, copy in box 41, NFPA Records.

52. H.J.R. 197, 85 Cong. 1, Jan. 28, 1957; H.R. 10410, 85 Cong. 2, Jan. 30, 1958; Johnston to Hornaday, Nov. 19, 1958, box 48, AFA Records; Kenneth Pomeroy to Hubert Humphrey, July 16, 1956, ibid.; Conrad Wirth to Howard Zahniser, March 19, 1956, ibid.

Commemoration in 1951 of the first industrial tree farm, established by the Weyerhaeuser Timber Company near Montesano, Washington, in 1941. Left to right: Chapin Collins, former managing director of the American Forest Products Industries; Mrs. C. H. Clemons; J. P. Weyerhaeuser, Jr.; William B. Greeley, chairman of the board, AFPI. Courtesy of the Forest History Society

Chief Richard E. McArdle (right) presents *Timber Resources for America's Future* to Secretary of Agriculture Ezra Taft Benson (left), with Assistant Chief Edward C. Crafts looking on (1958). Courtesy of the U.S. Forest Service

President Harry S. Truman signs proclamation creating the Gifford Pinchot National Forest, 1949. Courtesy of the U.S. Forest Service

Trail riders enter the Maroon Bells-Snowmass Wild Area, White River National Forest, Colorado, 1957. Courtesy of the U.S. Forest Service

Timber cruisers on the Eldorado National Forest, California, 1953. Courtesy of the U.S. Forest Service

better to leave well enough alone. McArdle, however, thought that the bill was vital and made what Crafts thought the "most crucial decision" of his administration—to pursue the multiple-use bill with all of the resources at his command. Typically, the ranks then closed, and McArdle received strong, unanimous support for the bill.[53]

As another draft of multiple-use legislation moved through Congress, Park Service opposition became more direct. Director Conrad Wirth saw the concept of multiple use as containing Forest Service criticism of the Park Service. In a speech Wirth referred to Park Service opponents as "rallying around and beating this old tom-tom called multiple use." Later Wirth stated to his staff that as expansion of the park system was an important part of Mission 66, a Park Service program, the "multiple use cure-all for land management problems" was an attempt to stigmatize the Park Service and head off increases. Over and over, Park Service opponents had been referring to recreation as a "locking up" of resources.[54]

Chief McArdle sent a memo to all Forest Service personnel, regretting Park Service opposition to legislation that their agency supported. He said that usually the two bureaus were able to iron out differences of opinion without resorting to name-calling. McArdle expressed his hope that the current Park Service expansion program would not cause a "lowering or cheapening of national-park standards." He advised Forest Service personnel to avoid involvement in the conflict and just to practice multiple use instead. He assured them that the Forest Service was in the recreation business to stay, even though the 1897 law had not specifically listed recreation as one of the resources that would justify creation of forest reserves.[55]

McArdle was not the only one who had noticed congressional failure to recognize recreation in the 1897 law. The National Lumber Manufacturers Association opposed multiple-use legislation, claiming it diluted the early mandate that national forests were to provide timber and water. Ralph Hodges of the NLMA was to meet with McArdle to discuss industrial opposition. Anticipating that McArdle would use the argument that multiple use was an emergency measure to fend off the Park Service and its expansion program, Hodges prom-

53. Crafts, "Saga of a Law," p. 16.

54. Conrad Wirth, speech to Visitor Services Conference, Williamsburg, Va., Nov. 30, 1959, box 1365A 1669, RFS, Washington, D.C., FRC; Wirth to staff, Dec. 11, 1960, ibid.

55. McArdle to Forest Service personnel, Feb. 12, 1960, box 1365A 1669, RFS, Washington, D.C., FRC.

ised his associates that he would be skeptical. Anyway, multiple use offered no advantage to the industry. The Forest Service already had authority to designate a wide range of recreation areas from which the industry was barred. The clincher, from Hodges' view, was that "by making all uses equal in priority the forest manager will probably have to act on the basis of public pressure. This doesn't give protection to the lumber industry."[56]

Of the conservation groups, only the Sierra Club opposed the multiple-use bill, feeling that it placed the wilderness bill in jeopardy. The October 1959 issue of the *Sierra Club Bulletin* carried an article by Grant McConnell, then associate professor of political science at the University of Chicago, which traced the history of multiple use, describing it as a useful administrative tool. He feared, however, that the Forest Service may have been granted too much discretion to use the tool wisely. To McConnell, multiple-use administration was flawed. Foresters were technically competent to identify resource problems and to propose an array of solutions, but unfortunately foresters lacked training that enabled them to choose between uses such as lumber and scenery. He noted the common erroneous assumption that expertise in one aspect of a subject automatically yielded competence in other aspects of the same field. McConnell then described what he saw as important deficiencies in forestry education. He ended the article with praise for past Forest Service achievements but stated flatly that "the Forest Service is not as well equipped administratively as it needs to be to deal with the problems of conflict in land use which it must face in the years to come."[57]

To help allay Sierra Club fears about multiple use, the next spring Assistant Chief Crafts met with the Sierra Club board of directors in San Francisco. Both sides agreed afterward that it had been a productive session. In a summary memo written for his files, Crafts noted that the Sierra Club had agreed not to oppose the multiple-use bill but wanted a clear statement that timber would not dominate. The *Sierra Club Bulletin* carried an account to its readers substantially the same as Crafts had recorded. Recreation should be equal to other uses, and the multiple-use bill would not prevent expansion of national parks. For the Forest Service, the result was not as good as Sierra

56. Hodges to Mortimer B. Doyle, March 30, 1960, box 97, NFPA Records.

57. Grant McConnell, "The Multiple Use Concept in Forest Service Policy," *Sierra Club Bulletin* 44 (October 1959):14–28.

Club support would have been; but at least by preventing its opposition, a major hurdle had been cleared.[58]

During the congressional hearings, 128 individuals representing 85 organizations testified, largely in support of multiple use. By now the NLMA had received necessary assurances, including incorporation of a phrase in the bill declaring the act to be "supplemental to, but not in derogation of" the 1897 Sundry Civil Act, which referred specifically to timber and water, and supported the measure. The Sierra Club declined to testify at all, accepting for the time being inclusion of the statement that wilderness was "consistent" with multiple use. When McArdle testified, he insisted that the measure was vital, otherwise "it would make impossible the continued management of the national forests for 'the greatest good of the greatest number in the long run.' " The all-important approval from the Bureau of the Budget, difficult to obtain at first because some in the bureau saw multiple use as a wedge toward an increase in Forest Service funding, was also obtained. Assistant Secretary of Agriculture E. L. Peterson did yeoman work in the bill's behalf, as he had for many Forest Service programs. Then the bill lodged in committee. McArdle enlisted Senator Humphrey to intervene, and the bill soon reached the Senate floor. As in the House, the measure carried by voice vote.[59]

On June 12, 1960, President Eisenhower signed the Multiple Use–Sustained Yield Act into law. McArdle described the law as "one of the most fundamental." For the first time the five major uses—timber, wildlife, range, water, outdoor recreation—were contained and specified in one law. No one use received priority. McArdle believed that the Multiple-Use Act "undoubtedly will be looked upon in years to come as the basic charter for the administration of the national forests."[60] As it turned out, the decade that followed sorely tested many American institutions—no less so the Multiple-Use Act.

58. Edward Crafts, memo to files, May 9, 1960, copy in box 1365A 1669, RFS, Washington, D.C., FRC; "Crafts Discusses Multiple Use Bill," *Sierra Club Bulletin* 45 (June 1960): 3.

59. Crafts to McArdle, Sept. 11, 1959, RFS, Washington, D.C., FRC; E. L. Peterson to President of the Senate, Feb. 5, 1960, ibid.; U.S. Congress, House, *National Forests—Multiple Use and Sustained Yield*, Subcommittee on Forests of the Committee on Agriculture, 86 Cong. 2, March 16, 18, 1960, p. 39; Charles A. Connaughton, retired regional forester, feels strongly that Assistant Secretary Peterson deserves much credit for the McArdle administration; he "called the shots in the operation of the Forest Service." See Connaughton, *Forty-three Years in the Field with the U.S. Forest Service*, Oral History Interview by Elwood R. Maunder, 1976, FHS, p. 54; McArdle to Elwood R. Maunder, Aug. 4, 1975, in McArdle, *Dr. Richard E. McArdle*, OHI, pp. 222–24.

60. McArdle to FS personnel, June 15, 1960, copy in box 1365A 1669, RFS, Washington, D.C., FRC. For a full analysis of the specific phraseology of the Multiple Use Act see Crafts, "Saga of a Law," *American Forests*, vol. 76, pt. 1 (June 1970): 12–19, 52–54; vol. 76, pt. 2 (July 1970): 28–35.

CHAPTER XII

Multiple Use Tested: An Environmental Epilogue

IN 1972, TWELVE YEARS after the Multiple Use–Sustained Yield Act became law, the Congressional Research Service sent a pessimistic report to the Senate Committee on Interior and Insular Affairs. It began: "For American forestry generally, this is a time of conflict and crisis. It is also a time of severe criticism of the policies carried out by the U.S. Forest Service." The report described a series of angry accusations leveled at various aspects of national forest administration, including clearcutting practices, timber supply, and wilderness protection.[1] But it was to be the public's burgeoning concern for environmental safeguards that prompted the severest criticism.

Two years earlier, Chief Edward P. Cliff had reported to all Forest Service personnel that their programs were "out of balance to meet public needs for the environmental 1970's." He noted "mounting criticism from all sides" and that the direction of the Forest Service must be and was being changed. On another occasion, the chief told regional foresters and experiment station directors that one of his "major disappointments" of the 1960s was the failure to attain wider support for multiple-use management.[2] Times were different. In Pinchot's day, the slogan "greatest good of the greatest number" sufficed. But McArdle and his successors served a more sophisticated and per-

1. U.S. Congress, Senate, "An Analysis of Forestry Issues in the First Session of the 92D Congress," Print for Committee on Interior and Insular Affairs, 92 Cong. 2, Apr. 1972, pp. 3–4.
2. Friday Newsletter no. 34, Sept. 18, 1970; Edward P. Cliff speech to Regional Foresters and Station Directors, Jan. 19, 1970.

haps more cynical public; multiple use failed to capture a broad following.

It is unlikely that anyone really thought that the 1960 Multiple Use–Sustained Yield Act would end disagreements over resource use on national forests. At the very least, however, the new law provided the sort of administrative housekeeping needed from time to time to place traditional policies into newer contexts. At its best, the Multiple-Use Act would enable the Forest Service to stave off unacceptable demands from any one sector, to resolve conflicts between user groups, and to prevent overuse of any of the national forest resources.

Chief McArdle, never one to be deluded by unfounded optimism, always portrayed the measure in positive terms. Multiple use was the theme of the Fifth World Forestry Congress, held in Seattle, Washington, during the late summer of 1960. As president of the congress, McArdle told a large, international audience of foresters that all American forest resources were to be used based upon the goal of the greatest good of the greatest number in the long run. "These instructions have constituted Forest Service doctrine from the beginning. They are the genesis of multiple use."

Fitting the multiple-use concept into worldwide forest management, McArdle explained that management required positive action and that multiple events occurring by chance did not constitute multiple use. This distinction was important, for otherwise forest managers might take credit for coincidental favorable conditions that did not result from conscious planning. He further explained that noneconomic considerations were valid, that economic returns were not the sole criterion upon which decisions were based. Although multiple use was not a panacea, it offered "overriding advantages" to the forest land manager. McArdle believed that multiple-use management would tend to reduce competition for resources by impeding any single use from attaining dominance over resources. "It offers balance."[3]

While McArdle was explaining Forest Service policy to the two thousand delegates from sixty-six nations, Patrick O. Goldsworthy, president of the North Cascades Conservation Council and an active Sierra Clubber, as a harbinger of protests to come, placed quantities of

3. Richard E. McArdle, "The Concept of Multiple Use of Forest and Associated Lands—Its Value and Limitations," in Fifth World Forestry Congress, *Proceedings*, 3 vols. (1962), 1:143–45.

an anti–multiple use newsletter on literature distribution tables outside. When he saw the pamphlet later, the chief was angered by what he considered an embarrassing and unethical ploy and protested to Sierra Club president Nathan C. Clark that the "statements contained therein are replete with untruths, half truths, slurs, innuendo and erroneous inferences." He labeled the brochure an "extreme publication." Sierra Club Executive Director David Brower answered McArdle's charges by pointing out that neither the chief nor Assistant Chief Edward C. Crafts had been available to discuss "some of the errors of your interpretation of our Newsletter" the previous month. Neither had the Forest Service responded yet to a Sierra Club statement on national forest policy that the club had developed over the years.[4]

The Society of American Foresters chastised the Sierra Club for making its newsletter available at the forestry congress on three counts: (1) common courtesy suggested that advance permission be obtained, (2) airing domestic conflicts at an international conference was a tactic that was in "shocking bad taste," and (3) the literature contained critical references to land conditions in Algeria and Greece, two of the nations that had sent delegates to the congress in response to an invitation from the American government. Brower rejected the charges in sequence: (1) prior approval of materials to be distributed is not typical of other conferences, (2) "preservation is not a domestic issue, but an international one", and (3) references to Algeria and Greece were in a context of ancient history and could offend only "oversensitive" land managers today. There seems to be no way to verify Brower's countercharge that a "Forest Service representative" confiscated the supply of newsletters, but it is obvious the earlier agreement not to oppose passage of the Multiple-Use Act was not a promise to refrain from acting when conditions seemed to merit.[5]

Another critic of multiple use also had been quick to react. In a lengthy and penetrating analysis of the multiple-use law, J. Michael McCloskey—eventually Brower's successor as executive director of the Sierra Club—questioned the law's intent and meaning. He disagreed with the official Forest Service position that the agency had long practiced multiple use, listing many examples of administrative

4. Sierra Club, *Outdoor Newsletter*, no. 6 (Aug. 22, 1960); McArdle to Clark, Oct. 28, 1960, Brower to McArdle, Nov. 1, 1960, Sierra Club Records, San Francisco, Calif.

5. Henry Clepper to Nathan C. Clark, Nov. 18, 1960, Brower to Clepper, Nov. 28, 1960, Sierra Club Records.

and legal decisions that had assigned priority to one use over another. Historically, then, all uses had not received equal consideration, which was a prime element inherent in the 1960 law. Of even greater concern to McCloskey was a host of ambiguities that seemed to increase the discretionary power of the Forest Service, even though the act supposedly reduced these powers. A key issue lay in the combining of multiple use with sustained yield, two concepts that were to him apparently philosophically inconsistent. Sustained yield, according to McCloskey, evoked high-yield images, but multiple use on the other hand suggested reducing yield when necessary. To the young attorney, the act compounded confusion: "resolution must await litigation stemming from the rising tempo of conflict among forest users."[6]

The litigation that McCloskey predicted has indeed occurred, testing and defining a flurry of new conservation laws and seeking to redefine the old. The case load from lawsuits challenging Forest Service land management decisions has increased from an average of one pending at the beginning of the 1960s to two dozen by the mid-1970s.[7] User groups for timber, water, range, wildlife, and outdoor recreation have expressed skepticism or dissatisfaction with the Multiple-Use Act, and some stridently so.

It was not McArdle, however, the champion of multiple use at the Seattle forestry congress, who would deal with these critics during the noisy 1960s and 1970s. He retired in 1962 after a thirty-nine-year career and was succeeded by Edward P. Cliff, his assistant chief for National Forest Administration. When Cliff retired in 1972, John R. McGuire, associate chief and with long experience in research, moved into the top spot. Both Cliff and McGuire would find their administrations during the era of environmental activism to be challenging, indeed.

Recreation

Forest Service recognition of the growing importance of recreation was evident throughout the 1950s. In 1957 Samuel T. Dana, with McArdle's support, studied the recreational scene and wrote a report

6. J. Michael McCloskey, "Natural Resources—National Forests—The Multiple Use–Sustained Yield Act of 1960," *Oregon Law Review* 41 (Dec. 1961): 49–78.

7. Clarence W. Brizec, "Judicial Review of Forest Service Land Management Decisions," 2 pts., *Journal of Forestry* 73 (July, Aug. 1975): 424.

advocating creation of a Division of Forest Recreation Research.[8] Dana's report was instrumental in gaining congressional approval for this substantive increase in recreational emphasis. Now the Forest Service could look at recreation more reflectively than in the past, with specialists freer from jangling telephones and other distractions so common to administrative life. The administrative counterpart to increased recreation research activity was the launching, also in 1957, of Operation Outdoors, a five-year program designed to upgrade existing facilities and to provide new construction to meet the ever-increasing demands.

During the latter part of the 1950s, it became obvious to many in and out of the government that there was need for an overall recreational survey. The Forest Service began to plan such a survey for the national forests, anticipating participation with either the Mission 66 program of the Park Service or with the proposed Outdoor Recreation Resource Review Commission (ORRRC). Basic planning was well underway in the Forest Service when Congress in 1958 established the commission to inventory and evaluate outdoor recreational resources, taking into account trends of population, leisure, and transportation. The commission was to estimate the amount and type of recreation facilities that would be needed to meet the demands of 1976 and 2000. To be completed by 1961, the report was to include data and recommendations on a state by state, region by region basis. In effect, it would be the recreational equivalent to the Timber Resources Review. Congress appropriated 2.5 million dollars to support the commission.

The Forest Service quickly moved to participate, trying wherever possible to adopt parameters that were mutually acceptable to the Park Service and other agencies. All land suitable for recreation was to be cataloged, classified, and mapped. All types of recreation activity would be considered. Numbers of visits would be the basic data, but it was clear that visits were a measure of use, not demand. The distinction between use, which might be relatively low because of inadequate facilities, and demand, an estimate of maximum use should all of the needed facilities be provided, was crucial when predicting future requirements and priorities.[9]

8. Samuel T. Dana, "Problem Analysis: Research in Forest Recreation" (Washington, D.C.: U.S. Forest Service, processed 1957).

9. George F. Burks and James C. Rettie to Edward C. Crafts, May 7, 1958, copy in James C. Rettie Papers, University of Oregon.

Not since the 1928 report of the National Conference on Outdoor Recreation had the topic received such thorough treatment. Predicting recreational demands through the year 2000, the ORRRC Report estimated that present facilities would have to be increased threefold to meet all future needs of public outdoor recreation. The report advocated creation of a federal recreation agency, and Congress responded by establishing the Bureau of Outdoor Recreation in the Department of the Interior in 1962. Edward C. Crafts, the outspoken Forest Service assistant chief, left his long career in the Department of Agriculture and became the new agency's first director.

Of all the recreational issues that the Forest Service had to consider, wilderness continued to dominate. The Multiple-Use Bill had moved through Congress with relative ease, but the Wilderness Bill—introduced in 1956 at the same time as the Multiple-Use Bill—took four controversial years longer to become law. Heavy opposition from industry and professional foresters delayed enactment. Also, the Forest Service had initially opposed the bill (as had the Park Service) until receiving assurances that it would not dilute or jeopardize the Multiple-Use Act.

The Wilderness Act of September 3, 1964, reflects lack of faith on the part of many recreationists in multiple use and the Forest Service. Although the Forest Service had invented the wilderness concept forty years earlier, and on its own initiative had created a multimillion-acre wilderness system, the agency could not rest on its past achievements. By the 1950s trust had begun to erode. Existing wilderness areas had been created during relatively noncontroversial times and when demands for publicly owned resources were modest. Now that competition for land was intensifying, many feared that the Forest Service would slack off on wilderness designation or even reclassify some areas to nonwilderness use.

Through negotiation, the Sierra Club had agreed not to oppose the Multiple-Use Bill and in return had received inclusion of the statement that "establishment and maintenance of areas of wilderness are consistent" with multiple use. The supporters of legislation to protect wilderness by law rather than through administrative decision (which could be later reversed at administrative discretion) stepped up their campaign during the 1960s. The last bastion of opposition was the mining industry. So powerful was this group that mining alone received special dispensation when Congress passed the Wilderness Act in 1964. Mining activities could continue for twenty additional years;

but after December 31, 1983, no new claims could be patented in wilderness areas.[10]

The Wilderness Act of 1964 provided Congress with a ten-year period in which to review proposed wilderness areas and to develop the wilderness system. Seemingly a great victory for recreationists—considering the nature and intensity of opposition—the law threatened to fall short of its potential, because congressional attention was diverted by the intense social issues of the late 1960s. Still, the Wilderness Act demonstrates the substantial political power that recreationists had achieved in only a decade, a power which would continue to grow.

Timber

Timber supplies were a major issue, both before and after the Multiple-Use Act was ratified. Mileage of roads in the national forests doubled to 160,000 miles between 1940 and 1960, with timber access the major justification for these new roads. The national forests had yielded only 1.5 billion board feet of timber in 1941, 4.4 billion in 1951, 8.3 billion in 1961, and 11.5 billion board feet in 1970.[11] With demands for public wood increasing still, it was imperative to have suitable timber policies.

Although the lumber industry had eventually offered support for multiple use, and continued to do so, the lumbermen remained vigilant to protect their interests. Like the recreationists, they, too, sought special legislative recognition. In the late 1960s, congressional supporters of the lumber industry introduced a bill that became known as the National Timber Supply Act. The bill proposed to create a revolving fund of timber sale receipts from the national forests to be used to support timber management, similar to the process that Pinchot had used until 1907, when Congress decided that receipts should be treated as general revenue. These designated monies would provide a

10. The literature on wilderness is enormous and growing daily. Most helpful on the Wilderness Act are J. Michael McCloskey, "The Wilderness Act of 1964: Its Background and Meaning," *Oregon Law Review* 45 (1966); Richard E. McArdle, *Dr. Richard E. McArdle: An Interview with the Former Chief, U.S. Forest Service, 1952–1962*, Oral History Interview by Elwood R. Maunder, 1975, FHS, pp. 144–65; Clinton P. Anderson, *Outsider in the Senate* (New York: World Publishing, 1970), chap. 8; Glen O. Robinson, *The Forest Service: A Study in Public Land Management* (Baltimore, Md.: Johns Hopkins Press for Resources for the Future, 1975), chap. 6; and Roderick Nash, *Wilderness and the American Mind*, rev. ed. (New Haven, Conn.: Yale University Press, 1973), chap. 12.

11. Senate, "Analysis of Forestry Issues," 1972, p. 5.

substantial increase in timber management funding, allowing a like increase in the national forest allowable cut. Lumbermen who a decade earlier had vigorously opposed the concept of earmarking funds—in that instance to expand recreation—now came out four-square in favor of special apportionments. Conversely, recreationists who had earlier advocated special consideration for their interests now protested that the Timber Supply Act would play favorites. Opponents of the bill had their way; massive opposition caused it to be withdrawn.

During the 1970s, fears about too much logging shifted dramatically to fears about one logging method—clearcutting. Clearcutting means that all of the trees are cut in an area regardless of size, quality, or species. Exceptions would be trees too small to be marked or those that are not merchantable for some other reason. For fire protection, regeneration, or other requirements, however, the falling of all trees in the area may be specified. Supporters of clearcutting pointed out that certain species, such as Douglas fir in the Far West, reproduce poorly in shade. Thus a clearing to provide sunlight is necessary for the regeneration of Douglas fir, an extremely valuable tree. Clearcutting was defended on other grounds, too. Difficult terrain often requires use of cable logging systems that would severely damage any standing timber. Cable systems also reduce the need for roads and for having heavy equipment moving over fragile soils; areas subject to erosion can often be protected more effectively by clearcutting than by logging selectively. Clearcutting is also prescribed at times as a treatment for timber stands infested with insects or disease. Another reason, of course, is economic. The cost of moving men and equipment to a logging area is the same to take one tree or many. It is obviously much more efficient to take many rather than few.

Conservationists began to insist that selective logging, that is, cutting of certain marked individuals, was preferable to clearcutting. The remaining trees reduce the adverse aesthetic impact and seemingly offer protection to the area. Additionally, selective logging develops an uneven-aged stand, thought by many to be ecologically superior to the even-aged forest produced by clearcutting. Conservationists acknowledged the need for logging because wood is a vital commodity, but they thought that loggers should use the selective method whenever possible.

In a historical context, clearcutting versus selective logging seems paradoxical. First it is necessary to distinguish between the vast, devastated areas that nineteenth-century lumbermen left in their wake and a modern clearcut prescribed by a professional forester as the optimum silvicultural method. In both situations all of the trees are removed, but with significant differences. The forester's clearcut is much smaller, with allowances for regeneration and soil protection. To a forester, clearcutting is not only a logging system but a regeneration system as well. Such systems are chosen according to growth characteristics of desired species. Clearcutting, seed tree, shelterwood, and selection are among the methods the forester considers. After taking species, terrain, soil type, and market into account, the forester picks the best method of removing the old stand—either en masse by clearcutting or in stages by one of the other techniques. Planning for the new forest, then, plays a major part in choosing the method to log the old.

During the twentieth century, as foresters gradually exerted influence over logging methods, selective logging was adopted in many areas as the most effective cure for destructive practices. But any system can be misused, and selective logging frequently deteriorated into "high-grading," taking only the most valuable individuals with little or no thought given to the future of the residual stand. There is evidence to suggest that clearcutting, at least in the Far West, was reinstituted as a means to stop the evils of selective logging. More recently, some insist that selective logging is the proper cure for the evils of clearcutting. One thing, at least, is clear: misapplication of any system can cause unacceptable damage.

However desirable clearcutting may be to wood production, its application brings criticism from many sectors. The most obvious objection to clearcutting is its distinctly unaesthetic aftermath. No one finds recent clearcuts to be attractive, although proponents urge patience, for a few years of regrowth greatly softens the stark and jumbled look. Opponents also offer loss of watershed, increased erosion, and destruction of wildlife habitat as compelling reasons to limit or even to ban clearcutting. Articles and books aimed at a broad audience appeared in steady succession during the late 1960s and 1970s, charging that clearcutting was a destructive logging method. Those who favored clearcutting generally restricted their rebuttals to technical outlets and house organs, which were read mainly by those who were already convinced.

Opponents of clearcutting adopted a tactic of claiming that the method was in conflict with multiple use. A series of articles in *Ameri-*

can Forests, in strongly criticizing Forest Service application of multiple use, focused on clearcutting practices in Montana. Resulting from a study by professors at the University of Montana Forestry School, which followed and overshadowed a comprehensive analysis of the situation by Forest Service staff, the Montana hubbub spawned bills in Congress to limit, prohibit, or at least declare a moratorium on clearcutting. In a very real sense, the regulation issue that had dropped from view in the early 1950s now reappeared. This time the clamor included the Forest Service among those needing regulation. The subsequent hearings provide a three-volume compendium of modern thought on logging methods. Foresters, through the Society of American Foresters and forestry schools, vigorously insisted upon retaining their full professional prerogatives when prescribing forest land treatments.[12]

The major event of the clearcutting controversy was a West Virginia District Court decision—one of McCloskey's predicted litigations—that judged the logging technique to be contrary to the intent of the 1897 Sundry Civil Act, by now renamed the Organic Act of 1897.[13] The decision was ironic, since the lumber industry had labored to have that act recognized in the 1960 Multiple-Use Act as a safeguard to its interests. But as often happens, the courts saw the law's intent from a different vantage point and perhaps established a precedent that placed clearcut logging in great jeopardy unless Congress enacts remedial legislation.

Ecological Awareness

Logging methods, timber supply, and recreation were far from the only issues under review following passage of the Multiple-Use Act, but they are representative of the value conflicts that became more and more evident during this period. Suddenly, another set of values

12. U.S. Congress, Senate. " 'Clearcutting' Practices on National Timberlands," Hearings, Subcommittee on Public Lands of the Committee on Interior and Insular Affairs, 92 Cong. 1, 3 pts., Apr.–June 1971. A summary of the issues is [Luke Popovich], "The Bitterroot—Remembrances of Things Past," *Journal of Forestry* 73 (Dec. 1975): 791–93. A book supportive of clearcutting is Eleanor C. J. Horowitz, *Clearcutting: A View from the Top* (Washington, D.C.: Acropolis Books, 1974). Opposed to the practice is Nancy Wood, *Clearcut: The Deforestation of America* (San Francisco, Calif.: Sierra Club, Battlebook Series no. 3, 1971). For an excellent summary of the history of clearcutting in the Pacific Northwest, see Ivan Doig, "The Murky Annals of Clearcutting," *Pacific Search* 10 (Dec./Jan. 1975/76): 12–14.

13. *Izaak Walton* v. *Butz*, 367 F. Supp. 422 (N.D. W. Va., 1973). In 1975, the U.S. Circuit Court of Appeals for the Fourth Circuit upheld the District Court decision.

began to push the big issues of the 1950s and 1960s into the background. The possibility of irreversible disruption of life systems—upsetting the ecological balance—began to seem likely. Preservationists became "environmentalists," with the Sierra Club adding "environmental survival" to its list of priorities. No one person, including a forester, could any longer handle on a professional level the full range of responsibilities thrust upon the Forest Service in recent times. As evidence of its increasing recognition of this fact, statistics show that foresters made up 90 percent of Forest Service professionals in 1958, but only 50 percent in 1973.[14] Landscape architects, engineers, economists and other social scientists now occupy a significant portion of Forest Service rosters.

The only thing new about ecology was the public awareness that mushroomed during the 1960s and 1970s. Forestry students have studied the relationship between plants and their environment for generations, while intellectual successors to George Perkins Marsh have for over a century lectured their readers on man's responsibility to the land and about the dangers of depletion and pollution. Yet, despite the wealth of published evidence, only specialists seemed aware that there was more to resource use than extraction and husbanding remaining inventories.

It remained for Rachel Carson to write *Silent Spring* in 1962 and produce that mysterious chemistry which finally penetrated public consciousness in a way her legion of predecessors had failed to do. Perhaps it was her man-centered focus on accumulation of pesticides in food chains that caused readers to see *Silent Spring* as a call to self-defense. Other writers who wrote from nature's point of view had repeatedly failed to capture public attention, but Carson, a scientist with exquisite literary expression, electrified a slowly growing, science-dominated environmental movement.[15]

Carson's success was reminiscent of the way that Pinchot had boosted the conservation movement begun by Fernow and others. Of course, Pinchot benefited from an aggressive and powerfully supportive president—and much simpler times; Carson did not have these advantages. Nonetheless, her impact was so great that future histo-

14. Minutes of Board of Directors, Sept. 20–21, 1969, Sierra Club Records; Robinson, *The Forest Service*, p. 34.

15. Rachel Carson, *Silent Spring* (Boston: Houghton Mifflin, 1962). For an excellent analysis, see Donald Fleming, "Roots of the New Conservation Movement," *Perspectives in American History* 6 (1972): 7–91.

rians may well rank Carson even higher in importance than have her contemporaries.[16]

Carson's attack on misuse or overuse of pesticides brought swift response from the highest levels of government. A presidential panel on pesticides saw its recommendation for more stringent safeguards implemented over strenuous protests of public agencies, including the Forest Service.[17] Foresters generally believed that only by extensively applying DDT would they be able to keep certain forest pests in check. The very qualities of the agent that made it potentially harmful were the same qualities that made it the most effective insecticide around.

The furor over a tussock moth infestation in eastern Oregon forests during the early 1970s offers an excellent case study of contrary points of view on insecticide use. Conventional wisdom dictated application of DDT to eliminate the insect menace, but by now permission from the Environmental Protection Agency (another result of increased public anxiety about the quality of life) was required. After complying with strict requirements, including preparation of a hefty environmental impact statement, the Forest Service received a temporary permit to spray. Opponents of DDT, meanwhile, insisted that natural forces would bring the epidemic under control and predicted that chemicals would cause greater environmental damage than the untreated insects. Imprecise monitoring technology, one of Rachel Carson's worries, prevents conclusive evaluation of the tussock moth spray project. Each side remains unconvinced by the other's data.[18]

The revitalized conservation movement of the 1960s and 1970s, with its environmental/ecological overlay, retrieved past spokesmen to add depth and substance to later prophets. A symposium at the University of Chicago in 1964 honored George Perkins Marsh and the centennial of *Man and Nature*. Forester-wildlife specialist Aldo Leopold, dead since 1948, was resurrected by a multitude of new

16. In response to an AFA questionnaire in 1975, 104 conservationists named Rachel Carson one of the ten most important contributors to conservation in American history. See Henry Clepper, "What Conservationists Think about Conservation," *American Forests* 81 (Aug. 1975): 29.

17. President's Science Advisory Committee, *Use of Pesticides* (Washington, D.C.: GPO, 1963).

18. An informative, four-part statement of the Forest Service position on use of DDT to control the tussock moth is "Insect Infestations: The Tussock Moth," in Washington State Forestry Conference, *Proceedings* (1974), pp. 79–105. For an insightful, opposing view, see Frank Graham, Jr., "Pest Control by Press Release: The Return of DDT," *Audubon* 76 (Sept. 1974): 65–71.

believers of land ethic—the eloquently simple concept of husbandry that Leopold had advanced in his *Sand County Almanac*.

Marsh and Leopold, and Pinchot, too, were now joined by biologists Barry Commoner, Paul Erlich, and others who extended the views and fears of Rachel Carson to greater, and maybe distorted, lengths in articles and monographs that emulated *Silent Spring*, at least to the extent that they became best-sellers. The American public of the 1960s purchased these environmental warnings by the millions. Evidence suggests that they were reading, too, and becoming aware that the old, traditional values that favored rapid growth and development cried out for re-examination.

Recognizing its need for wide-ranging information on land and resources in order to deal positively with the issues, Congress established the Public Land Law Review Commission (PLLRC) in 1964. The fourth of similar land commissions (the first being in 1880), it was to study and recommend changes or additions to the nation's land law. The voluminous commission report was summarized by 137 recommendations. Apparently out of touch with the rapidly changing times, the authors of the report were critical of the Forest Service for spending too much time and money on administering national forest resources. Such views, coupled with recommendations that mirrored the aborted National Timber Supply Act and the proposal that "dominant use" replace multiple use, lost the otherwise excellent and certainly valuable compendium much credibility with conservationists.[19] It seemed to many that a dominant use policy would result in timber demands displacing the noncommodity demands of recreationists. At a time when ecological and environmental safeguards were replacing recreation and wilderness as the number one conservation concern, it was most untimely for the commission to stress the importance of economic return to the federal treasury. At least in terms of prudence, protection of the environment ought to have been given prime display.

19. Public Land Law Review Commission, *One Third of the Nation's Land* (Washington, D.C.: GPO, 1970), p. 48. "Management of public lands should recognize the highest and best use of particular areas of land as dominant over other authorized use." The report offered wilderness areas as an example of land use that would receive support from a dominant use policy. There was fear in some quarters, however, that dominant use would be applied more often to timber, range, and other commodity resources. This fear reflects an ambivalent view toward multiple use; at first it was opposed, and then it became a policy to protect from encroachment by dominant use. For example, see Philip Berry and Michael McCloskey, "The Public Land Law Review Commission Report: An Analysis," *Sierra Club Bulletin* 55 (Oct. 1970): 18–30.

In response to this new public awareness and attitude, Congress passed the National Environmental Policy Act in 1969, establishing the Environmental Protection Agency and requiring systematic review of operations that might cause environmental deterioration.[20] In principle, the law required nothing new for Forest Service procedures; its activities and goals from the beginning had been heavily steeped in ecological terms, even if not precisely articulated. The campaign since the turn of the century to reduce livestock numbers in order to prevent overgrazing was really an effort to restore balance to range lands that were used beyond their capacity. Regulation of logging, a major issue for decades, revolved around the notion that lumbermen were destroying the capacity of cutover private forest lands to produce new forests. Fire was a menace, not only because of the obvious economic losses but because fire-blackened soil was exposed to the erosive powers of wind and rain. Predator eradication, pursued with single-minded vigor during the Pinchot years, was abandoned when Aldo Leopold and others demonstrated the near-fatal imbalances such programs could cause in wildlife populations.

All these pieces need merging into an administrative whole. Timber management affects water quality and quantity; domestic herds must be balanced with wild; and recreation with its aesthetic requirements is sensitive to the operations of all. As it traditionally did, the Forest Service continued to expend a significant portion of staff time planning programs to meet anticipated needs. After three years of preparation, the agency issued in 1974 the Environmental Program for the Future, a proposed ten-year plan for managing the national forest system. It offered three alternative levels of management and operation for six resource systems and eighteen research activities and indicated the volumes of national forest products and services which could be provided. Public comments were invited, received, and analyzed. The data gathered in this project served as a base for providing Congress with the more voluminous and complex information required by the Forest and Rangeland Renewable Resources Planning Act of 1974, also known as the Humphrey-Rarick Act. The act called for an even broader Forest Service assessment to include an analysis of the present and prospective demand and supply situation for all of the nation's

20. Respondents to the 1975 AFA survey ranked the National Environmental Policy Act of 1969, which became effective on January 1, 1970, as the "most constructively influential" conservation legislation of the twentieth century. The 1924 Clarke–McNary Act placed third. See Clepper, "What Conservationists Think," p. 29.

renewable resources (the national forest system encompasses only 8 percent of the nation's total area). In addition, the act required an immediate long-range management program for the Forest Service. As before, the public was invited to examine these reports and to recommend changes or additions.[21]

During earlier times, the forest officer, underpaid, understaffed, and often undertrained, had little opportunity to view his responsibilities in formal ecological terms. His crude, primary goal was to minimize damage to the vast resources under his jurisdiction. What more could a mere handful of men hope to do with a million or more sketchily mapped acres in a newly proclaimed national forest?

Impressive developments in cartography, inventory, training, procedures—and salary—have provided the modern counterpart to the horseback, *Use Book*–toting ranger with vastly greater abilities to achieve balance. These advances, however, were accompanied by a much larger agency, replete with bureaucratic inertia and commitments to institutional traditions. Pinchot's men were more nearly free to act, bound mainly by the limitations of their stamina. Today's ranger refers continuously to his "bible"—the manual—for direction and justification. Innovative rangers learn how to play off one section of the manual against another and thus achieve a measure of independence.

As late as the 1950s, the balance sought was still a balance of uses, leaving environmental safeguards unstated but inherent in planning. Then, in 1960, presidential ink was scarcely dry on the Multiple-Use Act, a congressional mandate for balanced use, when the environmental storm broke across the nation. Not only would a whole set of new criteria have to be reckoned with, but now the ranger felt scarcely able to move because of the bevy of watchdogs following at his heels. That the ranger was irritated by this close observation is ironic. Generations of foresters had lamented the lack of public interest in their work; now the public was interested—with a vengeance it seemed to some—but instead of catching the long-anticipated bouquets, the ranger found himself dodging brickbats.

Dismayed by this perverse turn of events, the ranger, with intense pride in his profession and institution, reacted. With defense mechanisms churning furiously, foresters rejected the Browers and

21. Forest and Rangeland Renewable Resources Planning Act of 1975, 88 Stat. 476 (Humphrey–Rarick Act).

the Carsons instead of recognizing them as forerunners of the rapidly approaching future. Scorned, these advocates of what to them was a new way of looking at our world then reacted in kind. Lashing out at resource agencies and professions, the ever-more militant environmentalists overstated their case, knowing that congressional compromise would inevitably trim their demands. Nose-to-nose confrontations were fashionable during the 1960s for many American institutions, no less so for those involved with natural resources.

Perhaps the 1970s shows a general softening of lines throughout all of society. The Forest Service staff and its antagonists are pulling back from a hard-line brink, one suspicious step at a time. We should hope that the decade of the federal forestry centennial and national bicentennial will see mutual trust and respect become a long-overdue element of forest resource planning and administration. Unanimity is an impossible and even undesirable goal in a democratic society, but there must be some middle ground that is acceptable to most.

APPENDIX 1

Chronological Summary of Events Important to the History of the U.S. Forest Service

1862 May 15: Department of Agriculture established (12 Stat. 387).

1873 August 22: American Association for the Advancement of Science resolves in favor of creating a federal forestry commission; Franklin B. Hough appointed chairman of a committee to implement resolution.

1875 September 10: American Forestry Association organized in Chicago.

1876 August 15: Congress appropriated $2,000 to employ federal forestry agent (19 Stat. 143, 167); Franklin B. Hough appointed.

1881 Division of Forestry established in the Department of Agriculture.

1882 April, August: First American Forest Congress.

1883 Nathaniel H. Egleston succeeds Hough as chief of the Division of Forestry.

1886 March 15: Bernhard E. Fernow succeeds Egleston as chief of the Division of Forestry.

June 30: Division of Forestry receives statutory permanence within the Department of Agriculture (24 Stat. 100, 103).

1889 February 9: Department of Agriculture receives Cabinet status.

1891 March 3: President authorized to set aside forest reserves from public domain (26 Stat. 1095).

1892 June 4: Sierra Club formed.

1896 National Academy of Sciences appoints special committee to investigate forest reserves; Charles S. Sargent is chairman.

1897 June 4: An amendment to the Sundry Civil Appropriations Act (30 Stat. 11, 34) specifies purposes for which forest reserves can be established and provides for their administration and protection (act now referred to as the Organic Act).

1898 July 1: Gifford Pinchot succeeds Fernow as chief of the Division of Forestry.

1900 November 19: Society of American Foresters formed.

1901 November 15: Division of Forestry created in the General Land Office with Filibert Roth at its head.

March 2: Division of Forestry in USDA becomes Bureau of Forestry (31 Stat. 922, 929).

1902 National Lumber Manufacturers Association formed.

1905 January: Second American Forest Congress.

February 1: Administration of the forest reserves is transferred from the Department of the Interior to the Department of Agriculture (33 Stat. 628).

March 3: Bureau of Forestry becomes U.S. Forest Service, effective July 1 (33 Stat. 861, 872–873).

1906 June 8: American Antiquities Act provides protection for objects of antiquity and authorizes presidential proclamations to create national monuments (34 Stat. 225).

June 11: Forest Homestead Act opens agricultural lands for entry within forest reserves (34 Stat. 233).

June 30: Ten percent of receipts from forest reserves to be returned to states or territories for benefit of public roads and schools (34 Stat. 669, 684).

1907 March 4: Forest reserves renamed national forests; further enlargement forbidden in six western states except by act of Congress (34 Stat. 1256, 1269).

1908 First Forest Service experiment station is established at Fort Valley, Arizona.

May 13–15: First conference of governors held in White House.

May 23: Payments to states for schools increased to 25 percent of national forest receipts (35 Stat. 251, 260).

December 1: Regional organization of Forest Service put into effect.

1909 Western Forestry and Conservation Association organized.

January 22: Report of National Conservation Commission sent to Congress.

1910 January 7: President William H. Taft fires Gifford Pinchot; Henry S. Graves named chief on January 12.

June: Forest Products Laboratory dedicated at Madison, Wisconsin.

1911 March 1: Weeks Law authorizes federal purchase of lands in watersheds of navigable streams, matching funds for state forestry agencies, and for other purposes (36 Stat. 961).

May 1, 3: Supreme Court holds that Congress has right to create national forests and that the secretary of agriculture has right to issue rules and regulations for the forests and to prescribe penalties for violation of these rules and regulations (220 US 506, 523).

1913-14 Bureau of Corporations issues report on the lumber industry.

1915 Branch of Research established in Forest Service.

1916 June 9: Chamberlain–Ferris Act revests federal title to unsold lands of the Oregon and California Railroad Company (39 Stat. 218).

August 25: National Park Service established in the Department of the Interior (39 Stat. 535).

1920 April 15: Henry S. Graves resigns as chief of the Forest Service; William B. Greeley named to succeed.

June 1: *Timber Depletion, Lumber Exports, and Concentration of Timber Ownership* (Capper Report) sent to Senate.

1922 March 20: Secretary of agriculture authorized to exchange land in national forests for private land of equal value within national forest boundaries (42 Stat. 465).

1924 Rachford Report on the grazing situation released.

May: President Calvin Coolidge convenes the National Conference on Outdoor Recreation.

June 3: First wilderness area established on the Gila National Forest in New Mexico.

June 7: Clarke–McNary Act expands or modifies many Weeks Law cooperative programs; purchase of forest lands no longer restricted to the watersheds of navigable streams (43 Stat. 653).

1925 January 1: Forest Service begins issuing ten-year grazing permits.

1928 April 30: McNary–Woodruff Act authorizes 8 million dollars to purchase land under Weeks Law (45 Stat. 468).

May 1: William B. Greeley resigns as chief; Robert Y. Stuart named as successor.

May 22: McSweeney–McNary Act authorizes forestry research program, including a forest survey (45 Stat. 699).

1930 June 9: Knutson–Vandenberg Act authorizes funds for reforestation of national forests and the creation of a revolving fund for reforestation or timber stand improvement on national forests (45 Stat. 527).

December 6: President Herbert Hoover appoints Timber Conservation Board.

1932 American Forest Products Industries incorporated as a subsidiary of the National Lumber Manufacturers Association.

1933 March 27: *National Plan for American Forestry* (Copeland Report) submitted to Senate.

April 5: Executive Order establishes Office of Emergency Conservation Work.

June 10: Executive Order places all national monuments under USDI.

June 16: National Industrial Recovery Act authorizes Codes of Fair Competition for the various industries (48 Stat. 195).

August 25: Soil Erosion Service established in Department of the Interior.

October 23: Chief Robert Y. Stuart dies; succeeded by Ferdinand A. Silcox on November 15.

1934 Prairie States Forestry Project (shelterbelt) started with emergency funds administered by the Forest Service.

June 28: Taylor Grazing Act authorizes secretary of the interior to establish 80 million acres of grazing districts in unreserved public domain (48 Stat. 1269).

1935 April 27: Soil Conservation Act establishes Soil Conservation Service in the Department of Agriculture, succeeding the Soil Erosion Service (49 Stat. 163).

May 27: Supreme Court invalidates National Industrial Recovery Act (295 US 495).

1936 April 28: The Forest Service transmits *The Western Range* to the Senate.

1937 May 18: Cooperative Farm Forestry Act (Norris–Doxey) authorizes 2.5 million dollars annually to promote farm forestry in cooperation with states (50 Stat. 188).

June 28: Civilian Conservation Corps officially succeeds Emergency Conservation Work (50 Stat. 319).

August 28: Secretary of the interior authorized to establish sustained yield units on the revested Oregon and California Railroad lands (50 Stat. 874).

1938 June 14: Concurrent resolutions create Joint Congressional Committee on Forestry (52 Stat. 1452).

1939 December 20: F. A. Silcox dies; Earle H. Clapp named acting chief.

1940 *Forest Outings*, a study of forest recreation, is published by the Forest Service.

1941 American Forest Products Industries (now American Forest Institute) launches campaign to explain the industrial position on public regulation of logging on private land.

March 24: *Forest Lands of the United States*, a report by the Joint Congressional Committee on Forestry, is released.

1943 January 8: Lyle F. Watts named chief; E. H. Clapp named associate chief.

1944 March 29: Sustained-Yield Forest Management Act authorizes the secretaries of agriculture and the interior to establish cooperative sustained yield units with private land owners who qualify, and federal units consisting of federal land (58 Stat. 132).

1946 Forest Service issues first of six *Reappraisal* reports.

May 16: General Land Office and Grazing Service are merged to form the Bureau of Land Management in the Department of the Interior (60 Stat. 1097, 1099).

October: Third American Forest Congress.

1947 June 25: Forest Pest Control Act makes protection of all forest lands in the United States against destructive insects and disease a federal policy (61 Stat. 177).

1949 November 7: U.S. Supreme Court affirms the constitutionality of state regulation of logging on private lands (338 US 863).

1950 April 24: Granger–Thye Act broadens authority of secretary of agriculture along several lines and adjusts range regulations (64 Stat. 82).

August 25: Cooperative Forest Management Act authorizes secretary of agriculture to cooperate with state foresters in assisting private landowners (64 Stat. 473).

1952 Lyle Watts retires as chief, succeeded by Richard E. McArdle on July 1.

1953 October: Fourth American Forest Congress.

1954 June 24: Controverted O&C lands disputed by the Departments of Agriculture and the Interior shall be administered as national forest lands (68 Stat. 270).

1955 July 23: Multiple Use Mining Act returns surface rights from mining claims to the United States, unless the claim is proven valid (69 Stat. 367, 375).

1958 *Timber Resources for America's Future* is published by the Forest Service.

June 28: Outdoor Recreation Review Commission created (72 Stat. 328).

1960 June 12: Multiple Use–Sustained Yield Act directs the Forest Service to give equal consideration to outdoor recreation, range, timber, water, and wildlife and fish (74 Stat. 215).

1962 January 31: ORRRC Report stresses the importance of outdoor recreation and recommends creation of an outdoor recreation bureau.

April 2: Bureau of Outdoor Recreation created by Secretarial Order no. 497.

July 1: Richard E. McArdle retires as chief; Edward P. Cliff named as successor.

October 10: Forestry Research, State Plans, Assistance Act (McIntire–Stennis) authorizes federal support for forestry research at land grant colleges (76 Stat. 806).

1963 May 28: The Outdoor Recreation Act improves coordination of the

eighteen federal agencies directly involved with outdoor recreation (77 Stat. 49).
October: Fifth American Forest Congress.

1964 September 3: Wilderness Act sets up ten-year congressional review program for wilderness designation (78 Stat. 890).
September 3: The Land and Water Conservation Fund Act authorizes establishment of a fund supported by sales of "golden eagle" windshield stickers, motorboat fuel taxes, and surplus property sales (78 Stat. 900). Partly used to match state expenditures, the fund is to support acquisition and development of state parks; sale of stickers discontinued in 1968.

1968 Public Land Law Review Commission releases its report.
October 2: The Wild and Scenic Rivers Act provides for the preservation of selected rivers in their natural state (82 Stat. 906).

1970 January 1: National Environmental Policy Act establishes the Council on Environmental Quality and requires evaluation of potential environmental impacts of pending federal legislation and agency programs (83 Stat. 852).
December 2: Environmental Protection Agency is created by executive reorganization with congressional concurrence (84 Stat. 2086). The agency is to enforce environmental standards, monitor conditions, and conduct relevant research.

1972 July 1: Edward P. Cliff retires as chief; John R. McGuire named as successor.
October 18: Federal Water Pollution Control Act, Amendment of 1972, defines standards for Environmental Protection Agency to control sources of water pollution (86 Stat. 816).
October 21: Federal Environmental Pesticide Act grants federal and state agencies increased control over pesticide use (86 Stat. 973).

1973 The U.S. District Court for the Northern District of West Virginia in *Izaak Walton* v. *Butz* decides that clearcut logging on the Monongahela National Forest in West Virginia is contrary to the Organic Act of 1897.
September 24: Report of President's Advisory Panel on Timber and the Environment is released.

1974 August 17: Forest and Rangeland Renewable Resources Planning Act (Humphrey–Rarick) directs the secretary of agriculture to assess all lands and to prepare a program (88 Stat. 476).

1975 August 21: U.S. Circuit Court of Appeals for the Fourth Circuit upholds the 1973 *Izaak Walton* v. *Butz* decision of the U.S. District Court, which results in a ban on clearcutting on national forests in West Virginia and other states subject to the District Court.
October: Sixth American Forest Congress.

APPENDIX 2

Chronology of Administrations

President			Secretary of Agriculture*		Forest Service Chief†	
Ulysses S. Grant	Republican	1869–77	Horace Capron	1867–71	Franklin B. Hough	1876–83
			Frederick Watts	1871–77		
Rutherford B. Hayes	Republican	1877–81	William G. LeDuc	1877–81	Franklin B. Hough	
James A. Garfield	Republican	1881	George B. Loring	1881–85	Franklin B. Hough	
Chester A. Arthur	Republican	1881–85	George B. Loring		Franklin B. Hough	
					Nathaniel H. Egleston	1883–86
Grover Cleveland	Democrat	1885–89	Norman J. Colman	1885–89	Nathaniel H. Egleston	
					Bernhard E. Fernow	1886–98
Benjamin Harrison	Republican	1889–93	Jeremiah M. Rusk	1889–93	Bernhard E. Fernow	
Grover Cleveland	Democrat	1893–97	J. Sterling Morton	1893–97	Bernhard E. Fernow	
William McKinley	Republican	1897–1901	James Wilson	1897–1913	Bernhard E. Fernow	
					Gifford Pinchot	1898–1910
Theodore Roosevelt	Republican	1901–9	James Wilson		Gifford Pinchot	
William H. Taft	Republican	1909–13	James Wilson		Gifford Pinchot	
					Henry S. Graves	1910–20

Woodrow Wilson	Democrat	1913–21	David F. Houston	1913–20	Henry S. Graves	
			Edwin T. Meredith	1920–21	William B. Greeley	1920–28
Warren G. Harding	Republican	1921–23	Henry C. Wallace	1921–24	William B. Greeley	
Calvin Coolidge	Republican	1923–29	Henry C. Wallace		William B. Greeley	
			Howard M. Gore	1924–25	Robert Y. Stuart	1928–33
			William M. Jardine	1925–29		
Herbert C. Hoover	Republican	1929–33	Arthur M. Hyde	1929–33	Robert Y. Stuart	
Franklin D. Roosevelt	Democrat	1933–45	Henry A. Wallace	1933–40	Robert Y. Stuart	
					Ferdinand A. Silcox	1933–39
			Claude R. Wickard	1940–45	Earle H. Clapp (Acting)	1939–43
					Lyle F. Watts	1943–52
Harry S. Truman	Democrat	1945–53	Claude R. Wickard		Lyle F. Watts	
			Clinton P. Anderson	1945–48		
			Charles F. Brannan	1948–53	Richard E. McArdle	1952–62
Dwight D. Eisenhower	Republican	1953–61	Ezra Taft Benson	1953–61	Richard E. McArdle	
John F. Kennedy	Democrat	1961–63	Orville L. Freeman	1963–69	Richard E. McArdle	
					Edward P. Cliff	1962–72
Lyndon B. Johnson	Democrat	1963–69	Orville L. Freeman		Edward P. Cliff	
Richard M. Nixon	Republican	1969–74	Clifford M. Hardin	1969–71	Edward P. Cliff	
			Earl L. Butz	1971–	John R. McGuire	1972–
Gerald R. Ford	Republican	1974–	Earl L. Butz		John R. McGuire	

*Commissioner of Agriculture until 1889.

†Agency became Forest Service in 1905. Head administrator was chief until 1898, forester from 1898 to 1935, and chief from 1935 to date.

APPENDIX 3

Organizational Charts of the Forest Service

THE FOREST SERVICE IN 1905

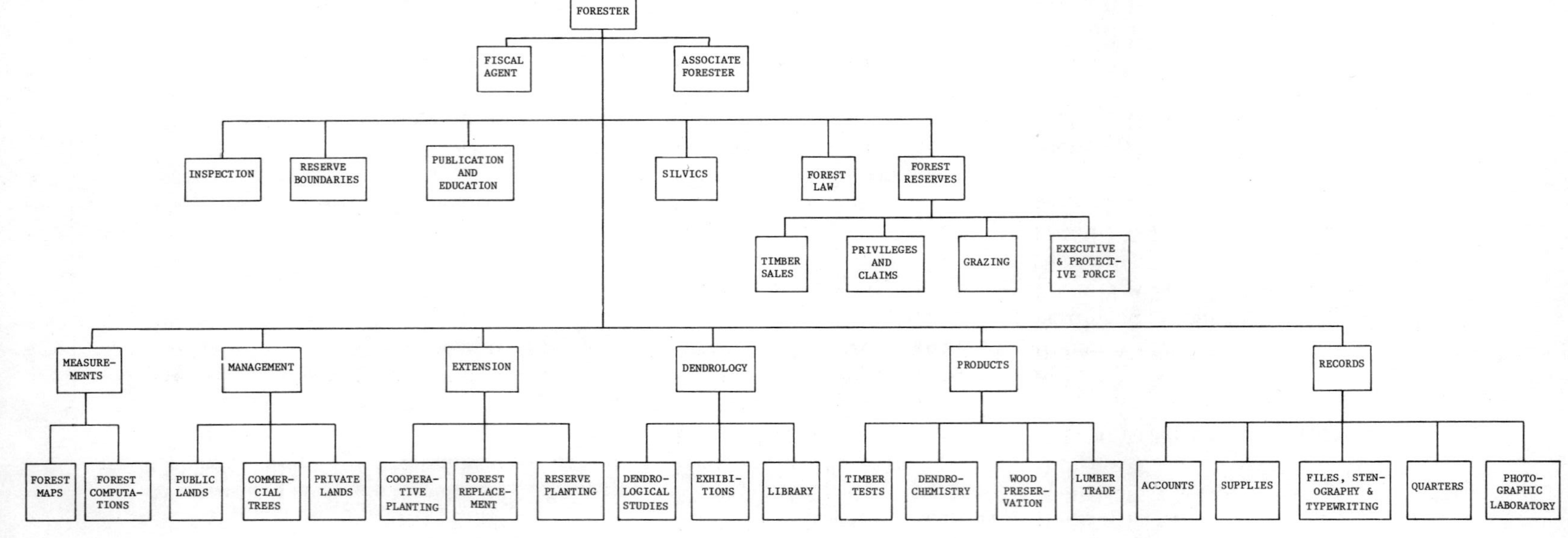

The Forest Service in 1908

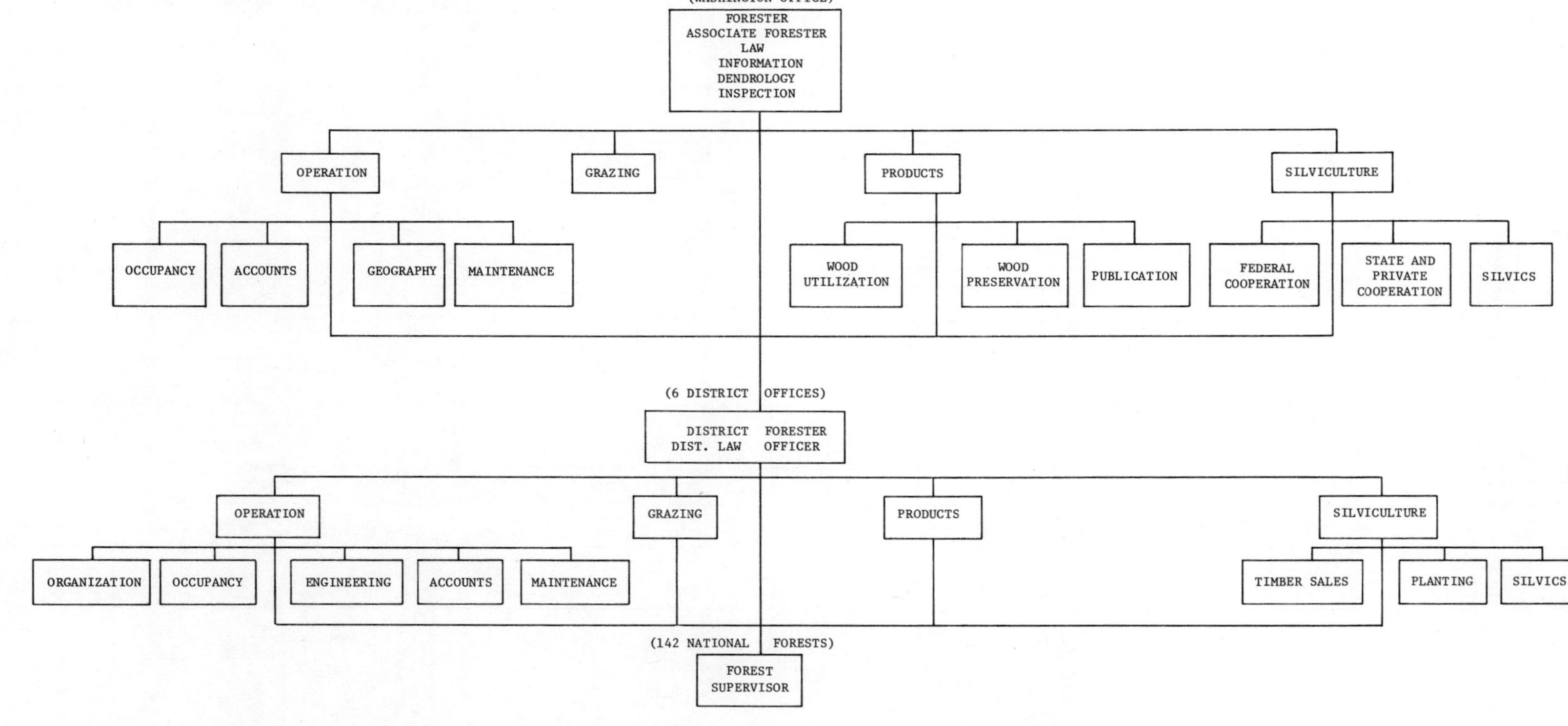

The Forest Service in 1917

FORESTER

ASSOCIATE FORESTER
EDITOR
ACCOUNTS

ACQUISITION
APPRAISAL
SURVEYS

LANDS
LAND CLASSIFICATION
SURVEYS
BOUNDARIES
TITLES
OCCUPANCY

GRAZING
RANGE RECONNAISSANCE
RANGE CONTROL
GRAZING STUDIES
FISH & GAME

OPERATION
ORGANIZATION AND PERSONNEL
MAINTENANCE
PERMANENT IMPROVEMENTS
ADMINISTRATIVE AUDIT
FIRE PROTECTION

ENGINEERING
ROADS
WATER POWER
MAPS AND SURVEYS

SILVICULTURE
WORKING PLANS AND TIMBER RECONNAISSANCE
TIMBER ADMINISTRATION
REFORESTATION
INSECT CONTROL
FEDERAL COOPERATION
STATE COOPERATION

RESEARCH
FOREST STUDIES
INDUSTRIAL INVESTIGATIONS
STATISTICAL INVESTIGATIONS
FOREST PRODUCTS LABORATORY

7 DISTRICT OFFICES

DISTRICT FORESTER

FISCAL AGENT
ENGINEERING

OPERATION
LANDS
GRAZING
SILVICULTURE
RESEARCH

152 NATIONAL FORESTS
8 EXPERIMENT STATIONS

FOREST SUPERVISOR

FOREST RANGER

The Forest Service in 1936

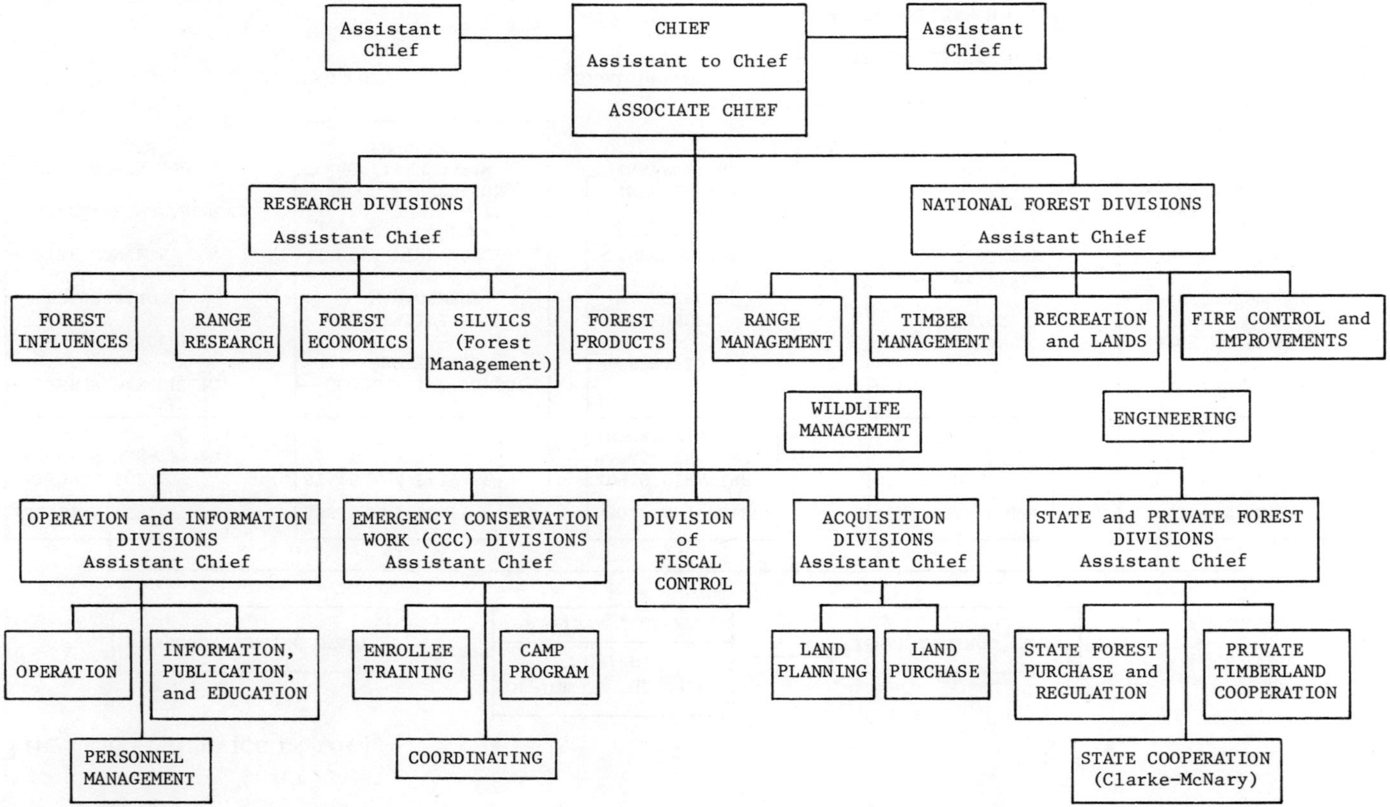

Assistant Chief

CHIEF
Assistant to Chief
ASSOCIATE CHIEF

Assistant Chief

RESEARCH DIVISIONS
Assistant Chief

FOREST INFLUENCES

RANGE RESEARCH

FOREST ECONOMICS

SILVICS (Forest Management)

FOREST PRODUCTS

NATIONAL FOREST DIVISIONS
Assistant Chief

RANGE MANAGEMENT

WILDLIFE MANAGEMENT

TIMBER MANAGEMENT

RECREATION and LANDS

ENGINEERING

FIRE CONTROL and IMPROVEMENTS

OPERATION and INFORMATION DIVISIONS
Assistant Chief

OPERATION

INFORMATION, PUBLICATION, and EDUCATION

PERSONNEL MANAGEMENT

EMERGENCY CONSERVATION WORK (CCC) DIVISIONS
Assistant Chief

ENROLLEE TRAINING

CAMP PROGRAM

COORDINATING

DIVISION of FISCAL CONTROL

ACQUISITION DIVISIONS
Assistant Chief

LAND PLANNING

LAND PURCHASE

STATE and PRIVATE FOREST DIVISIONS
Assistant Chief

STATE FOREST PURCHASE and REGULATION

PRIVATE TIMBERLAND COOPERATION

STATE COOPERATION (Clarke-McNary)

The Forest Service in 1954

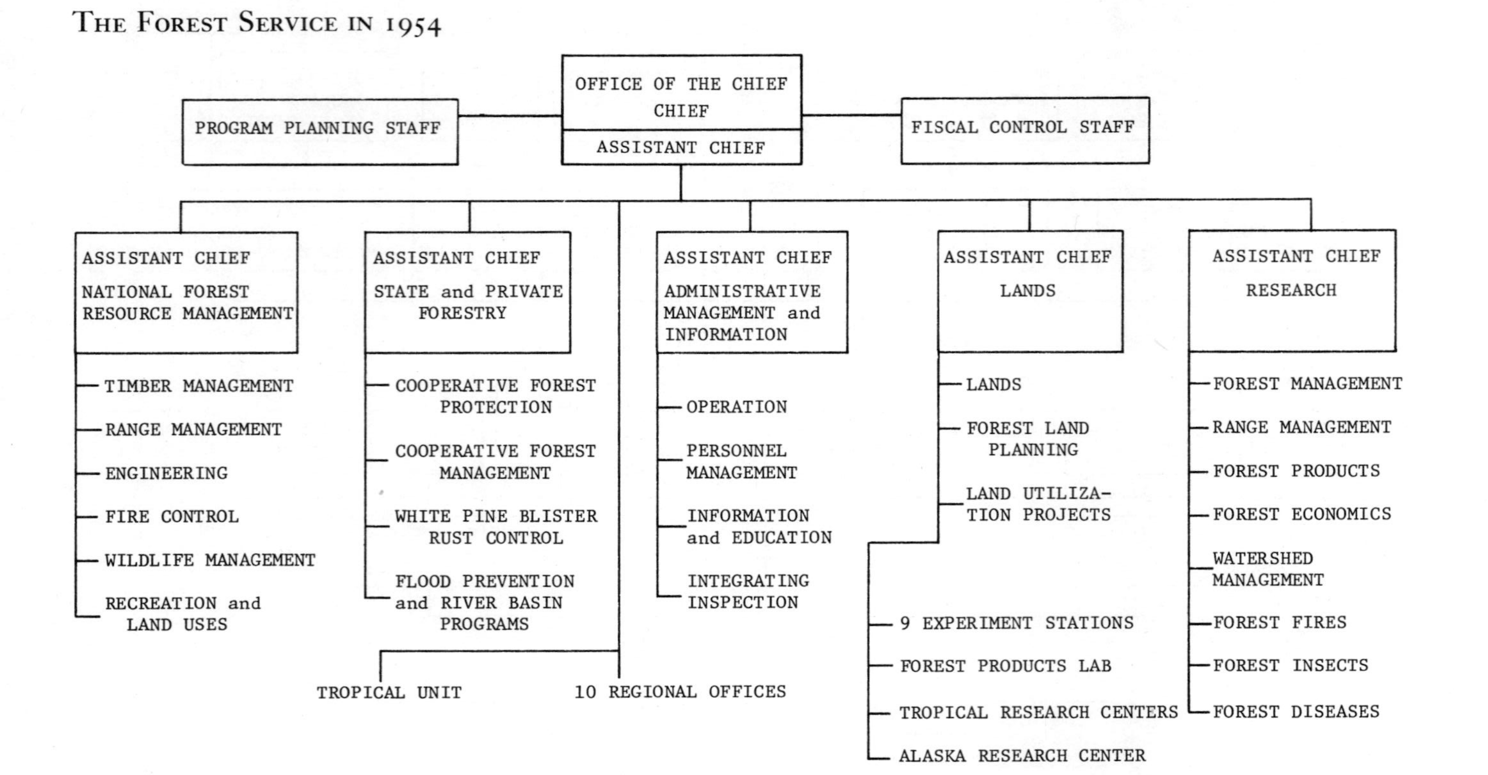

The Forest Service in 1975

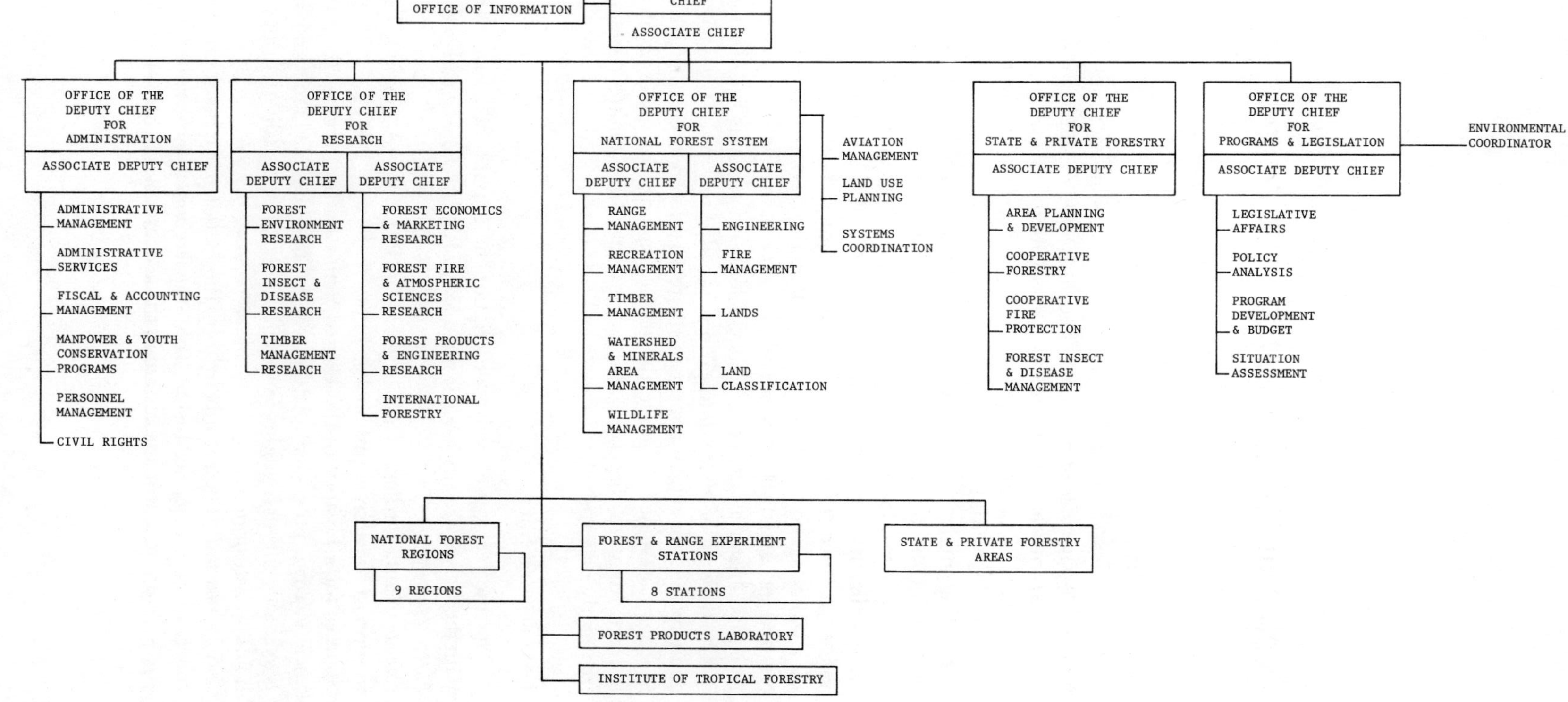

Notes on Sources of Forest Service History

THOSE EMBARKING on writing a history of the United States Forest Service will find no shortage of materials. The unpublished, and to a lesser extent the published, sources exist in quantity adequate to exhaust even the most tireless worker. Therefore, selectivity is essential or the project will never advance to the writing stage.

Existing guides, bibliographies, and advice from those who had tackled portions of the topic on other occasions were invaluable to me in setting priorities. The opportunity to examine the rich and previously unused manuscript collections and other library resources of the Forest History Society added a balance of views lacking in some studies. The Forest Service is, after all, a public agency, and the opinions of conservationists, professional foresters, and industrialists must be used in combination with its own evaluation. Adequate materials are readily available to support this broader view.

The description that follows is not intended to replace the documentation and explanations which appear in the textual footnotes; rather, this note on sources is designed to guide the researcher into the materials in a logical fashion. Many of the possible side trips have been omitted, since they are frequently of a specialized nature and best determined as the need arises.

UNPUBLISHED SOURCES

The most comprehensive description of unpublished materials dealing with the U.S. Forest Service is included in Richard C. Davis, *North American Forest History: A Guide to Archives and Manuscripts in the United States and Canada* (Santa Barbara, Calif.: ABC-Clio Press, 1976). Compiled especially to support research in this book is Judith A. Steen, "A Guide to Unpublished Sources for a History of the United States Forest Service" (Forest History Society, 1973, processed). Additional and more detailed archival information is available through guides to specific collections, usually distributed by the holding institution.

Governmental. The records held by the National Archives provide the key source of Forest Service history from the late nineteenth century to the early 1950s. More recent materials are in the agency's active files or are being held by the twelve federal records centers. Materials held by the federal records centers are presently under the custody of the Forest Service, and it is unclear which records will ultimately be transferred to the National Archives. The National Archives collection is well described by *Preliminary Inventory of the Records of the Forest Service*, compiled by Harold T. Pinkett and revised by Terry W. Good, National Archives Publication no. 69-10. Of greatest value to this study were the several categories of chiefs' correspondence; minutes of the Service Committee (chief and staff), which allow weekly or biweekly views of major and minor developments up to 1935 (very recently, the chief has reinstated minute-taking during staff meetings); and records of the Research Compilation File, which has little to do with research functions but instead was designed to aid those engaged in research on the Forest Service. Divisional or functional files tended to be less useful because of the immense quantities of unsynthesized, routine, statistical information contained therein. In most instances, summaries of divisional activity appear in the chiefs' files.

Non–Forest Service federal records of interest include those generated at the secretarial level within the Department of Agriculture. Most pertinent materials or copies have been transferred to the agency files. Records of the General Land Office and its Division R within the Department of the Interior are relevant, particularly for the years prior to the 1905 transfer of the forest reserves. Records of presidential commissions, such as the National Conference on Outdoor Recreation (1924–29), provide much information on specialized aspects of Forest Service history. Of a less precise but still useful nature are materials held by the several presidential libraries. References to routine operations seldom appear in the White House files, but many clues and other supportive data crop up in a range of presidential categories.

Institutional. The activities of many forestry-related institutions must be examined in addition to official records to gain a broader view of Forest Service history. The following collections proved to be of great value. The records of the American Forestry Association, the oldest forest conservation group in America, and the Society of American Foresters, representing the profession of forestry, contain numerous references to events and individuals important to Forest Service history. Officers and general membership of the two associations overlap in a major way with the agency's roster. The National Forest Products Association (formerly the National Lumber Manufacturers Association) and the American Forest Institute (formerly American Forest Products Industries) have worked to achieve industrial goals. The records of these industrial organizations illuminate the relations between the Forest Service and the private sector. The Forest History Society, Santa Cruz, California, holds the records of these four major organizations.

Similarly, the records of the American National Cattlemen's Association (formerly the American National Live Stock Association) (University of

Wyoming, Laramie), the Western Forestry and Conservation Association (Oregon Historical Society, Portland), and the National Forestry Program Committee (Cornell University, Ithaca, New York) show how industrial interests are advanced. Of equal importance, in order to balance external views of the Forest Service, are the records of the National Association of State Foresters (Denver Public Library), the Natural Resources Council of America (Forest History Society), and the Sierra Club (University of California, Berkeley).

Individuals. Papers of individuals who participated in some aspect of Forest Service history frequently broaden the record left in official files. Papers of chiefs were of particular value. Of the Franklin B. Hough materials (New York State Archives, Albany), his personal diary was the most useful for the period 1873–84. The Bernhard E. Fernow Papers at Cornell University were helpful but usually only supplemented his official correspondence as part of the Forest Service Records relating to the latter decades of the nineteenth century. The voluminous Gifford Pinchot Papers (Library of Congress) are undoubtedly the most valuable personal collection on the history of conservation, forestry, and the Forest Service, especially from 1890 until 1910, when he left federal employ. The diary of Henry S. Graves (Yale University, New Haven, Connecticut) illuminates his personality and relates his inner thoughts on key forestry events during the second decade of the twentieth century; the remainder of his collection is less useful for Forest Service history, although of great value for other forestry affairs. The William B. Greeley Papers (University of Oregon, Eugene) show many aspects of his private life during his tenure as chief, 1920–28. The Earle H. Clapp Papers (National Archives) is largely a collection of official memoranda assembled to produce a personal memoir. Nonetheless, the Clapp collection, partially closed, is extremely valuable for the New Deal period and for the development of research policy.

Personal papers of individuals who were not chief or cannot be classified as Forest Service employees also add a large measure of understanding to agency history. Foresters like Raphael Zon (Minnesota Historical Society, Saint Paul), Tom Gill (Forest History Society), and E. T. Allen (Oregon Historical Society) provide a range of perceptions and interpretations about day-to-day Forest Service activities. Pieces of history were also found in the Harold L. Ickes Papers (Library of Congress), James C. Rettie Papers (University of Oregon), and Corydon Wagner Papers (University of Washington, Seattle).

Oral History Sources and Dissertations

Oral history interviews and doctoral dissertations are a genre of quasi-published materials that many historians fail to utilize. Both have added to this study in a substantive way.

Oral interviews—personal statements of the past—contain the biases and inaccuracies common to memoirs but are replete with clues for further search and observations about men and events that are difficult if not impossible to obtain through conventional methods. The Regional Oral History Office of the Bancroft Library and the Forest History Society have produced a lengthy list of interviews with foresters, conservationists, and lumbermen whose recollections have a bearing on Forest Service history. Three interviews proved especially valuable for this study, although many others made some contribution. Arthur C. Ringland, *Conserving Human and Natural Resources*, Oral History Interview (OHI) by Amelia R. Fry et al., 1970, Regional Oral History Office, Bancroft Library, University of California, Berkeley, provides many insights into early years. Edward C. Crafts, *Forest Service Researcher and Congressional Liaison: An Eye to Multiple Use*, OHI by Susan R. Schrepfer, 1972, Forest History Society, and Richard E. McArdle, *Dr. Richard E. McArdle: An Interview with the Former Chief, U.S. Forest Service, 1952–1962*, OHI by Elwood R. Maunder, 1975, Forest History Society, provide much information and observations about the 1950s and 1960s.

The Forest History Society library contains microfilm copies of most doctoral dissertations pertinent to Forest Service history. Of particular value were those that filled gaps in conventional literature, such as biographies of secretaries of agriculture and the interior. Others contained useful interpretations of specific issues, but frequently dissertations served best as guides to specialized literature.

Published Sources

Existing monographic literature tends to treat the history of the Forest Service in only a fragmented way. Certain aspects or episodes have been covered in detail, while others have been largely neglected. This is not to say, however, that there are not many valuable works available for examination. It must also be kept in mind that this project is national in scope. There are many, excellent published or processed volumes that deal with regional aspects of Forest Service history.

The most comprehensive bibliography of Forest Service history is included in Ronald J. Fahl, *North American Forest and Conservation History: A Bibliography* (Santa Barbara, Calif.: ABC-Clio Press, 1976). Gerald R. Ogden, "The United States Forest Service: A Historical Bibliography, 1876–1972," Forest History Society, 1973, processed, contains seven thousand citations to books and articles relating to Forest Service history. Many other bibliographies are also available but generally treat only narrow or technical features of forestry or the Forest Service.

The most valuable, single reference on the history of forestry and the Forest Service is Samuel T. Dana, *Forest and Range Policy* (New York: McGraw-Hill, 1956). Accurate, well indexed, and based largely on published sources, the Dana volume is in effect a one-volume encyclopedia of the subject. Of a different style and focus and drawing more heavily on unpublished materials is Henry Clepper, *Professional Forestry in the United States* (Baltimore, Md.: Johns Hopkins Press for Resources for the Future, 1971). Dated but still useful is Darrell H. Smith, *The Forest Service: Its History, Activities and Organization* (Washington, D.C.: Brookings Institution, 1930). A more recent work by Michael Frome, *The Forest Service* (New York: Praeger Publishers, 1971), treats history only incidentally. Glen O. Robinson, *The Forest Service: A Study in Public Land Management* (Baltimore, Md.: Johns Hopkins Press for Resources for the Future, 1975), offers some historical views but primarily deals with the modern Forest Service. Of great utility for Forest Service history is Frank E. Smith, ed., *Conservation in the United States: A Documentary History*, 5 vols. (New York: Chelsea House Publishers in association with Van Nostrand Reinhold Company, 1971), in which pertinent legislation, reports, and speeches are excerpted, or printed whole, with annotation. Two other general works must be cited: John Ise, *The United States Forest Policy* (New Haven, Conn.: Yale University Press, 1920), and Jenks Cameron, *The Development of Governmental Forest Control in the United States* (Baltimore, Md.: Johns Hopkins Press, 1928). Ise places heavy emphasis on interpretations of the *Congressional Record*, while Cameron looks to more diverse sources. Although dated, both have yet to be replaced; in 1972, Arno Press reprinted Ise and DaCapo Press reprinted Cameron.

Specific topics of Forest Service history have been explored. Most influential, because of its seminal interpretations, is Samuel P. Hays, *Conservation and the Gospel of Efficiency: The Progressive Conservation Movement, 1890–1920* (Cambridge, Mass.: Harvard University Press, 1959). Of narrower focus for the same time period is Elmo R. Richardson, *The Politics of Conservation: Crusades and Controversies, 1897–1913* (Berkeley: University of California Press, 1962). A moderate revision of Gifford Pinchot's battle with Secretary of the Interior Richard A. Ballinger is offered in James Penick, Jr., *Progressive Politics and Conservation: The Ballinger–Pinchot Affair* (Chicago: University of Chicago Press, 1968). Donald C. Swain, *Federal Conservation Policy, 1921–1933* (Berkeley: University of California Press, 1963), places the Forest Service into context with other federal conservation programs and sets the stage for study of the New Deal years.

The most valuable single source of Forest Service and conservation history during the New Deal is Edgar B. Nixon, ed., *Franklin D. Roosevelt and Conservation, 1911–1945*, 2 vols. (Hyde Park, N.Y.: General Services Administration, 1957), which is an indexed, annotated collection of the most important presidential correspondence concerning conservation. A full calendar of FDR's conservation letters is also available from the presidential library at Hyde Park. One of the most popular New Deal programs has been examined, with emphasis on administration, by John A. Salmond in *The*

Civilian Conservation Corps, 1933–42 (Durham, N.C.: Duke University Press, 1967). Tree planting on the American prairies also captured the public's imagination and has been treated comprehensively by Wilmon H. Droze, *Trees, Prairies, and People: Field Shelterbelt Planting in the Trans-Mississippi West* (Denton: Texas Woman's University Press, 1976).

Political storms common to the New Deal period at times involved the Forest Service. A chapter in Richard Polenberg, *Reorganizing Roosevelt's Government: The Controversy over Executive Reorganization, 1936–1939* (Cambridge, Mass.: Harvard University Press, 1966), looks closely at the successful efforts by Forest Service supporters to keep the agency in the Department of Agriculture. Management of the public range was a major conservation issue of the New Deal period. Wesley Calef, *Private Grazing and Public Lands: Studies of the Local Management of the Taylor Grazing Act* (Chicago: University of Chicago Press, 1960), which focuses on the central Rocky Mountain region, and the flawed but useful Phillip O. Foss, *Politics and Grass: The Administration of Grazing on the Public Domain* (Seattle: University of Washington Press, 1960), describe the enactment and administration of one of the most important federal range statutes. For a quasi-official discussion of early Forest Service range policies, see Will C. Barnes, *Western Grazing Grounds and Forest Ranges* (Chicago: The Breeder's Gazette, 1913). His chapter 11 is especially pertinent. Another useful, personal account of grazing policies and practices prior to the New Deal may be found in Paul H. Roberts, *Hoof Prints on Forest Ranges: The Early Years of National Forest Range Administration* (San Antonio, Tex.: Naylor Company, 1963).

Two monographs relevant to the recent past provide distinctly different views of the Forest Service. Herbert Kaufman, *The Forest Ranger: A Study in Administrative Behavior* (Baltimore, Md.: Johns Hopkins Press for Resources for the Future, 1960), offers valuable insights by a political scientist into how the decentralized Forest Service maintains control of its field personnel. Presenting a historian's analysis of conservation issues of the 1950s, Elmo R. Richardson, *Dams, Parks and Politics: Resource Development and Preservation in the Truman–Eisenhower Era* (Lexington: University Press of Kentucky, 1973), is invaluable for measuring the tone of the times and contains extensive references to pertinent archival collections.

Research is an important facet of Forest Service history. Charles A. Nelson, *History of the U.S. Forest Products Laboratory, 1910–1963* (Madison, Wis.: Forest Products Laboratory, 1971), presents a very useful study of this key research facility. Brief historical views of two Forest Service experiment stations are Susan R. Schrepfer et al., *A History of the Northeastern Forest Experiment Station* (Forest Service Technical Report no. NE-7, 1973), and Ivan Doig, *Early Forestry Research: A History of the Pacific Northwest Forest and Range Experiment Station, 1925–1975* (Portland, Ore.: PNW Forest and Range Experiment Station, 1976). Verne L. Harper, *A Forest Service Research Scientist and Administrator Views Multiple Use*, OHI by Elwood R. Maunder, 1972, Forest History Society, is of particular value for recent research policies and programs. Ashley L. Schiff, *Fire and Water: Scientific Heresy in the Forest Service* (Cambridge, Mass.: Harvard University Press, 1962), provides a critical view of the relationship between Forest Service administrators and researchers. T. Swann Harding, *Two Blades of Grass: A History of Scientific Developments in the U.S. Department of Agriculture* (Norman: University of Oklahoma Press, 1947), contains many references to Forest Service research.

Histories of other institutions and agencies parallel that of the Forest Service. Ralph R. Widner, ed., *Forest and Forestry in the American States: A Reference Anthology* (Washington, D.C.: National Association of State Foresters, 1968), provides a beginning for the understudied history of state forestry activities. Henry Clepper, *Crusade for Conservation: The Centennial History of the American Forestry Association* (Washington, D.C.: American Forestry Association, 1975), which also appears in an unindexed version in the October 1975 issue of *American Forests*, shows the importance of the oldest forest conservation group to the Forest Service and to conservation history generally. Monographs on the National Park Service abound. The most useful are John Ise, *Our National Park Policy: A Critical History* (Baltimore, Md.: Johns Hopkins Press for Resources for the Future, 1961), and William C. Everhart, *The National Park Service* (New York: Praeger Publishers, 1972). The formative years of the Sierra Club are unevenly treated in Holway R. Jones, *John Muir and the Sierra Club: The Battle for Yosemite* (San Francisco, Calif.: Sierra Club, 1965). Marion Clawson, *The Bureau of Land Management* (New York: Praeger Publishers, 1971), provides some historical treatment for the major resource bureau in the Department of the Interior. For a wide-ranging, insiders' view of all aspects of forestry, useful anthologies are Robert K. Winters, ed., *Fifty Years of Forestry in the U.S.A.* (Washington, D.C.: Society of American Foresters, 1950), and Henry Clepper and Arthur B. Meyer, eds., *American Forestry: Six Decades of Growth* (Washington, D.C.: Society of American Foresters, 1960).

Biographical accounts and memoirs make a substantive and valuable contribution to the history of the Forest Service. A useful collection of biographical references is Henry Clepper, ed., *Leaders of American Conservation* (New York: Ronald Press Company, 1971). A brief biographical view of the first head of the newly formed federal forestry agency may be found in Edna L. Jacobsen, "Franklin B. Hough, A Pioneer in Scientific Forestry in America," *New York History* 15 (July 1934). Andrew Denny Rodgers III, *Bernhard Eduard Fernow: A Story of North American Forestry* (Princeton, N.J.: Princeton University Press, 1951), offers ample rewards for those who stay with it. Gifford Pinchot, unquestionably the central figure in Forest Service history, has been treated by several authors. His lengthy autobiography, *Breaking New Ground* (New York: Harcourt Brace, 1947; reprinted 1972 by

University of Washington Press, Seattle), is invaluable for an understanding of the man and his agency, even though his account ends in 1910. M. Nelson McGeary, *Gifford Pinchot: Forester–Politician* (Princeton, N.J.: Princeton University Press, 1968), is a revealing account of the man who molded the agency in its formative years and aspects of his life after he was fired as chief. Harold T. Pinkett, *Gifford Pinchot: Private and Public Forester* (Urbana: University of Illinois Press, 1970), carefully focuses on Pinchot's forestry achievements. Hazel Dawson, *Gifford Pinchot: A Bio-Bibliography* (Washington, D.C.: Department of the Interior, 1971), shows the extent to which Pinchot has captured the attention of authors. The only other chief to receive biographical treatment is William Greeley, in George T. Morgan, Jr., *William B. Greeley: A Practical Forester* (Saint Paul, Minn.: Forest History Society, 1961). Greeley's autobiographical account of Forest Service history is *Forests and Men* (Garden City, N.Y.: Doubleday and Company, 1951). McArdle, *An Interview with the Chief*, offers a personal account of his administration, from 1952 to 1962, and observations about forestry events since 1962.

S. B. Sutton, *Charles Sprague Sargent and the Arnold Arboretum* (Cambridge, Mass.: Harvard University Press, 1970), offers early views of Pinchot's career. Coert DuBois, *Trail Blazers* (Stonington, Conn.: Stonington Publishing Company, 1951), is a delightful but brief personal account of early Forest Service adventures. Gordon B. Dodds, *Hiram Martin Chittenden: His Public Career* (Lexington: University Press of Kentucky, 1973),contains a chapter on Chittenden's conflict with Pinchot over the effect of forests on flooding. Donald L. Baldwin, *The Quiet Revolution: The Grass Roots of Today's Wilderness Preservation Movement* (Boulder, Colo.: Pruett Publishing Company, 1972), presents an uneven biographical view of Arthur H. Carhart's contribution to wilderness preservation.

Another leading participant in the wilderness movement may be seen in Susan L. Flader, *Thinking Like a Mountain: Aldo Leopold and the Evolution of an Ecological Attitude toward Deer, Wolves, and Forests* (Columbia: University of Missouri Press, 1974). Evolution of sustained yield and the National Recovery Administration Lumber Code are only two of the topics touched on in Rodney C. Loehr, ed., *Forests for the Future: The Story of Sustained Yield as Told in the Diaries of David T. Mason, 1907–1950* (Saint Paul: Forest Products History Foundation, Minnesota Historical Society, 1952).

Articles pertinent to Forest Service history are scattered throughout history journals. Only the *Journal of Forest History* (formerly *Forest History*) contains a concentration of relevant articles and oral history excerpts. Fahl's bibliography provides the best access to periodical literature written as history.

There is an immense amount of published source material available in contemporary periodical literature and agency reports. The *Journal of Forestry* and *American Forests* are especially rich for technical information and policy statements, frequently by Forest Service staff. Both journals are indexed annually and the *Journal of Forestry* also can be approached through a series of decennial, cumulative indexes. Trade journals of the forest industries, particularly for the earlier years, contain articles, reprints of speeches and newspaper pieces, and statements by industrial spokesmen that are most useful to document the positions and attitudes of the private sector. Access to contemporary periodical and technical literature may be facilitated by E. N. Munns, *A Selected Bibliography of North American Forestry*, 2 vols., U.S. Department of Agriculture Miscellaneous Publication 364, 1940; *Forestry Abstracts*; and the *Bibliography of Agriculture*. Annual reports of the Forest Service and its predecessors are also an excellent source of data on policies, activities, and statistics.

Newspapers per se were not consulted for this study. The Forest History Society archives do contain a newspaper clipping file that was inherited from the U.S. Forest Service via the National Agricultural Library. This collection is cited in footnotes as the FHS clipping file. The file begins with the Bernhard Fernow administration of the 1890s and continues through the Pinchot years to approximately 1910. Most of the clippings are mounted and provide accounts of Forest Service activities from newspapers across the nation. Also randomly included in this file are an assortment of Fernow correspondence and memos, and Forest Service news releases and similar items up to the 1920s. Despite its rather heterogeneous nature, the clipping file proved very valuable to this study. The Denver Public Library has a portion of the same collection that encompasses later years.

There is an excellent photographic record of Forest Service history. Hundreds of thousands of official photographs have recorded daily activities. The vast collection that the agency began accumulating in the 1890s is currently being sorted, cataloged, and transferred to the National Archives. A sophisticated index provides ready access to those pictures as they are accessioned. The collection and the cataloging project is described in Leland J. Prater, "Historical Forest Service Photo Collection," *Journal of Forest History* 18 (April 1974): 28–31.

Index